Automated Machine Learning

Hyperparameter optimization, neural architecture search, and algorithm selection with cloud platforms

Adnan Masood, PhD

BIRMINGHAM—MUMBAI

Automated Machine Learning

Group Product Manager: Kunal Parikh

Publishing Product Manager: Ali Abidi

Senior Editor: Mohammed Yusuf Imaratwale

Content Development Editor: Nazia Shaikh

Technical Editor: Sonam Pandey

Copy Editor: Safis Editing

Project Coordinator: Aparna Ravikumar Nair

Proofreader: Safis Editing

Indexer: Pratik Shirodkar

Production Designer: Vijay Kamble

First published: February 2021

Production reference: 1180221

Published by Packt Publishing Ltd.

Livery Place

35 Livery Street

Birmingham

B3 2PB, UK.

ISBN 978-1-80056-768-9

www.packt.com

Foreword

There are moments in your life that stick with you no matter the circumstances. For me, it was the moment that I first met Dr. Adnan Masood. It was not at a tech conference or at a work function. It was at a Sunday school event which both of our children attended. He introduced himself and asked me what I did. I usually just give a generic canned response as most folks I speak with outside of my field of work don't really get what I do. Instead, his eyes lit up when I told him that I work with data. He kept asking me deeper and deeper questions about some of the most obscure Machine Learning and Deep Learning Algorithms that even I had not heard in a long time. It is a nice realization when you find out that you are not alone in this world and that there are others who have the same passion as you.

It is this passion that I see Dr. Masood bringing to a quickly growing and often misunderstood field of Automated Machine Learning. As a Data Scientist working at Microsoft, I often hear from organizational leads that Automated Machine Learning will lead to the end of the need for data science expertise. This is truly not the case and Automated Machine Learning should not be treated as a *"black-box"* or a *"One-size-fits-all"* approach to feature engineering, data pre-processing, model training, and model selection. Rather, Automated Machine Learning can help cut down the time and cost affiliated with work that takes away from the overall beauty of Data Science, Machine Learning, and Artificial Intelligence.

The great news about the current publication you hold in your hand or read on your tablet is that you now have a nuanced understanding of the benefits of applying Automated Machine Learning with every current and future project in your organization. Additionally, you will get hands-on expertise leveraging AutoML with open-source packages as well as cloud solutions offered by Azure, Amazon Web Services, and Google Cloud Platform. Whether you are a seasoned data scientist, a budding data scientist, a data engineer, an ML engineer, a DevOps engineer, or a data analyst, you will find that AutoML can help get you to the next level in your Machine Learning journey.

Ahmed Sherif

Cloud Solution Architect, AI & Analytics – Microsoft Corporation

Contributors

About the author

Adnan Masood, PhD is an artificial intelligence and machine learning researcher, visiting scholar at Stanford AI Lab, software engineer, Microsoft MVP (Most Valuable Professional), and Microsoft's regional director for artificial intelligence. As chief architect of AI and machine learning at UST Global, he collaborates with Stanford AI Lab and MIT CSAIL, and leads a team of data scientists and engineers building artificial intelligence solutions to produce business value and insights that affect a range of businesses, products, and initiatives.

About the reviewer

Jamshaid Sohail is passionate about data science, machine learning, computer vision, and natural language processing and has more than 2 years of experience in the industry. He has worked at a Silicon Valley-based start-up named FunnelBeam, the founders of which are from Stanford University, as a data scientist. Currently, he is working as a data scientist at Systems Limited. He has completed over 66 online courses from different platforms. He authored the book *Data Wrangling with Python 3.X* for Packt Publishing and has reviewed multiple books and courses. He is also developing a comprehensive course on data science at Educative and is in the process of writing books for multiple publishers.

Table of Contents

2

Automated Machine Learning, Algorithms, and Techniques

3

Automated Machine Learning with Open Source Tools and Libraries

Section 2: AutoML with Cloud Platforms

4

Getting Started with Azure Machine Learning

5

Automated Machine Learning with Microsoft Azure

6

Machine Learning with AWS

7

Doing Automated Machine Learning with Amazon SageMaker Autopilot

8

Machine Learning with Google Cloud Platform

9

Automated Machine Learning with GCP

Section 3: Applied Automated Machine Learning

10

AutoML in the Enterprise

Preface

Every machine learning engineer deals with systems that have hyperparameters, and the most basic task in **automated machine learning** (**AutoML**) is to automatically set these hyperparameters to optimize performance. The latest deep neural networks have a wide range of hyperparameters for their architecture, regularization, and optimization, which can be customized effectively to save time and effort.

This book reviews the underlying techniques of automated feature engineering, model and hyperparameter tuning, gradient-based approaches, and more. You'll explore different ways of implementing these techniques in open source tools. Next, you'll focus on enterprise tools, learning about different ways of implementing AutoML in three major cloud service providers: **Microsoft Azure**, **Amazon Web Services** (**AWS**), and **Google Cloud Platform** (**GCP**). As you progress, you'll explore the features of cloud AutoML platforms by building machine learning models using AutoML. Later chapters will show you how to develop accurate models by automating time-consuming and repetitive tasks involved in the machine learning development life cycle.

By the end of this book, you'll be able to build and deploy AutoML models that are not only accurate, but that also increase productivity, allow interoperability, and minimize featuring engineering tasks.

Who this book is for

Citizen data scientists, machine learning developers, AI enthusiasts, or anyone looking to automatically build machine learning models using the features offered by open source tools, Microsoft Azure Machine Learning, AWS, and Google Cloud Platform will find this book useful.

What this book covers

Chapter 1, A Lap around Automated Machine Learning, presents a detailed overview of AutoML methods by both providing a solid overview for novices and serving as a reference for experienced machine learning practitioners. This chapter starts with the machine learning development life cycle and navigates the problem of hyperparameter optimization that AutoML solves.

Chapter 2, Automated Machine Learning, Algorithms, and Techniques, allows citizen data scientists to build AI solutions without extensive experience. In this chapter, we review the current developments of AutoML in terms of three categories: **automated feature engineering** (**AutoFE**), **automated model and hyperparameter learning** (**AutoMHL**), and **automated deep learning** (**AutoDL**). State-of-the-art techniques adopted in these three categories are presented, including Bayesian optimization, reinforcement learning, evolutionary algorithms, and gradient-based approaches. In this chapter, we'll summarize popular AutoML frameworks and conclude with the current open challenges of AutoML.

Chapter 3, Automated Machine Learning with Open Source Tools and Libraries, teaches you about AutoML **open source software** (**OSS**) tools and libraries that automate the entire life cycle of the ideation, conceptualization, development, and deployment of predictive models. From data preparation through model training to validation as well as deployment, these tools do everything with almost zero human intervention. In this chapter, we'll review the major OSS tools, including TPOT, AutoKeras, Auto-Sklearn, Featuretools, H2O AutoML, Auto-PyTorch, Microsoft NNI, and Amazon AutoGluon, and help you to understand the different value propositions and approaches used in each of these libraries.

Chapter 4, Getting Started with Azure Machine Learning, covers Azure Machine Learning, which helps accelerate the end-to-end machine learning life cycle using the power of the Windows Azure platform and services. In this chapter, we will review how to get started with an enterprise-grade machine learning service to build and deploy models that empower developers and data scientists for building, training, and deploying machine learning models faster. With examples, we will set up the groundwork to build and deploy AutoML solutions.

Chapter 5, Automated Machine Learning with Microsoft Azure, reviews in detail and with code examples, how can we automate time-consuming and iterative tasks of model development using an Azure machine learning stack and perform operations such as regression, classification, and time series analysis using Azure AutoML. This chapter will enable you to perform hyperparameter tuning to find the optimal parameters and find the optimal model with Azure AutoML.

Chapter 6, Machine Learning with Amazon Web Services, covers Amazon SageMaker Studio, Amazon SageMaker Autopilot, Amazon SageMaker Ground Truth, and Amazon SageMaker Neo, along with the other AI services and frameworks offered by AWS. As well as hyperscalers (cloud offerings), AWS offers one of the broadest and deepest sets of machine learning services and supporting cloud infrastructure, putting machine learning in the hands of every developer, data scientist, and expert practitioner. AWS offers machine learning services, AI services, deep learning frameworks, and learning tools to build, train, and deploy machine learning models fast.

Chapter 7, Doing Automated Machine Learning with Amazon SageMaker Autopilot, takes us on a deep dive into Amazon SageMaker Studio, using SageMaker Autopilot to run several candidates to figure out the optimal combination of data preprocessing steps, machine learning algorithms, and hyperparameters. The chapter provides a hands-on, illustrative overview of training an inference pipeline, for easy deployment on a real-time endpoint or batch processing.

Chapter 8, Machine Learning with Google Cloud Platform, reviews Google's AI and machine learning offerings. Google Cloud offers innovative machine learning products and services on a trusted and scalable platform. These services include AI Hub, AI building blocks such as sight, language, conversation, and structured data services, and AI Platform. In this chapter, you will become familiar with these offerings and understand how AI Platform supports Kubeflow, Google's open source platform, which lets developers build portable machine learning pipelines with access to cutting-edge Google AI technology such as TensorFlow, TPUs, and TFX tools to deploy your AI applications to production.

Chapter 9, Automated Machine Learning with GCP Cloud AutoML, shows you how to train custom business-specific machine learning models, with minimum effort and machine learning expertise. With hands-on examples and code walk-throughs, we will explore the Google Cloud AutoML platform to create customized deep learning models in natural language, vision, unstructured data, language translation, and video intelligence, without any knowledge of data science or programming.

Chapter 10, AutoML in the Enterprise, presents AutoML in an enterprise setting as a system to automate data science by generating fully automated reports that include an analysis of the data, as well as predictive models and a comparison of their performance. A unique feature of AutoML is that it provides natural-language descriptions of the results, suitable for non-experts in machine learning. We emphasize the operationalization of an MLOps pipeline with a discussion on approaches that perform well on practical problems and determine the best overall approach. The chapter details ideas and concepts behind real-world challenges and provides a journey map to address these problems.

To get the most out of this book

This book is an introduction to AutoML. Familiarity with data science, machine learning, and deep learning methodologies will be helpful to understand how AutoML improves upon existing methods.

Software/hardware covered in the book	OS requirements
Python 3	Windows, Mac OS X, or Linux (any)
The Jupyter Notebook/Anaconda	
A modern web browser (preferably Chrome or Edge)	

If you are using the digital version of this book, we advise you to type the code yourself or access the code via the GitHub repository (link available in the next section). Doing so will help you avoid any potential errors related to the copying and pasting of code.

Download the color images

We also provide a PDF file that has color images of the screenshots/diagrams used in this book. You can download it here: `https://static.packt-cdn.com/downloads/9781800567689_ColorImages.pdf`.

Conventions used

There are a number of text conventions used throughout this book.

`Code in text`: Indicates code words in text, database table names, folder names, filenames, file extensions, pathnames, dummy URLs, user input, and Twitter handles. Here is an example: "Open the `autopilot_customer_churn` notebook from the `amazonsagemaker-examples/autopilot` folder."

A block of code is set as follows:

```
[37]: import sagemaker
      prefix = 'sagemaker/automlbook-bankds/input'
      sess   = sagemaker.Session()
      uri = sess.upload_data(path="automl-train.csv", key_prefix=prefix)
      print(uri)

      s3://sagemaker-us-east-1-385578370913/sagemaker/automlbook-bankds/input/automl-train.csv
```

Bold: Indicates a new term, an important word, or words that you see onscreen. For example, words in menus or dialog boxes appear in the text like this. Here is an example: "From Amazon SageMaker Studio, start a data science notebook by clicking on the **Python 3** button."

> **Tips or important notes**
> Appear like this.

Get in touch

Feedback from our readers is always welcome.

General feedback: If you have questions about any aspect of this book, mention the book title in the subject of your message and email us at `customercare@packtpub.com`.

Errata: Although we have taken every care to ensure the accuracy of our content, mistakes do happen. If you have found a mistake in this book, we would be grateful if you would report this to us. Please visit `www.packtpub.com/support/errata`, selecting your book, clicking on the Errata Submission Form link, and entering the details.

Piracy: If you come across any illegal copies of our works in any form on the Internet, we would be grateful if you would provide us with the location address or website name. Please contact us at `copyright@packt.com` with a link to the material.

If you are interested in becoming an author: If there is a topic that you have expertise in and you are interested in either writing or contributing to a book, please visit `authors.packtpub.com`.

Reviews

Please leave a review. Once you have read and used this book, why not leave a review on the site that you purchased it from? Potential readers can then see and use your unbiased opinion to make purchase decisions, we at Packt can understand what you think about our products, and our authors can see your feedback on their book. Thank you!

For more information about Packt, please visit `packt.com`.

Section 1: Introduction to Automated Machine Learning

This part provides a detailed introduction to the landscape of automated machine learning, its pros and cons, and how it can be applied using open source tools and libraries. In this section, you will come to understand, with the aid of hands-on coding examples, that automated machine learning techniques are diverse, and there are different approaches taken by different libraries to address similar problems.

This section comprises the following chapters:

1
A Lap around Automated Machine Learning

"All models are wrong, but some are useful."

– George Edward Pelham Box FRS

"One of the holy grails of machine learning is to automate more and more of the feature engineering process."

– Pedro Domingos, A Few Useful Things to Know about Machine Learning

This chapter will provide an overview of the concepts, tools, and technologies surrounding automated **Machine Learning** (**ML**). This introduction hopes to provide both a solid overview for novices and serve as a reference for experienced ML practitioners. We will start by introducing the ML development life cycle while navigating through the product ecosystem and the data science problems it addresses, before looking at feature selection, neural architecture search, and hyperparameter optimization.

It's very plausible that you are reading this book on an e-reader that's connected to a website that recommended this manuscript based on your reading interests. We live in a world today where your digital breadcrumbs give telltale signs of not only your reading interests, but where you like to eat, which friend you like most, where you will shop next, whether you will show up to your next appointment, and who you would vote for. In this age of big data, this raw data becomes information that, in turn, helps build knowledge and insights into so-called wisdom.

Artificial Intelligence (**AI**) and its underlying implementations of ML and deep learning help us not only find the metaphorical needle in the haystack, but also to see the underlying trends, seasonality, and patterns in these large data streams to make better predictions. In this book, we will cover one of the key emerging technologies in AI and ML; that is, **automated ML**, or **AutoML** for short.

In this chapter, we will cover the following topics:

- The ML development life cycle
- Automated ML
- How automated ML works
- Democratization of data science
- Debunking automated ML myths
- Automated ML ecosystem (open source and commercial)
- Automated ML challenges and limitations

Let's get started!

The ML development life cycle

Before introducing you to automated ML, we should first define how we operationalize and scale ML experiments into production. To go beyond Hello-World apps and *works-on-my-machine-in-my-Jupyter-notebook* kinds of projects, enterprises need to adapt a robust, reliable, and repeatable model development and deployment process. Just as in a **software development life cycle** (**SDLC**), the ML or data science life cycle is also a multi-stage, iterative process.

The life cycle includes several steps – the process of problem definition and analysis, building the hypothesis (unless you are doing exploratory data analysis), selecting business outcome metrices, exploring and preparing data, building and creating ML models, training those ML models, evaluating and deploying them, and maintaining the feedback loop:

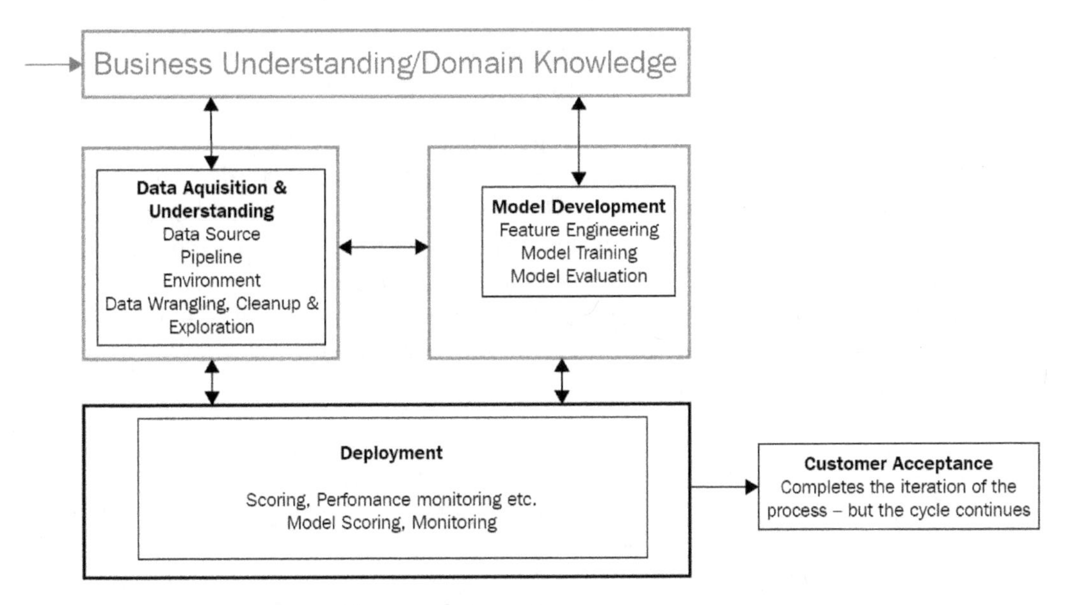

Figure 1.1 – Team data science process

A successful data science team has the discipline to prepare the problem statement and hypothesis, preprocess the data, select the appropriate features from the data based on the input of the **Subject-Matter Expert** (**SME**) and the right model family, optimize model hyperparameters, review outcomes and the resulting metrics, and finally fine-tune the models. If this sounds like a lot, remember that it is an iterative process where the data scientist also has to ensure that the data, model versioning, and drift are being addressed. They must also put guardrails in place to guarantee the model's performance is being monitored. Just to make this even more interesting, there are also frequent champion challenger and A/B experimentations happening in production – may the best model win.

In such an intricate and multifaceted environment, data scientists can use all the help they can get. Automated ML extends a helping hand with the promise to take care of the mundane, the repetitive, and the intellectually less efficient tasks so that the data scientists can focus on the important stuff.

Automated ML

"How many members of a certain demographic group does it take to perform a specified task?"

"A finite number: one to perform the task and the remainder to act in a manner stereotypical of the group in question." <insert your light bulb joke here>

This is meta humor – the finest type of humor for ensuing hilarity for those who are quantitatively inclined. Similarly, automated ML is a class of meta learning, also known as learning to learn – the idea that you can apply the automation principles to themselves to make the process of gaining insights even faster and more elegant.

Automated ML is the approach and underlying technology of applying certain automation techniques to accelerate the model's development life cycle. Automated ML enables citizen data scientists and domain experts to train ML models, and helps them build optimal solutions to ML problems. It provides a higher level of abstraction for finding out what the best model is, or an ensemble of models suitable for a specific problem. It assists data scientists by automating the mundane and repetitive tasks of feature engineering, including architecture search and hyperparameter optimization. The following diagram represents the ecosystem of automated ML:

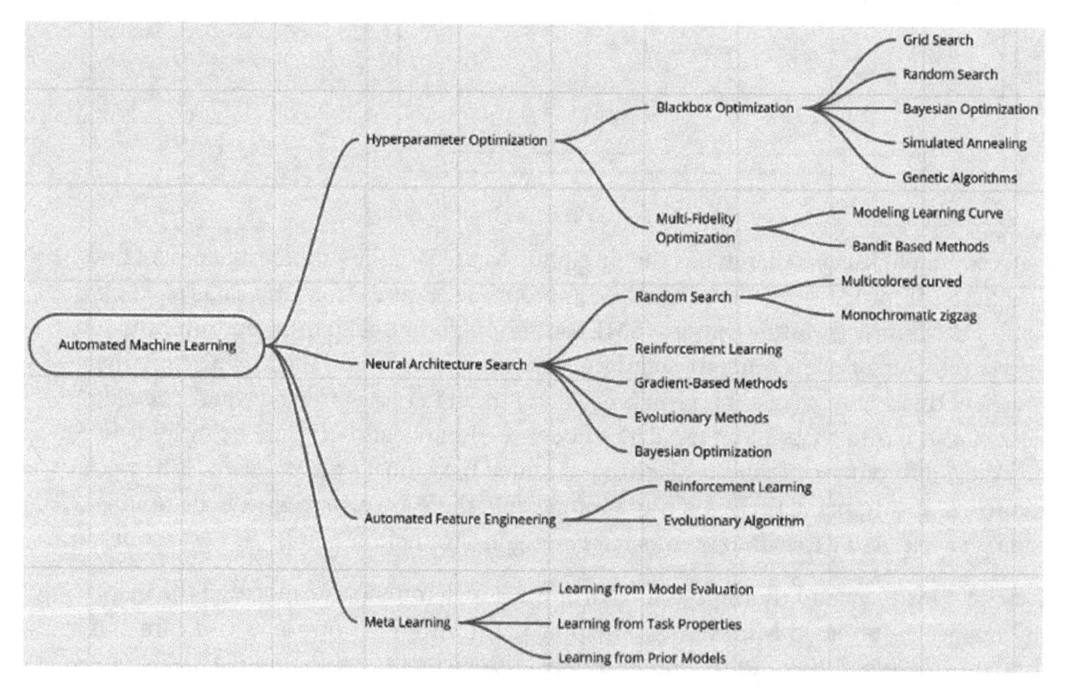

Figure 1.2 – Automated ML ecosystem

These three key areas – feature engineering, architecture search, and hyperparameter optimization – hold the most promise for the democratization of AI and ML. Some automated feature engineering techniques that are finding domain-specific usable features in datasets include expand/reduce, hierarchically organizing transformations, meta learning, and reinforcement learning. For architectural search (also known as neural architecture search), evolutionary algorithms, local search, meta learning, reinforcement learning, transfer learning, network morphism, and continuous optimization are employed.

Last, but not least, we have hyperparameter optimization, which is the art and science of finding the right type of parameters outside the model. A variety of techniques are used here, including Bayesian optimization, evolutionary algorithms, Lipchitz functions, local search, meta learning, particle swarm optimization, random search, and transfer learning, to name a few.

In the next section, we will provide a detailed overview of these three key areas of automated ML. You will see some examples of them, alongside code, in the upcoming chapters. Now, let's discuss how automated ML really works in detail by covering feature engineering, architecture search, and hyperparameter optimization.

How automated ML works

ML techniques work great when it comes to finding patterns in large datasets. Today, we use these techniques for anomaly detection, customer segmentation, customer churn analysis, demand forecasting, predictive maintenance, and pricing optimization, among hundreds of other use cases.

A typical ML life cycle is comprised of data collection, data wrangling, pipeline management, model retraining, and model deployment, during which data wrangling is typically the most time-consuming task.

Extracting meaningful features out of data, and then using them to build a model while finding the right algorithm and tuning the parameters, is also a very time-consuming process. Can we automate this process using the very thing we are trying to build here (meta enough?); that is, should we automate ML? Well, that is how this all started – with someone attempting to print a 3D printer using a 3D printer.

A typical data science workflow starts with a business problem (hopefully!), and it is used to either prove a hypothesis or to discover new patterns in the existing data. It requires data; the need to clean and preprocess the data, which takes an awfully large amount of time – almost as much as 80% of your total time; and "data munging" or wrangling, which includes cleaning, de-duplication, outlier analysis and removal, transforming, mapping, structuring, and enriching. Essentially, we're taming this unwieldy vivacious raw real-world data and putting it in a tame desired format for analysis and modeling so that we can gain meaningful insights from it.

Next, we must select and engineer features, which means figuring out what features are useful, and then brainstorming and working with **SMEs** on the importance and validity of these features. Validating how these features would work with your model, the fitness from both a technical and business perspective, and improving these features as needed is also a critical part of the feature engineering process. The feedback loop to the SME is often very important, albeit being the least emphasized part of the feature engineering pipeline. The transparency of models stems from clear features – if features such as race or gender give higher accuracy regarding your loan repayment propensity model, this does not mean it's a good idea use them. In fact, an SME would tell you – if your conscious mind hasn't – that it's a terrible idea and that you should look for more meaningful and less sexist, racist, and xenophobic features. We will discuss this further in *Chapter 10, AutoML in the Enterprise*, when we discuss operationalization.

Even though the task of "selecting a model family" sounds like a reality show, that is what data scientists and ML engineers do as part of their day-to-day job. Model selection is the task of picking the right model that best describes the data at hand. This involves selecting a ML model from a set of candidate models. Automated ML can give you a helping hand with this.

Hyperparameters

You will hear about hyperparameters a lot, so let's make sure you understand what they are.

Each model has its own internal and external parameters. Internal parameters (also known as model parameters, or just parameters) are the ones intrinsic to the model, such as its weight and predictor matrix, while external parameters or hyperparameters are "outside" the model itself, such as the learning rate and its number of iterations. An intuitive example can be derived from k-means, a well-understood unsupervised clustering algorithm known for its simplicity.

The *k* in k-means stands for the number of clusters required, and epochs (pronounced epics, as in *Doctor Who is an epic show!*) are used to specify the number of passes that are done over the training data. Both of these are examples of hyperparameters – that is, the parameters that are not intrinsic to the model itself. Similarly, the learning rate for training a neural network, C and sigma for support vector machines, the *k* number of leaves or depth of a tree, the latent factors in a matrix factorization, and the number of hidden layers in a deep neural network are all examples of hyperparameters.

Selecting the right hyperparameters has been called tuning your instrument, which is where the magic happens. In ML tribal folklore, these elusive numbers have been brandished as "*nuisance parameters*", to the point where proverbial statements such as "*tuning is more of an art than a science*" and "*tuning models is like black magic*" tend to discourage newcomers in the industry. Automated ML is here to change this perception by helping you choose the right hyperparameters; more on this later. Automated ML enables citizen data scientists to build, train, and deploy ML models, thus possibly disrupting the status quo.

> **Important note**
>
> Some consider the term "citizen data scientists" as a euphuism for non-experts, but SME and people who are curious about analytics are some of the most important people – and don't let anyone tell you otherwise.

In conclusion, from building the correct ensembles of models to preprocessing the data, selecting the right features and model family, choosing and optimizing model hyperparameters, and evaluating the results, automated ML offers algorithmic solutions that can programmatically address these challenges.

The need for automated ML

At the time of writing, Open AI's GPT-3 model has recently been announced, and it has an incredible 175 billion parameters. Due to this ever-increasing model complexity, which includes big data and an exponentially increasing number of features, we now have a necessity to not only be able to tune these parameters, but also have sophisticated, repeatable procedures in place to tweak these proverbial knobs so that they can be adjusted. This complexity makes it less accessible for citizen data scientists, business subject matter experts, and domain experts – which might sound like job security, but it is not good for business, nor for the long-term success of the field.

Also, this isn't just about the hyperparameters, but the entire pipeline and the reproducibility of the results becoming harder as the model's complexity grows, which curtails AI democratization.

Democratization of data science

To nobody's surprise, data scientists are in high demand! As a LinkedIn Workforce Report found in August 2018, there were more than 151,000 data scientist jobs going unfilled across the US (`https://economicgraph.linkedin.com/resources/linkedin-workforce-report-august-2018`). Due to this disparity in supply and demand, the notion of democratization of AI, which is enabling people who are not formally trained in math, statistics, computer science, and related quantitative fields to design, develop, and use predictive models, has become quite popular. There are arguments on both sides regarding whether an SME, a domain SME, a business executive, or a program manager can effectively work as a citizen data scientist – which I consider to be a layer of abstraction argument. For businesses to gain meaningful actionable insights in a timely manner, there is no other way than to accelerate the process of raw data to insight, and insights to action. It is quite evident to anyone who has served in the analytics trenches. This means that no citizen data scientists are left behind.

As disclaimers and caveats go, like everything else, automatic ML is not the proverbial silver bullet. However, automated methods for model selection and hyperparameter optimization bear the promise of enabling non-experts and citizen data scientists to train, test, and deploy high quality ML models. The tooling around automated ML is shaping up and hopefully, this gap will be reduced, allowing for increased participation. Now, let's review some of the myths surrounding automated ML and debunk them, MythBusters style!

Debunking automated ML myths

Much like the moon landing, when it comes to automated ML, there are more than a few conspiracy theories and myths surrounding it. Let's take a look at a few that have been debunked.

Myth #1 – The end of data scientists

One of the most frequently asked questions around automated ML is, *"Will automated ML be a job killer for data scientists?"*

The short answer is, not anytime soon – and the long answer, as always, is more nuanced and boring.

The data science life cycle, as we discussed previously, has several moving parts where domain expertise and subject matter insights are critical. The data scientists collaborate with businesses to build a hypothesis, analyze the results, and decide on any actionable insights that may create business impact. The act of automating mundane and repeatable tasks in data science, does not take away from the cognitively challenging task of discovering insights. If anything, instead of spending hours sifting through data and cleaning up features, it frees up data scientists to learn more about the underlying business. A large variety of real-world data science applications need dedicated human supervision, as well as the steady gaze of domain experts to ensure the fine-grained actions that come out of these insights reflect the desired outcome.

One of the proposed approaches, *A Human-in-the-Loop (HITL) Perspective on AutoML: Milestones and the Road Ahead* by *Doris Jung-Lin Lee et al.*, builds upon the notion of keeping humans in the loop. HITL suggests three different level of automation in data science workflows: user-driven, cruise control, and autopilot. As you progress through the maturity curve and the confidence of specific models increases, the user-driven flows move to cruise control and eventually to the autopilot stage. By leveraging different areas of expertise by building a talent pool, automated ML can help in multiple stages of the data science life cycle by engaging humans.

Myth #2 – Automated ML can only solve toy problems

This is a frequent argument from the skeptics of automated ML – that it can only be used to solve well-defined, controlled toy problems in data science and does not bode well for any real-world scenario.

The reality is quite the contrary – but I think the confusion arises from an incorrect assumption that we can just take a dataset, throw it to an automated ML model, and we will get meaningful insights. If we were to believe the hype around automated ML, then it should be able to look at messy data, perform a magical cleanup, figure out all the important features (including target variables), find the right model, tune its hyperparameters, and voila – it's built a magical pipeline!

Even though it does sound absurd when spoken out loud, this is exactly what you see in carefully crafted automated ML product demos. Then, there's the hype cycle, which has the opposite effect of diminishing the real value of automated ML offerings. The technical approaches powering automated ML are robust, and the academic rigor that's put into bringing these theories and techniques to life is like any other area of AI and ML.

In future chapters, we will look at several examples of hyperscalar platforms that benefit from automated ML, including – but not limited to – Google Cloud Platform, AWS, and Azure. These testimonials lead us to believe that real-world automated ML is not limited to eking out better accuracy in Kaggle championships, but rather poised to disrupt the industry in a big way.

Automated ML ecosystem

It almost feels redundant to point out that automated ML is a rapidly growing field; it's far from being commoditized – existing frameworks are constantly being evolved and new offerings and platforms are becoming mainstream. In the upcoming chapters, we will discuss some of these frameworks and libraries in detail. For now, we will provide you with a breadth-first introduction to get you acquainted with the automated ML ecosystem before we do a deep dive.

Open source platforms and tools

In this section, we will briefly review some of the open source automated ML platforms and tools that are available. We will deep dive into some of these platforms in *Chapter 3, Automated Machine Learning with Open Source Tools and Libraries.*

Microsoft NNI

Microsoft **Neural Network Intelligence** (**NNI**) is an open source platform that addresses the three key areas of any automated ML life cycle – automated feature engineering, architectural search (also referred to as **neural architectural search** or **NAS**), and **hyperparameter tunning** (**HPI**). The toolkit also offers model compression features and operationalization via **KubeFlow**, **Azure ML**, **DL Workspace** (**DLTS**), and **Kubernetes** over AWS.

The toolkit is available on GitHub to be downloaded: `https://github.com/microsoft/nni`.

auto-sklearn

Scikit-learn (also known as **sklearn**) is a popular ML library for Python development. As part of this ecosystem and based on *Efficient and Robust Automated ML* by *Feurer et al.*, **auto-sklearn** is an automated ML toolkit that performs algorithm selection and hyperparameter tuning using Bayesian optimization, meta-learning, and ensemble construction.

The toolkit is available on GitHub to be downloaded: `github.com/automl/auto-sklearn`.

Auto-Weka

Weka, short for **Waikato Environment for Knowledge** Analysis, is an open source ML library that provides a collection of visualization tools and algorithms for data analysis and predictive modeling. Auto-Weka is similar to **auto-sklearn** but is built on top of Weka and implements the approaches described in the paper for model selection, hyperparameter optimization, and more.

The developers describe Auto-WEKA as going beyond selecting a learning algorithm and setting its hyperparameters in isolation. Instead, it implements a fully automated approach. The author's intent is for Auto-WEKA "*to help non-expert users to more effectively identify ML algorithms*" – that is, democratization for SMEs – via "*hyperparameter settings appropriate to their applications*".

The toolkit is available on GitHub to be downloaded: `github.com/automl/autoweka`.

Auto-Keras

Keras is one of the most widely used deep learning frameworks and is an integral part of the TensorFlow 2.0 ecosystem. **Auto-Keras**, based on the paper by Jin et al., proposes that it is "*a novel method for efficient neural architecture search with network morphism, enabling Bayesian optimization*". This helps the neural architectural search "*by designing a neural network kernel and algorithm for optimizing acquisition functions in a tree-structured space*". Auto-Keras is the implementation of this deep learning architecture search via Bayesian optimization.

The toolkit is available on GitHub to be downloaded: `github.com/jhfjhfj1/autokeras`.

TPOT

The **Tree-based Pipeline Optimization Tool**, or **TPOT** for short (nice acronym, eh!), is a product of University of Pennsylvania, Computational Genetics Lab. TPOT is an automated ML tool written in Python. It helps build and optimize ML pipelines with genetic programming. Built on top of scikit-learn, TPOT helps automate feature selection, preprocessing, construction, model selection, and parameter optimization by "*exploring thousands of possible pipelines to find the best one*". It is just one of the many toolkits with a small learning curve.

The toolkit is available on GitHub to be downloaded: `github.com/EpistasisLab/tpot`.

Ludwig – a code-free AutoML toolbox

Uber's automated ML tool, Ludwig, is an open source deep learning toolbox used for experimentation, testing, and training ML models. Built on top of TensorFlow, Ludwig enables users to create model baselines and perform automated ML-style experiments with different network architectures and models. In its latest release (at the time of writing), **Ludwig** now integrates with **CometML** and supports **BERT** text encoders.

The toolkit is available on GitHub to be downloaded: `https://github.com/uber/ludwig`.

AutoGluon – an AutoML toolkit for deep learning

From AWS Labs, with the goal of democratization of ML in mind, **AutoGluon** has been developed to enable "*easy-to-use and easy-to-extend AutoML with a focus on deep learning and real-world applications spanning image, text, or tabular data*". AutoGluon, an integral part of AWS's automated ML strategy, enables both junior and seasoned data scientists to build deep learning models and end-to-end solutions with ease. Like other automated ML toolkits, AutoGluon offers network architecture search, model selection, and custom model improvements.

The toolkit is available on GitHub to be downloaded: `https://github.com/awslabs/autogluon`.

Featuretools

Featuretools is an excellent Python framework that helps with automated feature engineering by using deep feature synthesis. Feature engineering is a tough problem due to its very nuanced nature. However, this open source toolkit, with its excellent timestamp handling and reusable feature primitives, provides an excellent framework you can use to build and extract a combination of features and look at what impact they have.

The toolkit is available on GitHub to be downloaded: `https://github.com/FeatureLabs/featuretools/`.

H2O AutoML

H2O's AutoML provides an open source version of H2O's commercial product, with APIs in R, Python, and Scala. This is an open source, distributed (multi-core and multi-node) implementation for automated ML algorithms and supports basic data preparation via a mix of grid and random search.

The toolkit is available on GitHub to be downloaded: `github.com/h2oai/h2o-3`.

Commercial tools and platforms

Now, let's go through some commercial tools and platforms that are used for automated ML.

DataRobot

DataRobot is a proprietary platform for automated ML. As one of the leaders in the automated ML space, Data Robot claims to "*automate the end-to-end process for building, deploying, and maintaining AI at scale*". Data Robot's model repository contains open source as well as proprietary algorithms and approaches for data scientists, with a focus on business outcomes. Data Robot's offerings are available for both the cloud and on-premises implementations.

The platform can be accessed here: `https://www.datarobot.com/platform/`.

Google Cloud AutoML

Integrated in the **Google Cloud Compute platform**, the Google Cloud AutoML offering aims to help train high-quality custom ML models with minimal effort and ML expertise. This offering provides AutoML Vision, AutoML Video Intelligence, AutoML Natural Language, AutoML Translation, and AutoML Tables for structured data analysis. We will discuss this Google offering in more detail in *Chapter 8, Machine Learning with Google Cloud Platform,* and *Chapter 9, Automated Machine Learning with GCP Cloud AutoML* of this book.

Google Cloud AutoML can be accessed at `https://cloud.google.com/automl`.

Amazon SageMaker Autopilot

AWS offers a wide variety of capabilities around AI and ML. SageMaker Autopilot is among one of these offerings and helps to "*automatically build, train, and tune models*" as part of the AWS ecosystem. **SageMaker Autopilot** provides an end-to-end automated ML life cycle that includes automatic feature engineering, model and algorithm selection, model tuning, deployment, and ranking based on performance. We will discuss AWS SageMaker Autopilot in *Chapter 6, Machine Learning with Amazon Web Services,* and *Chapter 7, Doing Automated Machine Learning with Amazon SageMaker Autopilot.*

Amazon SageMaker Autopilot can be accessed at `https://aws.amazon.com/sagemaker/autopilot/`.

Azure Automated ML

Microsoft Azure provides automated ML capabilities to help data scientists build ML models with speed and at scale. The platform offers automated feature engineering capabilities such as missing value imputation, transformations and encodings, drop ping high cardinality, and no variance features. Azure's automated ML also supports time series forecasting, algorithm selection, hyperparameter tunning, guardrails to keep model bias in check, and a model leaderboard for ranking and scoring. We will discuss the Azure ML and AutoML offerings in *Chapter 4, Getting Started with Azure Machine Learning*, and *Chapter 5, Automated Machine Learning with Microsoft Azure.*

Azure's automated ML offering can be accessed at `https://azure.microsoft.com/en-us/services/machine-learning/automatedml/`.

H2O Driverless AI

H2O's open source offerings were discussed earlier in the *Open source platforms and books* section. The commercial offering of H2O Driverless AI is an automated ML platform that addresses the needs of feature engineering, architecture search, and pipeline generation. The "bring your own recipe" feature is unique (even though it's now being adapted by other vendors) and is used to integrate custom algorithms. The commercial product has extensive capabilities and a feature-rich user interface for data scientists to get up to speed.

H2O Driverless AI can be accessed at `https://www.h2o.ai/products/h2o-driverless-ai/`.

Other notable frameworks and tools in this space include **Autoxgboost**, **RapidMiner Auto Model**, **BigML**, **MLJar**, **MLBox**, **DATAIKU**, and **Salesforce Einstein** (powered by Transmogrif AI). The links to their toolkits can be found in this book's *Appendix*. The following table is from Mark Lin's Awesome AutoML repository and outlines some of the most important automated machine learning toolkits, along with their corresponding links:

Project	Type	License
Auto-Keras	NAS	Custom
AutoML Vision	NAS	Commercial
AutoML Video Intelligence	NAS	Commercial
AutoML Natural Language	NAS	Commercial
AutoML Translation	NAS	Commercial
AutoML Tables	AutoFE, HPO	Commercial
auto-sklearn	HPO	Custom
auto_ml	HPO	MIT
BayesianOptimization	HPO	MIT
comet	HPO	Commercial
DataRobot	HPO	Commercial
Driverless AI	AutoFE	Commercial
H2O AutoML	HPO	Apache-2.0
Katib	HPO	Apache-2.0
MLJAR	HPO	Commercial
NNI	HPO, NAS	MIT
TPOT	AutoFE, HPO	LGPL-3.0
TransmogrifAI	HPO	BSD-3-Clause
MLBox	AutoFE, HPO	BSD-3 License
AutoAI Watson	AutoFE, HPO	Commercial

Figure 1.3 – Automated ML projects from Awesome-AutoML-Papers by Mark Lin

The classification type column specifies whether the library supports **Network Architecture Search** (**NAS**), **Hyperparameter Optimization** (**HPO**), and **Automated Feature Engineering** (**AutoFE**).

The future of automated ML

As the industry makes significant investments in the area surrounding automated ML, it is poised to become an important part of our enterprise data science workflows, if it isn't already. Serving as a valuable assistant, this apprentice will help data scientists and knowledge workers focus on the business problem and take care of any thing unwieldy and trivial. Even though the current focus is limited to automated feature engineering, architecture search, and hyperparameter optimization, we will also see that meta-learning techniques will be introduced in other areas to help automate this automation process.

Due to the increasing demand of democratization of AI and ML, we will see automated ML become mainstream in the industry – with all the major tools and **hyperscaler** platforms providing it as an inherent part of their ML offerings. This next generation of automated ML equipped tools will allow us to perform data preparation, domain customized feature engineering, model selection and **counterfactual** analysis, **operationalization**, **explainability**, **monitoring**, and create **feedback loops**. This will make it easier for us to focus on what's important in the business, including business insights and impact.

The automated ML challenges and limitations

As we mentioned earlier, data scientists aren't getting replaced, and automated ML is not a job killer – for now. The job of data scientists will evolve as the toolsets and their functions continue to change.

The reasons for this are twofold. Firstly, automated ML does not automate data science as a discipline. It is definitely a time saver for performing automated feature engineering, architecture search, hyperparameter optimization, or running multiple experiments in parallel. However, there are various other essential parts of the data science life cycle that cannot be easily automated, thus providing the current state of automated ML.

The second key reason is that being a data scientist is not a homogenous role – the competencies and responsibilities related to it vary across the industry and organizations. In lieu of democratizing data science with automated ML, the so-called junior data scientists will gain assistance from automated feature engineering capabilities, and this will speed up their data munging and wrangling practices. Meanwhile, senior engineers will have more time to focus on improving their business outcomes by designing better KPI metrices and enhancing the model's performance. As you can see, this will help all tiers of data science practitioners gain familiarity with the business domain and explore any cross-cutting concerns. Senior data scientists also have the responsibility of monitoring model and data quality and drift, as well as maintaining versioning, auditability, governance, lineage, and other **MLOps** (**Machine Learning Operations**) cross-cutting concerns.

Enabling the explainability and transparency of models to address any underlying bias is also a critical component for regulated industries across the world. Due to its highly subjective nature, there is limited functionality to address this automatically in the current toolsets; this is where a socially aware data scientist can provide a tremendous amount of value to stop the perpetuation of algorithmic bias.

A Getting Started guide for enterprises

Congratulations! You have almost made it to the end of the first chapter without dozing off – kudos! Now, you must be wondering: this automated ML thing sounds rad, but how do I go about using it in my company? Here are some pointers.

First, read the rest of this book to familiarize yourself with the concepts, technology, tools, and platforms. It is important to understand the landscape and understand that automated ML is a tool in your data science toolkit – it does not replace your data scientists.

Second, use automated ML as a democratization tool across the enterprise when you're dealing with analytics. Build a training plan for your team to become familiar with the tools, provide guidance, and chart a path to automation in data science workflows.

Lastly, due to the large churn in the feature sets, start with a smaller footprint, probably with an open source stack, before you commit to an enterprise framework. Scaling up this way will help you understand your own automation needs and give you time to do comparison shopping.

Summary

In this chapter, we covered the ML development life cycle and then defined automated ML and how it works. While building a case for the need for automated ML, we discussed the democratization of data science, debunked the myths surrounding automated ML, and provided a detailed walk-through of the automated ML ecosystem. Here, we reviewed the open source tools and then explored the commercial landscape. Finally, we discussed the future of automated ML, commented on the challenges and limitations of it, and finally provided some pointers on how to get started in an enterprise.

In the next chapter, we'll look under the hood of the technologies, techniques, and tools that are used to make automated ML possible. We hope that this chapter has introduced you to the automated ML fundamentals and that you are now ready to do a deeper dive into the topics that we discussed.

Further reading

For more information on the topics that were covered in this chapter, please take a look at the following suggested books and links:

- *Automated ML: Methods, Systems, Challenges*, by Frank Hutter (Editor), Lars Kotthoff (Editor), and Joaquin Vanschoren (Editor)
- *The Springer Series on Challenges in ML*
- *Hands-On Automated ML: A beginner's guide to building automated ML systems using AutoML and Python*, bby Sibanjan Das and Umit Mert Cakmak, *Packt*
- *Auto XGBoost*: `https://github.com/ja-thomas/autoxgboost`
- *RapidMiner*: `https://rapidminer.com/products/auto-model/`
- *BigML*: `https://bigml.com/`
- *MLJar*: `https://mljar.com/`
- *MLBOX*: `https://github.com/AxeldeRomblay/MLBox`
- *DataIKU*: `https://www.dataiku.com/`
- *Awesome-AutoML-Papers by Mark Lin*: `https://github.com/hibayesian/awesome-automl-papers`
- *Auto-WEKA 2.0: Automatic model selection and hyperparameter optimization in WEKA*: `https://www.cs.ubc.ca/labs/beta/Projects/autoweka/`
- *Auto-Keras: An Efficient Neural Architecture Search System*: `https://arxiv.org/pdf/1806.10282.pdf`
- *A Human-in-the-loop Perspective on AutoML: Milestones and the Road Ahead. Doris Jung-Lin Lee et al.*: `dorisjunglinlee.com/files/MILE.pdf`
- *What is Data Wrangling and Why Does it Take So Long? by Mike Thurber*: `https://www.elderresearch.com/blog/what-is-data-wrangling`
- *Efficient and Robust Automated ML*: `http://papers.nips.cc/paper/5872-efficient-and-robust-automated-machine-learning.pdf`
- LinkedIn Workforce Report: `https://economicgraph.linkedin.com/resources/linkedin-workforce-report-august-2018`

2
Automated Machine Learning, Algorithms, and Techniques

"Machine intelligence is the last invention that humanity will ever need to make."

– Nick Bostrom

"The key to artificial intelligence has always been the representation."

– Jeff Hawkins

"By far, the greatest danger of artificial intelligence is that people conclude too early that they understand it."

– Eliezer Yudkowsky

Automating the automation sounds like one of those wonderful Zen meta ideas, but learning to learn is not without its challenges. In the last chapter, we covered the **Machine Learning** (**ML**) development life cycle, and defined automated ML, with a brief overview of how it works.

In this chapter, we will explore under-the-hood technologies, techniques, and tools used to make automated ML possible. Here, you will see how **AutoML** actually works, the algorithms and techniques of automated feature engineering, automated model and hyperparameter turning, and automated deep learning. You will learn about meta-learning as well as state-of-the-art techniques, including Bayesian optimization, reinforcement learning, evolutionary algorithms, and gradient-based approaches.

In this chapter, we will cover the following topics:

- Automated ML – Opening the hood
- Automated feature engineering
- Hyperparameter optimization
- Neural architecture search

Automated ML – Opening the hood

To oversimplify, a typical ML pipeline comprises data cleaning, feature selection, pre-processing, model development, deployment, and consumption steps, as seen in the following workflow:

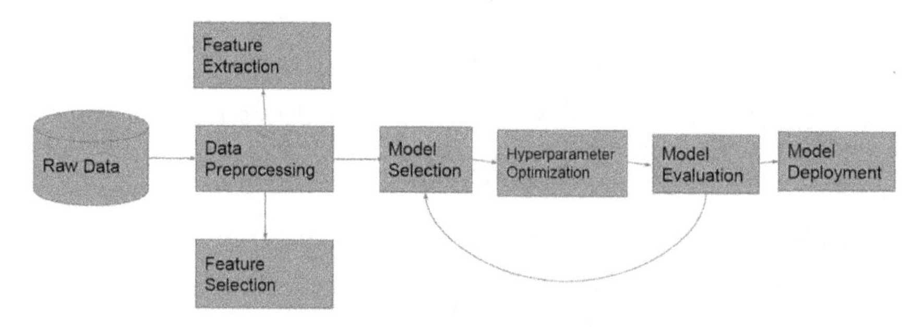

Figure 2.1 – The ML life cycle

The goal of automated ML is to simplify and democratize the steps of this pipeline so that it is accessible by citizen data scientists. Originally, the key focus of the automated ML community was model selection and hyperparameter tuning, that is, finding the best-performing model for the job and the corresponding parameters that work best for the problem. However, in recent years, it has been shifted to include the entire pipeline as shown in the following diagram:

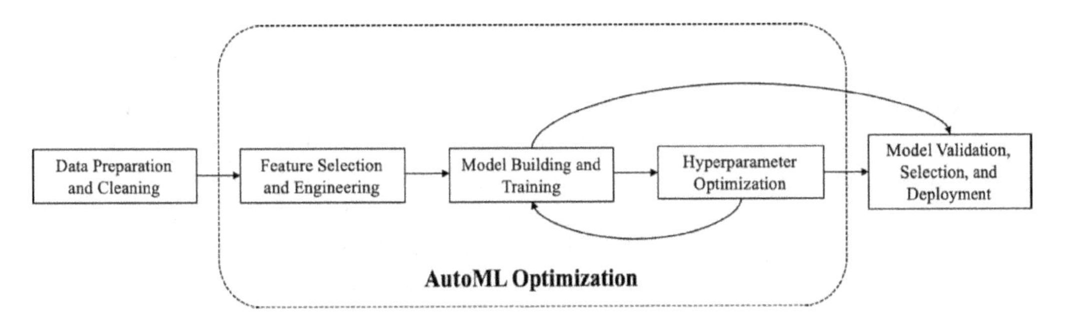

Figure 2.2 – A simplified AutoML pipeline by Waring et al.

The notion of meta-learning, that is, learning to learn, is an overarching theme in the automated ML landscape. Meta-learning techniques are used to learn optimal hyperparameters and architectures by observing learning algorithms, similar tasks, and those from previous models. Techniques such as learning task similarity, active testing, surrogate model transfer, Bayesian optimization, and stacking are used to learn these meta-features to improve the automated ML pipeline based on similar tasks; essentially, a warm start. The automated ML pipeline function does not really end at deployment – an iterative feedback loop is required to monitor the predictions that arise for drift and consistency. This feedback loop ensures that the outcome distribution of prediction matches the business metrics, and that there are anomalies in terms of hardware resource consumption. From an operational point of view, the logs of errors and warnings, including custom error logs, are audited and monitored in an automated manner. All these best practices also apply to the training cycle, where the concept drift, model drift, or data drift can wreak havoc on your predictions; heed the caveat emptor warning.

Now, let's explore some of the key automated ML terms you will see in this and future chapters.

The taxonomy of automated ML terms

For newcomers to automated ML, one of the biggest challenges is to become familiar with the industry jargon – large numbers of new or overlapping terminologies can overwhelm and discourage those exploring the automated ML landscape. Therefore, in this book we try to keep things simple and generalize as much as possible without losing any depth. You will repeatedly see in this book, as well as in other automated ML literature, the emphasis being placed on three key areas – namely, automated feature engineering, automated hyperparameter turning, and automated neural architecture search methods.

Automated feature engineering is further classified into feature extraction, selection, and generation or construction. Automated hyperparameter tuning, or the learning of hyperparameters for a specific model, sometimes gets bundled with learning the model itself, and hence becomes part of a larger neural architecture search area. This approach is known as the **Full Model Selection (FMS)** or **Combined Algorithm Selection and Hyperparameter (CASH)** optimization problem. Neural architecture search is also known as **automated deep learning** (abbreviated as **AutoDL**), or simply architecture search. The following diagram outlines how **data preparation**, **feature engineering**, **model generation**, and **evaluation**, along with their subcategories, become part of the larger ML pipeline:

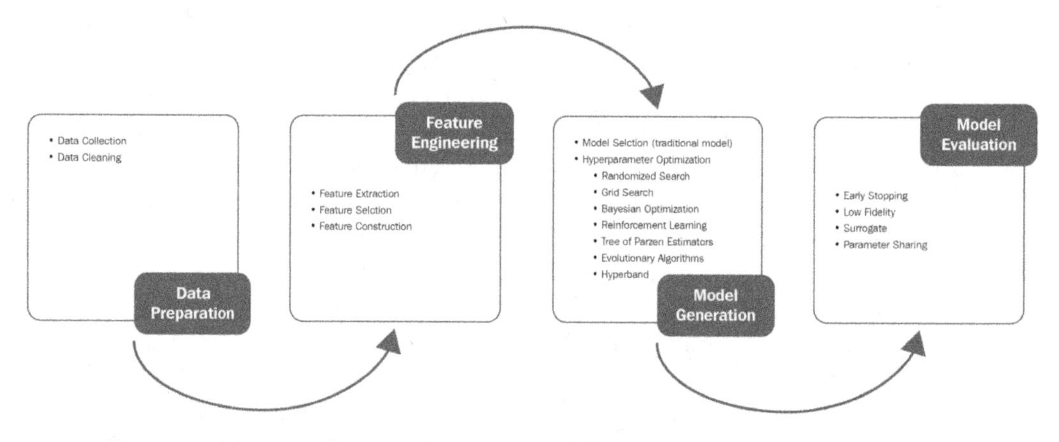

Figure 2.3 – Automated ML pipeline via state-of-the-art AutoML survey, He, et al., 2019

The techniques used to perform these three key tenets of automated ML have a few things in common. Bayesian optimization, reinforcement learning, evolutionary algorithms, gradient-free, and gradient-based approaches are used in almost all these different areas, with variations as shown in the following diagram:

	Bayesian Optimization	Reinforcement Learning	Evolutionary Algorithms	Gradient - Based Approaches	Frameworks
Automated Feature Engineering		FeatureRL	GP (Genetic Programming) for Feature Engineering		FeatureTools
Automated Model and Hyper Parameter Search	TPE - Tree of Parzen Estimators SMAC (Sequential Model-Based Optimization for General Algorithm Configuration) Auto-SKLearn FABOLAS Fast Bayesian Optimization of Machine Learning Hyperparameters on Large Datasets BOHB: Robust and Efficient Hyperparameter Optimization at Scale	APRL (Autonomous Predictive Modeler via Reinforcement Learning) Hyperband: A Novel Bandit-Based Approach to Hyperparameter Optimization	TPOT – Tree-based pipeline optimization. AutoStacker - Automatic Evolutionary Hierarchical Machine Learning System DarwinML - Graph-based Evolutionary Algorithm for Automated Machine Learning.		Hyperopt: Distributed Asynchronous Hyper-Parameter Optimization SMAC (Sequential Model-Based Optimization for General Algorithm Configuration) Auto-Sklearn TPOT – Tree based pipeline optimization.
Automated Deep Learning or Neural Architecture Search	AutoKeras NASBot	NAS – Neural Architecture Search NASNET (Neural Architecture Search Network) ENAS - Efficient Neural Architecture Search via Parameter Sharing		DARTS: Differentiable Architecture Search ProxylessNAS: Direct Neural Architecture Search on Target Task and Hardware NAONet (Neural Architecture Optimization NET)	AutoKeras AdaNet Neural Network Intelligence (NNI)

Figure 2.4 – Automated ML techniques

So, you may get perplexed looks if you refer to using genetic programming in automated feature engineering, while someone considers evolutionary hierarchical ML systems as a hyperparameter optimization algorithm. That is because you can apply the same class of techniques, such as reinforcement learning, evolutionary algorithms, gradient descent, or random search, to different parts of automated ML pipelines, and that works just fine.

We hope that the information provided between *Figure 2.2* and *Figure 2.4* help you to understand the relationship between ML pipelines, automated ML salient traits, and techniques/algorithms used to achieve those three key characteristic traits. The mental model you will build in this chapter will go a long way, especially when you encounter preposterous terms coined by marketing (yes Todd,
I am talking about you!), such as deep-learning-based-hyperparameter-optimization-product-with-bitcoins-and-hyperledger.

The next stop is automated feature engineering, the first pillar of the automated ML pipeline.

Automated feature engineering

Feature engineering is the art and science of extracting and selecting the right attributes from the dataset. It is an art because it not only requires subject matter expertise, but also domain knowledge and an understanding of ethical and social concerns. From a scientific perspective, the importance of a feature is highly correlated with its resulting impact on the outcome. Feature importance in predictive modeling measures how much a feature influences the target, hence making it easier in retrospect to assign ranking to attributes with the most impact. The following diagram explains how the iterative process of automated feature generation works, by generating candidate features, ranking them, and then selecting the specific ones to become part of the final feature set:

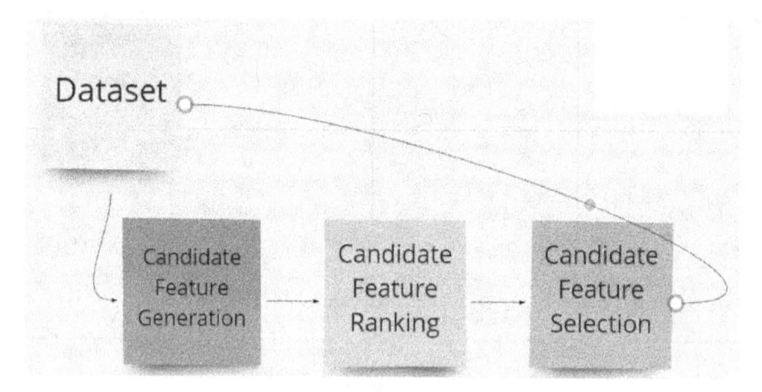

Figure 2.5 – Iterative feature generation process by Zoller et al. Benchmark and survey of automated ML frameworks, 2020

Extracting a feature from the dataset requires the generation of categorical binary features based on columns with multiple possible values, scaling the features, eliminating highly correlated features, adding feature interactions, substituting cyclic features, and handling data/time scenarios. Date fields, for instance, result in several features, such as year, month, day, season, weekend/weekday, holiday, and enrollment period. Once extracted, selecting a feature from a dataset requires the removal of sparse and low variance features, as well as applying dimensionality reduction techniques such as **Principal Component Analysis** (**PCA**) to make the number of features manageable. We will now investigate hyperparameter optimization, which used to be a synonym for automated ML, and is still a fundamental entity in the space.

Hyperparameter optimization

Due to its ubiquity and ease of framing, hyperparameter optimization is sometimes regarded as being synonymous with automated ML. Depending on the search space, if you include features, hyperparameter optimization, also dubbed hyperparameter tuning and hyperparameter learning, is known as automated pipeline learning. All these terms can be bit daunting for something as simple as finding the right parameters for a model, but graduating students must publish, and I digress.

There are a couple of key points regarding hyperparameters that are important to note as we look further into these constructs. It is well established that the default parameters are not optimized. Olson et al., in their NIH paper, demonstrated how the default parameters are almost always a bad idea. Olson mentions that "*Tuning often improves an algorithm's accuracy by 3–5%, depending on the algorithm.... In some cases, parameter tuning led to CV accuracy improvements of 50%.*" This was observed in *Cross-validation accuracy improvement – Data-driven advice for applying ML to bioinformatics problems*, by Olson et al.: `https://www.ncbi.nlm.nih.gov/pmc/articles/PMC5890912/`.

The second important point is that a comparative analysis of these models leads to greater accuracy; as you will see in forthcoming chapters, the entire pipeline (model, automated features, hyperparameters) are all key to getting the best accuracy trade-off. The *Heatmap for comparative analysis of algorithms* section in *Data-driven advice for applying ML to bioinformatics problems*, by Olson et al. (`https://www.ncbi.nlm.nih.gov/pmc/articles/PMC5890912/`) shows the experiment performed by Olson et al., where 165 datasets were used against multiple different algorithms to determine the best accuracy, ranked from top to bottom based on performance. The takeaway from this experiment is that no single algorithm can be considered best-performing across all the datasets. Therefore, there is a definite need to consider different ML algorithms when solving these data science problems.

Let's do a quick recap of what the hyperparameters are. Each model has its internal and external parameters. Internal parameters or model parameters are intrinsic to the model, such as weight or the predictor matrix, while external parameters also known as hyperparameters, are "outside" the model; for example learning rate and the number of iterations. For instance in k-means, k stands for the number of clusters required and epochs are used to specify the number of passes done over the training data. Both of these are examples of hyperparameters, that is, parameters that are not intrinsic to the model itself. Similarly, the learning rate for training a neural network, C and sigma for **Support Vector Machines** (**SVMs**), *k* number of leaves or depth of a tree, latent factors in a matrix factorization, the number of hidden layers in a deep neural network, and so on are all examples of hyperparameters.

To find the correct hyperparameters, there are a number of approaches, but first let's see what different types of hyperparameters there are. Hyperparameters can be continuous, for example:

- The learning rate of a model
- The number of hidden layers
- The number of iterations
- Batch size

Hyperparameters can also be categorical, for example, the type of operator, activation function, or the choice of algorithm. They can also be conditional, for example, selecting the convolutional kernel size if a convolutional layer is used, or the kernel width if a **Radial Basis Function** (**RBF**) kernel is selected in an SVM. Since there are multiple types of hyperparameters, there are also a variety of hyperparameter optimization techniques. Grid, random search, Bayesian optimization, evolutionary techniques, multi-arm bandit approaches, and gradient descent-based techniques are all used for hyperparameter optimization:

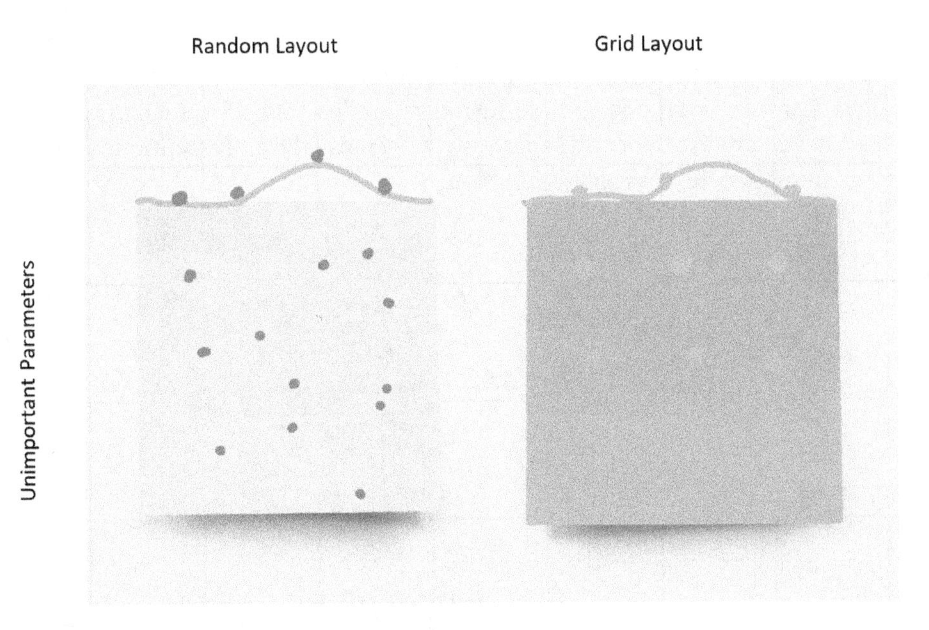

Figure 2.6 – Grid and random search layout. Bergstra and Bengio – JMLR 2012

The simplest techniques for hyperparameter tuning are manual, grid, and random search. Manual turning, as the name suggests, is based on intuition and guessing based on past experiences. Grid search and random search are slightly different, as you pick a set of hyperparameters either for each combination (grid), or randomly and iterate through to keep the best performing ones. However, as you can imagine, this can get computationally out of hand quickly as the search space gets bigger.

The other prominent technique is Bayesian optimization, in which you start with a random combination of hyperparameters and use it to construct a surrogate model. Then you use this surrogate model to predict how other combinations of hyperparameters would work. As a general principle, Bayesian optimization builds a probability model to minimize the objective's function, using past performance to select the future values, and that is exactly what's Bayesian about it. As known in the Bayesian universe, your observations are less important than your prior belief.

The greedy nature of Bayesian optimization is controlled by exploration and exploitation trade-off (expected improvement), allocating fixed-time evaluations, setting thresholds, and suchlike. There are variations of these surrogate models that exist, such as random forest surrogate and gradient boosting surrogate, which use the aforementioned techniques to minimize the surrogate's function:

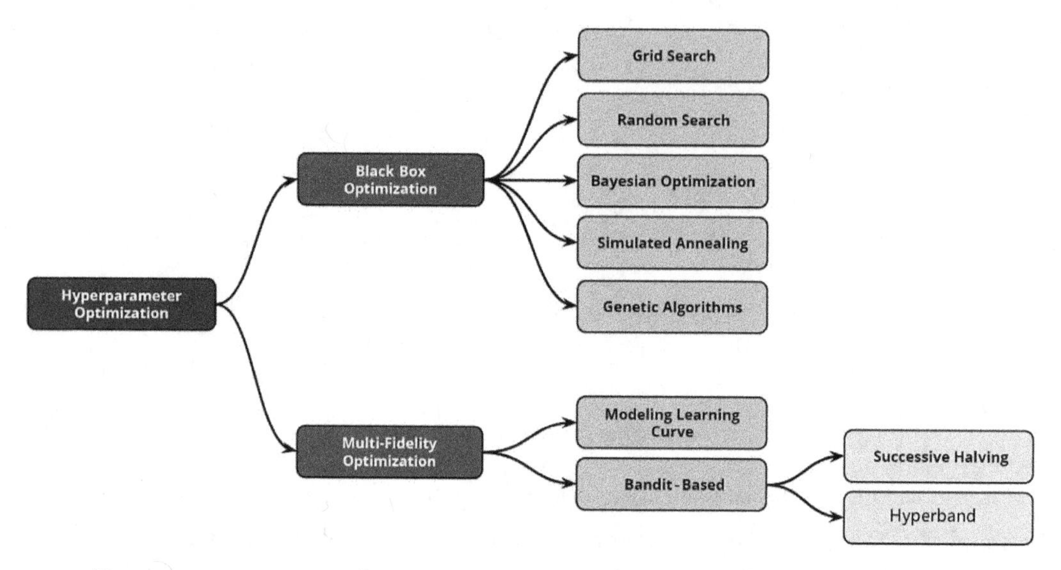

Figure 2.7 – A taxonomy of hyperparameter optimization techniques, Elshawi et al., 2019

The class of population-based methods (also called meta-heuristic techniques or optimization from samples methods) is also widely used to perform hyperparameter tuning, with genetic programming (evolutionary algorithms) being the most popular, where hyperparameters are added, mutated, selected, crossed over, and tuned. A particle swarm moves toward the best individual configurations when the configuration space is updated at each iteration. On the other hand, evolutionary algorithms work by maintaining a configuration space, and improve it by making smaller changes and combining individual solutions to build a new generation of hyperparameter configuration. Let's now explore the final piece of the automated ML puzzle – neural architecture search.

Neural architecture search

Selecting models can be challenging. In the case of regression, that is, predicting a numerical value, you have a choice of linear regression, decision trees, random forest, lasso versus ridge regression, k-means elastic net, gradient boosting methods, including **XGBoost**, and SVMs, among many others.

For classification, that in other words, separating out things by classes, you have **logistic regression**, **random forest**, **AdaBoost**, **gradient boost**, and **SVM-based classifiers** at your disposal.

Neural architecture has the notion of search space, which defines which architectures can be used in principle. Then, a search strategy must be defined that outlines how to explore using the exploration-exploitation trade-off. Finally, there has to be a performance estimation strategy, which estimates the candidate's performance. This includes training and validation of the architecture.

There are several techniques for performing the exploration of search space. The most common ones include chain structured neural networks, multi-branch networks, cell-based search, and optimizing approaches using existing architecture. Search strategies include random search, evolutionary approaches, Bayesian optimization, reinforcement learning, and gradient-free versus gradient-based optimization approaches, such as **Differentiable Architecture Search** (**DARTS**). The search strategy to hierarchically explore the architectural search spaces, using Monte Carlo tree search or hill climbing, is popular as it helps discover high-quality architectures by rapidly approaching better performing architectures. These are the gradient "free" methods. In gradient-based methods, the underlying assumption of a continuous search space facilitates DARTS, which, unlike traditional reinforcement learning or evolutionary search approaches, explores the search space using gradient descent. A visual taxonomy of neural architectural search can be seen in the following diagram:

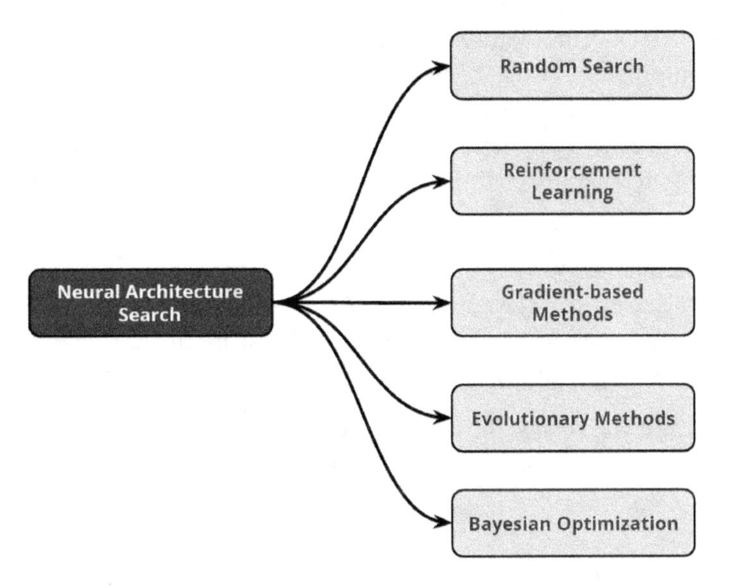

Figure 2.8 – A taxonomy of neural architecture search techniques, Elshawi et al., 2019

To evaluate which approach works best for the specific dataset, the performance estimation strategies have a spectrum of simple to more complex (albeit optimized) approaches. Simplest among the estimation strategies is to just train the candidate architecture and evaluate its performance on test data – if it works out, great. Otherwise, toss it out and try a different architectural combination. This approach can quickly become prohibitively resource-intensive as the number of candidate architectures grows; hence, the low-fidelity strategies, such as shorter training times, subset training, and fewer filters per layer are introduced, which are not nearly as exhaustive. Early stopping, in other words, estimating an architecture's performance by extrapolating its learning curve, is also a helpful optimization for such an approximation. Morphing a trained neural architecture, and one-short searches treating all architectures as a subgraph of a super graph, are also effective approaches as regards one-shot architecture search.

Several surveys have been conducted in relation to automated ML that provide an in-depth overview of these techniques. The specific techniques also have their own publications, with well-articulated benchmark data, challenges, and triumphs – all of which is beyond the scope of this manuscript. In the next chapter however, we will use the libraries that utilize these techniques, so you will get better hands-on exposure vis-à-vis their usability.

Summary

Today, the success of ML within an enterprise largely depends on human ML experts who can construct business-specific features and workflows. Automated ML aims to change this, as it aims to automate ML so as to provide off-the-shelf ML methods that can be utilized without expert knowledge. To understand how automated ML works, we need to review the underlying four subfields, or pillars, of automated ML: hyperparameter optimization; automated feature engineering; neural architecture search; and meta-learning.

In this chapter, we explained what is under the hood in terms of the technologies, techniques, and tools used to make automated ML possible. We hope that this chapter has introduced you to automated ML techniques and that you are now ready to do a deeper dive into the implementation phase.

In the next chapter, we will review the open source tools and libraries that implement these algorithms to get a hands-on overview of how to use these concepts in practice, so stay tuned.

Further reading

For more information on the following topics, refer to the suggested resources and links:

- *Automated ML: Methods, Systems, Challenges*: Frank Hutter (Editor), Lars Kotthoff (Editor), and Joaquin Vanschoren (Editor). The Springer Series on Challenges in ML
- *Hands-On Automated ML*: A Beginner's Guide to Building Automated ML Systems Using AutoML and Python, by Sibanjan Das and Umit Mert Cakmak, Packt
- *Neural Architecture Search with Reinforcement Learning*: `https://arxiv.org/pdf/1611.01578.pdf`
- *Learning Transferable Architectures for Scalable Image Recognition*: `https://arxiv.org/pdf/1707.07012.pdf`
- *Progressive Neural Architecture Search*: `https://arxiv.org/pdf/1712.00559.pdf`
- *Efficient Neural Architecture Search via Parameter Sharing*: `https://arxiv.org/pdf/1802.03268.pdf`
- *Efficient Architecture Search by Network Transformation*: `https://arxiv.org/pdf/1707.04873.pdf`
- *Network Morphism*: `https://arxiv.org/pdf/1603.01670.pdf`
- *Efficient Multi-Objective Neural Architecture Search via Lamarckian Evolution*: `https://arxiv.org/pdf/1804.09081.pdf`
- *Auto-Keras: An Efficient Neural Architecture Search System*: `https://arxiv.org/pdf/1806.10282.pdf`
- *Convolutional Neural Fabrics*: `https://arxiv.org/pdf/1606.02492.pdf`
- *DARTS: Differentiable Architecture Search*: `https://arxiv.org/pdf/1806.09055.pdf`
- *Neural Architecture Optimization*: `https://arxiv.org/pdf/1808.07233.pdf`
- *SMASH*: One-Shot Model Architecture Search through HyperNetworks: `https://arxiv.org/pdf/1708.05344.pdf`
- *DARTS in PyTorch*: `https://github.com/quark0/darts`

- Hyperparameter Tuning Using Simulated Annealing: `https://santhoshhari.github.io/2018/05/18/hyperparameter-tuning-using-simulated-annealing.html`
- Bayesian Optimization: `http://krasserm.github.io/2018/03/21/bayesian-optimization/`
- Neural Architecture Search: A Survey: `https://www.jmlr.org/papers/volume20/18-598/18-598.pdf`
- Data-driven advice for applying ML to bioinformatics problems: `https://www.ncbi.nlm.nih.gov/pmc/articles/PMC5890912/`

3

Automated Machine Learning with Open Source Tools and Libraries

"Empowerment of individuals is a key part of what makes open source work since, in the end, innovations tend to come from small groups, not from large, structured efforts."

– Tim O'Reilly

"In open source, we feel strongly that to really do something well, you have to get a lot of people involved."

– Linus Torvalds

In the previous chapter, you looked under the hood of automated **Machine Learning** (**ML**) technologies, techniques, and tools. You learned how AutoML actually works – that is, the algorithms and techniques of automated feature engineering, automated model and hyperparameter turning, and automated deep learning. You also explored Bayesian optimization, reinforcement learning, the evolutionary algorithm, and various gradient-based approaches by looking at their use in automated ML.

However, as a hands-on engineer, you probably don't get the satisfaction of understanding something fully until you get your hands dirty by trying it out. This chapter will give you the very opportunity to do this. AutoML **open source software** (**OSS**) tools and libraries automate the entire life cycle of ideating, conceptualizing, developing, and deploying predictive models. From data preparation through model training to validation as well as deployment, these tools do everything with almost zero human intervention.

In this chapter, we will review the major OSS tools, including **TPOT**, **AutoKeras**, **auto-sklearn**, **Featuretools**, and **Microsoft NNI**, to help you understand the differential value propositions and approaches that are used in each of these libraries.

In this chapter, we will cover the following topics:

- The open source ecosystem for AutoML
- Introducing TPOT
- Introducing Featuretools
- Introducing Microsoft NNI
- Introducing auto-sklearn
- Introducing AutoKeras

Let's get started!

Technical requirements

The technical requirements for this chapter are as follows:

- TPOT installation: `github.com/EpistasisLab/tpot`
- Featuretools installation: `https://pypi.org/project/featuretools/`
- Microsoft NNI installation: `https://github.com/microsoft/nni`
- auto-sklearn installation: `https://automl.github.io/auto-sklearn/master/installation.html`

- AutoKeras installation: `https://autokeras.com/install/`
- MNIST download: `https://www.kaggle.com/c/digit-recognizer`

The open source ecosystem for AutoML

By reviewing the history of automated ML, it is evident that, in the early days, the focus had always been on **hyperparameter** optimization. The earlier tools, such as **AutoWeka** and **HyperoptSkLearn**, and later **TPOT**, had an original focus on using Bayesian optimization techniques to find the most suitable **hyperparameters** for the model. However, this trend shifted left to include model selection, which eventually engulfed the entire pipeline by including feature selection, preprocessing, construction, and data cleaning. The following table shows some of the prominent automated ML tools that are available, including **TPOT**, **AutoKeras**, **auto-sklearn**, and **Featuretools**, along with their optimization techniques, ML tasks, and training frameworks:

	Language	Automated Machine Learning Technique	Automated Feature Extraction	Meta Learning	Link
AutoWeka	Java	Bayesian Optimization	Yes	No	https://github.com/automl/autoweka
AutoSklearn	Python	Bayesian Optimization	Yes	Yes	https://automl.github.io/auto-sklearn/master/
TPOT	Python	Genetic Algorithm	Yes	No	http://epistasislab.github.io/tpot/
Hyperopt-Sklearn	Python	Bayesian Optimization & Random Search	Yes	No	https://github.com/hyperopt/hyperopt-sklearn
AutoStacker	Python	Genetic Algorithm	Yes	No	https://arxiv.org/abs/1803.00684
AlphaD3M	Python	Reinforcement Learning	Yes	Yes	https://www.cs.columbia.edu/~idrori/AlphaD3M.pdf
OBOE	Python	Collaborative Filtering	No	Yes	https://github.com/udellgroup/oboe
PMF	Python	Collaborative Filtering & Bayesian Optimization	Yes	Yes	https://github.com/rsheth80/pmf-automl

Figure 3.1 – Features of automated ML frameworks

For several of the examples in this chapter, we will be using the **MNIST** database of handwritten digits. We will be using the **scikit-learn** `datasets` package since it has already taken care of data loading and preprocessing **MNIST** 60,000 training examples and 10,000 test examples. Most data scientists are ML enthusiasts and are very familiar with the **MNIST** database, which makes it a great candidate for teaching you how to use this library:

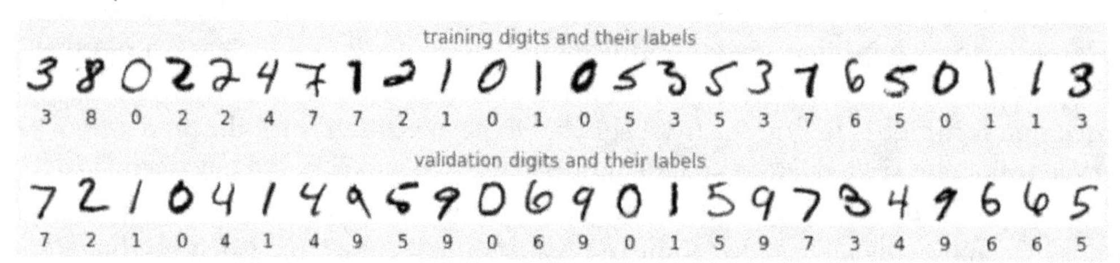

Figure 3.2 – MNIST database of handwritten digits – visualization

The preceding image shows what the MNIST dataset looks like. The dataset is available as part of all major ML and deep learning libraries, and can be downloaded from `https://www.kaggle.com/c/digit-recognizer`.

Introducing TPOT

The **Tree-based Pipeline Optimization Tool**, or **TPOT** for short, is a product of the University of Pennsylvania's, Computational Genetics Lab. TPOT is an automated ML tool written in Python. It helps build and optimize ML pipelines with genetic programming. Built on top of scikit-learn, TPOT helps automate the feature selection, preprocessing, construction, model selection, and parameter optimization processes by "*exploring thousands of possible pipelines to find the best one*". It is one of the only toolkits with a short learning curve.

The toolkit is available on GitHub to be downloaded: `github.com/EpistasisLab/tpot`.

To explain the framework, let's start with a minimal working example. For this example, we will be using the **MNIST** database of handwritten digits:

1. Create a new **Colab** notebook and run `pip install TPOT`. TPOT can be directly used from the command line or via Python code:

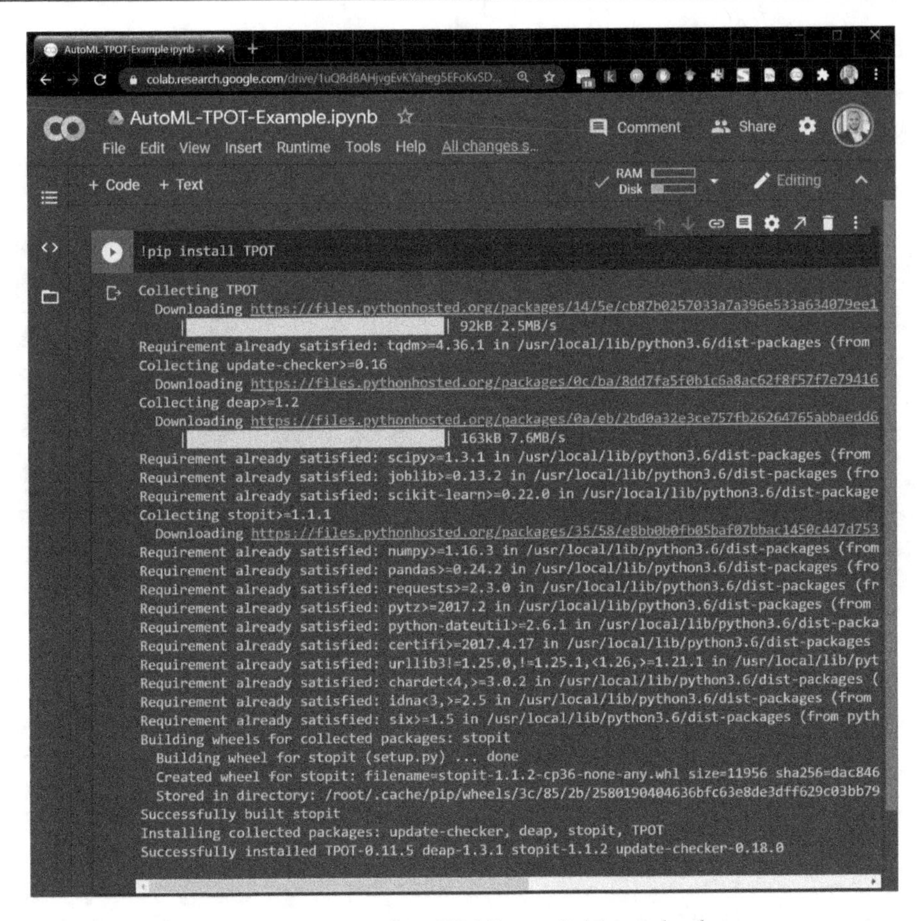

Figure 3.3 – Installing TPOT on a Colab notebook

2. Import `TPOTClassifier`, the scikit-learn `datasets` package, and the model selection libraries. We will use these libraries to load the data that we will be using for classification within TPOT:

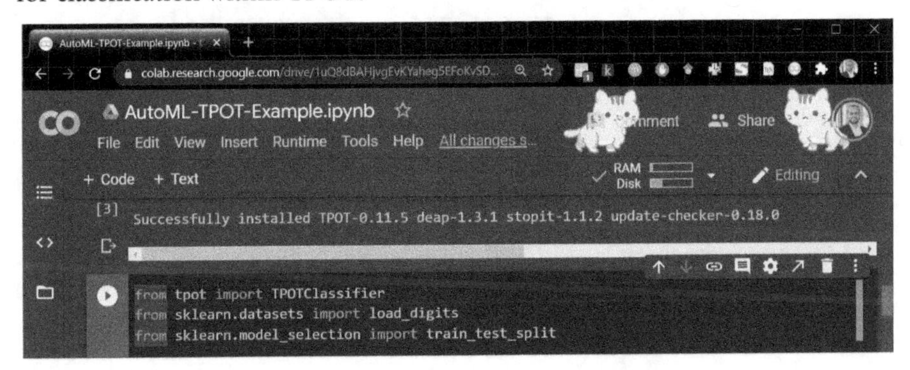

Figure 3.4 – AutoML TPOT example – import statement

3. Now, proceed by loading the **MNIST** digits dataset. The following `train_test_split` method returns a list containing a train-test split of given inputs. In this case, the inputs are digits data and digits target arrays. Here, you can see that the training size is **0.75** and that the test size is **0.25**, which signifies a standard 75-25 split in training and testing data:

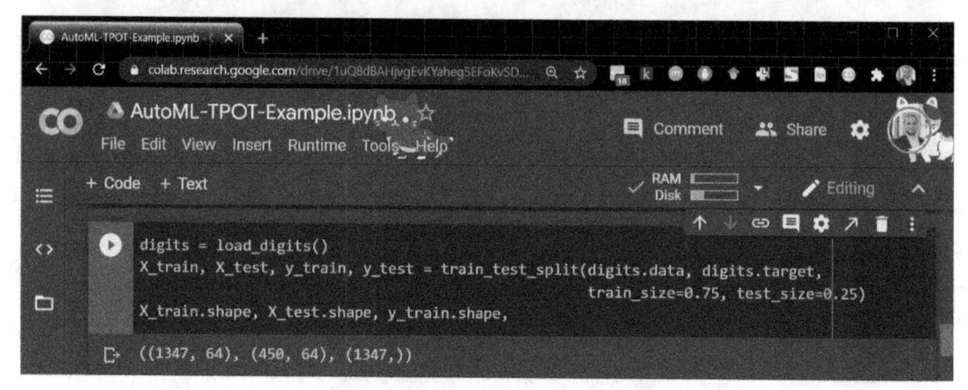

Figure 3.5 – AutoML TPOT example – loading the digits dataset

4. In a typical scenario, this is where we will choose a model, assign **hyperparameters**, and then try to fit it on the given data. However, since we are using automated ML as our virtual assistant, let's ask **TPOT** to do this for us. It's actually pretty easy.

   To find the right classifier for the job, you must instantiate a `TPOTClassifier`. This class is parametrically quite extensive, as shown in the following screenshot, but we will only be using three key parameters; that is, `verbosity`, `max_time_mins`, and `population_size`:

```
class tpot.TPOTClassifier(generations=100, population_size=100,
                          offspring_size=None, mutation_rate=0.9,
                          crossover_rate=0.1,
                          scoring='accuracy', cv=5,
                          subsample=1.0, n_jobs=1,
                          max_time_mins=None, max_eval_time_mins=5,
                          random_state=None, config_dict=None,
                          template=None,
                          warm_start=False,
                          memory=None,
                          use_dask=False,
                          periodic_checkpoint_folder=None,
                          early_stop=None,
                          verbosity=0,
                          disable_update_check=False,
                          log_file=None
                          )
```

Figure 3.6 – AutoML TPOT example – instantiating the TPOTClassifier object

A quick note about the arguments being passed to `Classifier` – setting `Verbosity` to `2` will make TPOT print information alongside a progress bar. The `max_time_mins` parameter sets the time allocation in minutes for TPOT to optimize the pipeline, while the `population_size` parameter is the number of individuals in the genetic programming population for every generation.

Upon starting the experiment, we will set the maximum time to only 1 minute:

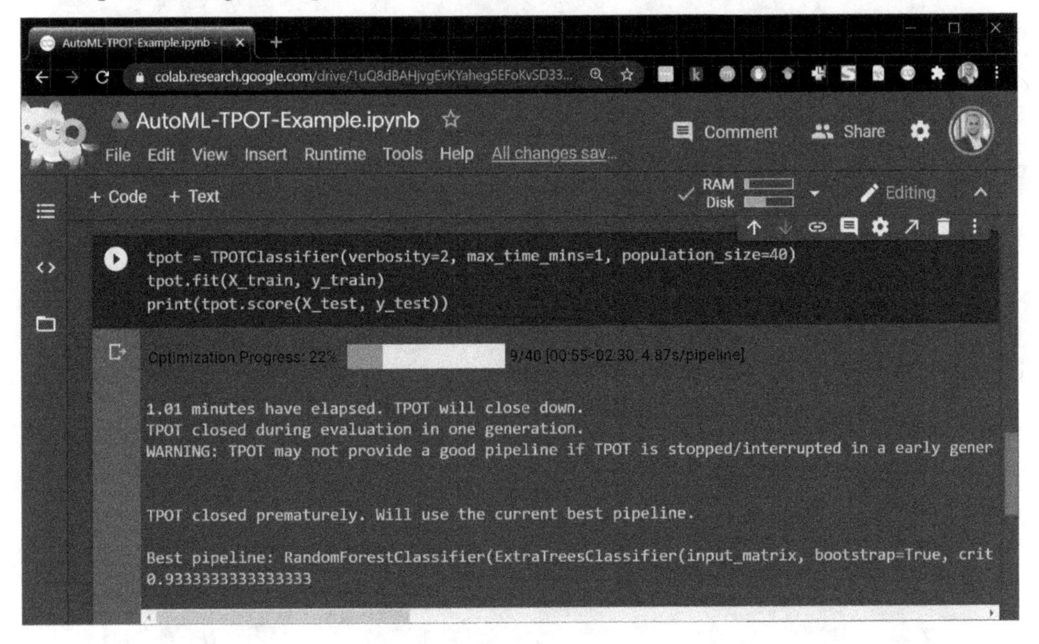

Figure 3.7 – AutoML TPOT example – optimization run of TPOTClassifier

You will see that the optimization progress isn't quite as good; it's at 22% since only 9 out of the 40 individuals have been processed in this generation. In this case, the best recommended pipeline is based on `RandomForestClassifier`.

5. Now, let's increase this to 5 minutes and check the resulting pipeline. At this point, it seems like the recommended classifier is the Gradient Boosting classifier. This is quite interesting:

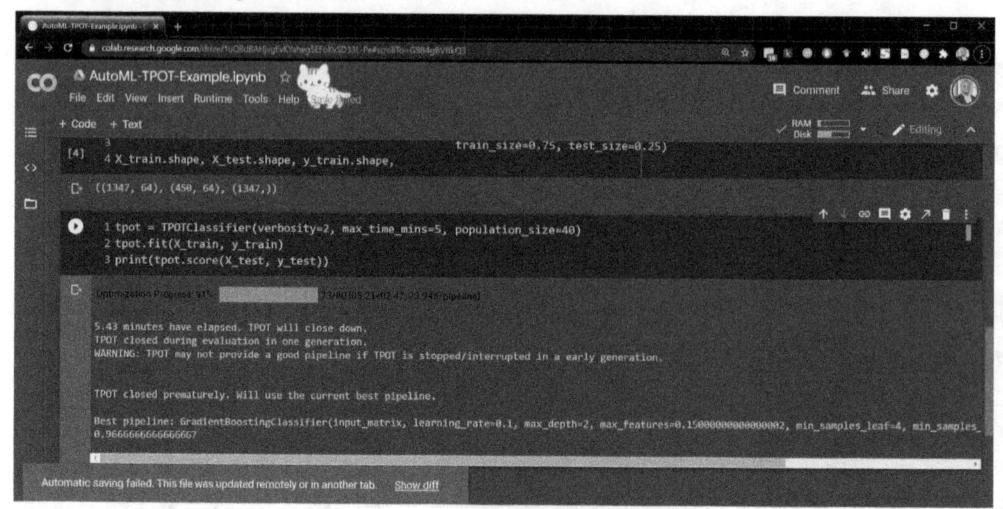

Figure 3.8 – AutoML TPOT example – executing TPOTClassifier

6. This time, we will gradually increase the time to `15` minutes, in which case the best pipeline will turn out to be from the **k-nearest neighbours** (KNN) classifier:

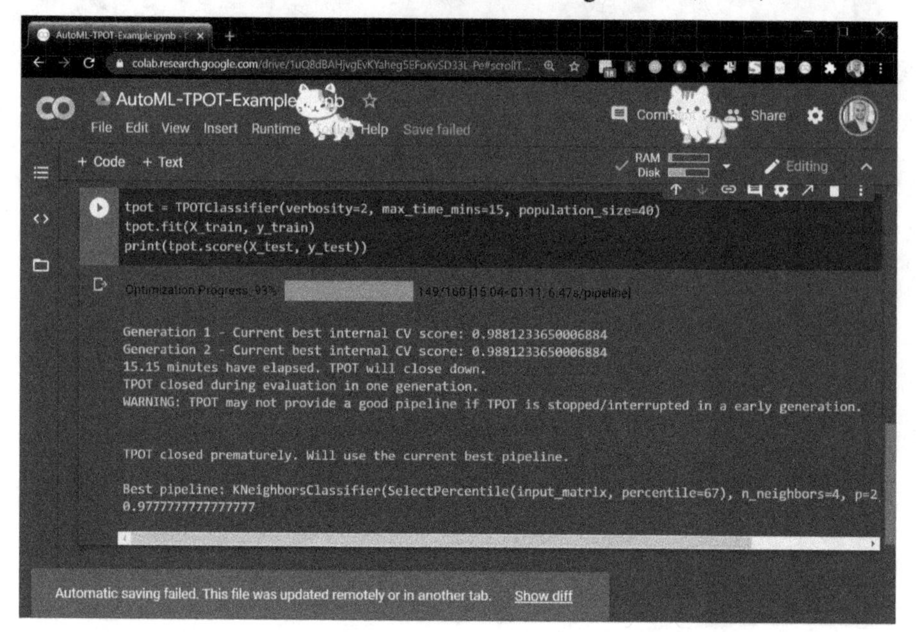

Figure 3.9 – AutoML TPOT classifier – TPOTClassifier fit to get the predictions

7. Increasing the time to `25` minutes does not change the algorithm, but other **hyperparameters** (number of neighbors) and their accuracy are increased:

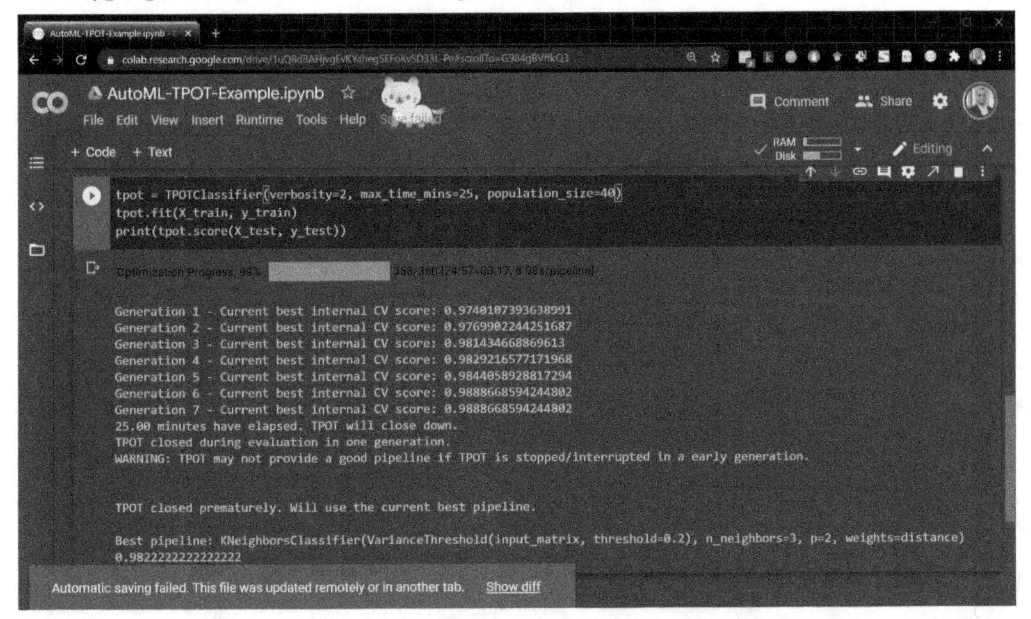

Figure 3.10 – AutoML TPOT example – running multiple generations and scores

8. Finally, let's run the experiment for an entire hour:

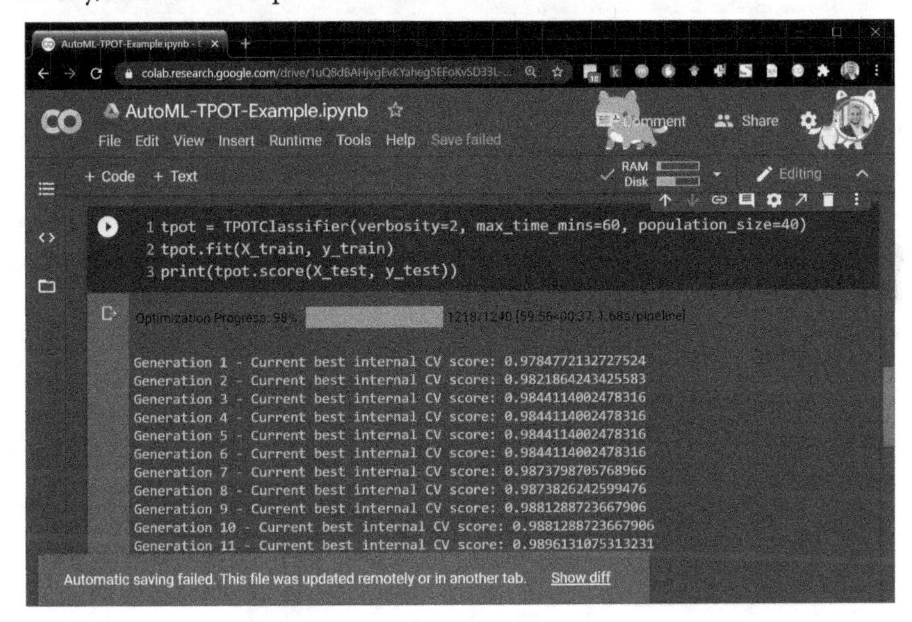

Figure 3.11 – AutoML TPOT example – TPOT generations and cross-validation scores

The resulting best pipeline is `KNeighborsClassifier` using feature ranking with recursive feature elimination. Other hyperparameters include `max_features` and `n_estimators`, and the pipeline has an accuracy of `0.98666`:

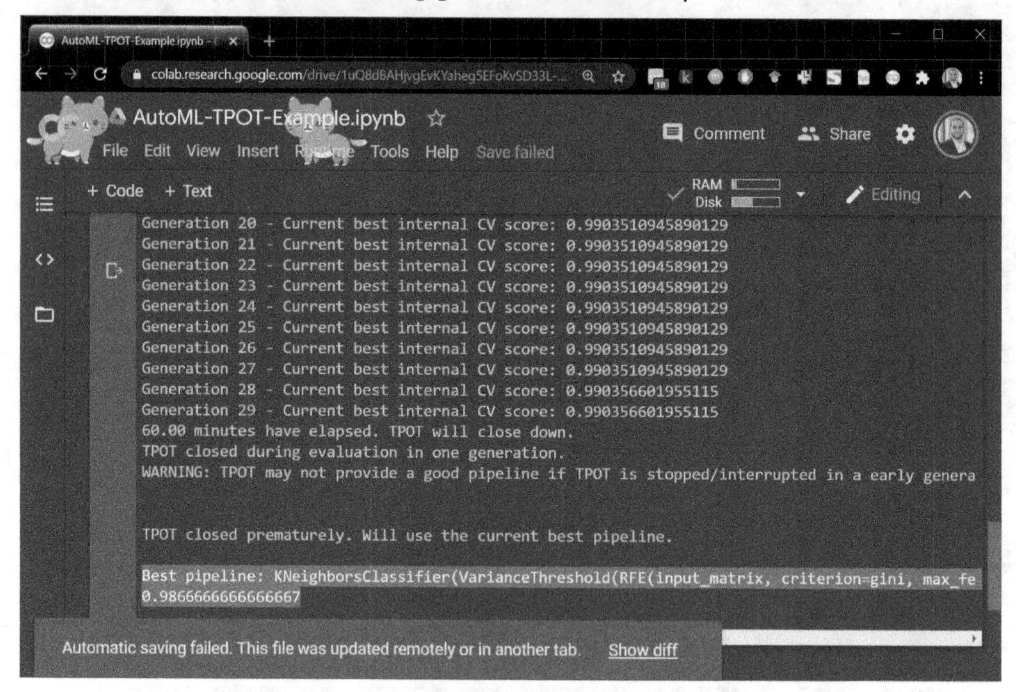

Figure 3.12 – AutoML TPOT example – the best pipeline

This reminds me – if 666 is considered an evil number, then 25.8069758011 is the root of all evil.

9. Also, as you have probably observed, the amount of time TPOT had to run its **cross-validation** (**CV**) for multiple generations, the pipeline changes, and not only the algorithm but the **hyperparameters** have evolved. There are also diminishing returns. The improvements in CV scores become smaller and smaller to where, at a certain point, these refinements doesn't make much difference.

Now, you can export the actual model from TPOT by calling the `export` method:

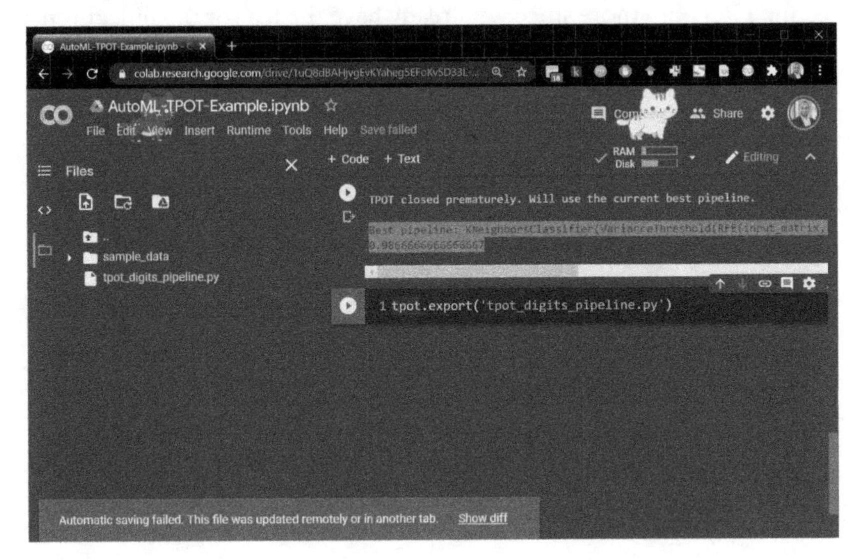

Figure 3.13 – AutoML TPOT example – exploring the digits pipeline

Once the model has been exported, you will be able to see the file in the left-hand pane of **Google Colab**, as shown in the following screenshot:

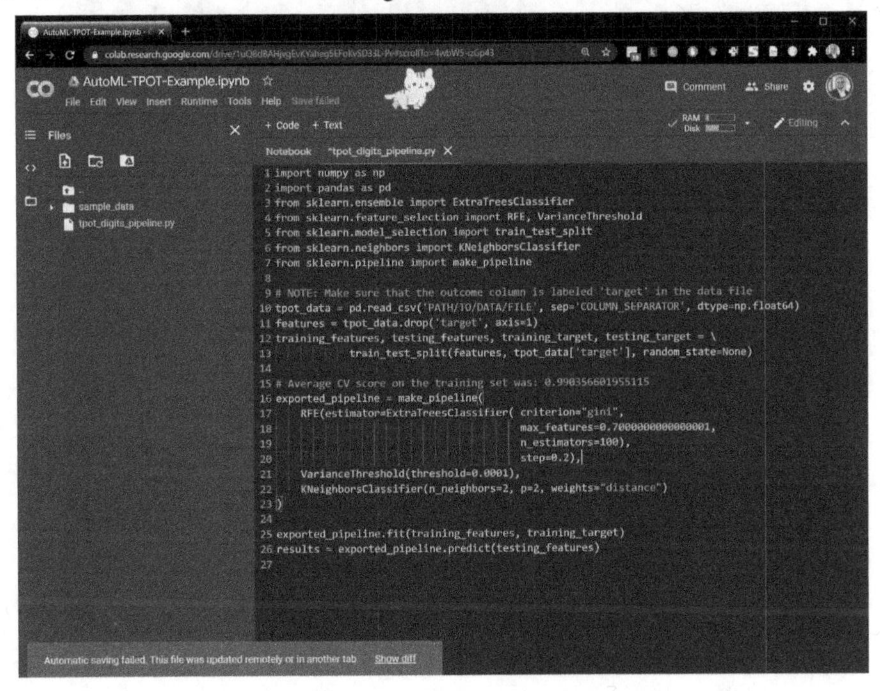

Figure 3.14 – AutoML TPOT example – visualizing the TPOT digits pipeline

Now that we know that this pipeline works the best, let's try this out. Notice how we don't have any need for TPOT anymore since we already have the tea (or pipeline, in this case):

```
import numpy as np
import pandas as pd
import numpy as np
from sklearn.ensemble import ExtraTreesClassifier
from sklearn.feature_selection import RFE, VarianceThreshold
from sklearn.model_selection import train_test_split
from sklearn.neighbors import KNeighborsClassifier
from sklearn.pipeline import make_pipeline
from sklearn.datasets import load_digits
from sklearn.externals import joblib

exported_pipeline = make_pipeline(
    RFE(estimator=ExtraTreesClassifier( criterion="gini",
                                        max_features=0.7000000000000001,
                                        n_estimators=100),
                                        step=0.2),
    VarianceThreshold(threshold=0.0001),
    KNeighborsClassifier(n_neighbors=2, p=2, weights="distance")
)
best_model = exported_pipeline._final_estimator
print("best model:\n", best_model)
```

Figure 3.15 – AutoML TPOT example – exporting the pipeline with ExtraTreesClassifier

Now that we've created the exported pipeline, let's load up the dataset. Instead of reading it from the CSV file, I can just use the `sklearn` datasets to expedite things. Also, I chose digit **1** here (`target [10]`) in the array, and voila – the prediction was right:

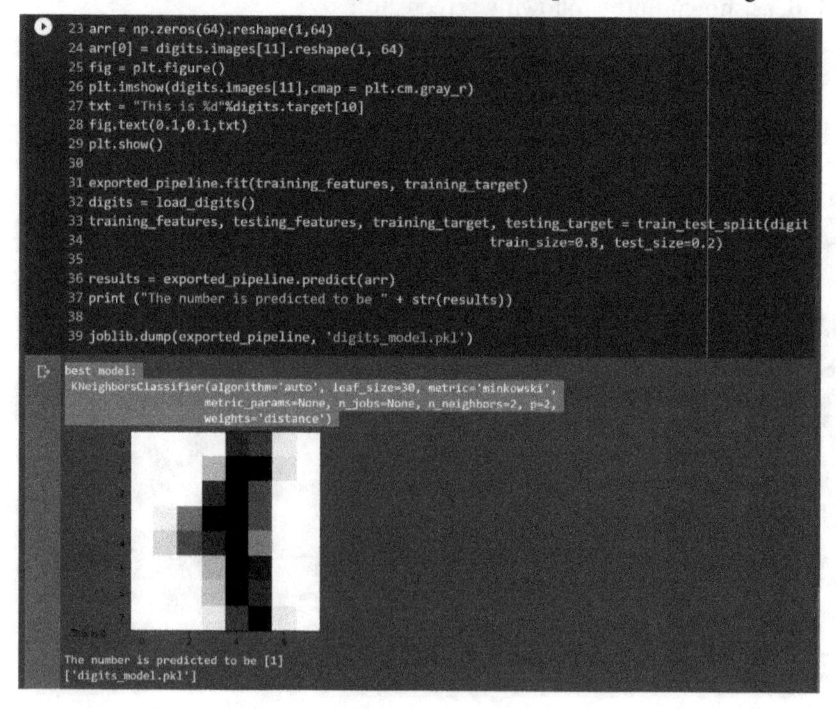

```
arr = np.zeros(64).reshape(1,64)
arr[0] = digits.images[11].reshape(1, 64)
fig = plt.figure()
plt.imshow(digits.images[11],cmap = plt.cm.gray_r)
txt = "This is %d"%digits.target[10]
fig.text(0.1,0.1,txt)
plt.show()

exported_pipeline.fit(training_features, training_target)
digits = load_digits()
training_features, testing_features, training_target, testing_target = train_test_split(digit
                                                        train_size=0.8, test_size=0.2)

results = exported_pipeline.predict(arr)
print ("The number is predicted to be " + str(results))

joblib.dump(exported_pipeline, 'digits_model.pkl')
```

```
best model:
 KNeighborsClassifier(algorithm='auto', leaf_size=30, metric='minkowski',
                     metric_params=None, n_jobs=None, n_neighbors=2, p=2,
                     weights='distance')
The number is predicted to be [1]
['digits_model.pkl']
```

Figure 3.16 – AutoML TPOT example – results from the exported pipeline

How does TPOT do this?

This looks great, but you didn't buy this book just to learn how to use an API - you want to understand a bit more about what is going on under the hood. Well, here is the scoop: TPOT has automated the key components of the pipeline using genetic programming; it tried different approaches, as you saw, and then eventually settled on using KNN as the best classifier:

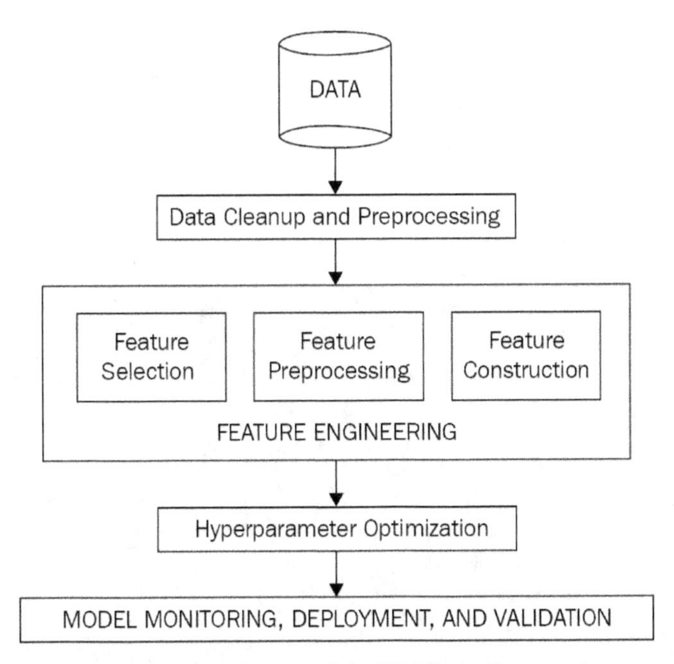

Figure 3.17 – Overview of the TPOT pipeline search

Behind the scenes, TPOT uses genetic programming constructs (selection, crossover, and mutation) to optimize transformation, which helps maximize classification accuracy. The following is a list of operators provided by TPOT:

Supervised Classification Operators	Feature Preprocessing Operators	Feature Selection Operators
Decision Tree, RandomForest, eXtreme Gradient Boosting Classifier, LogisticRegression, and KNearestNeighborClassifier.	StandardScaler, RobustScaler, MinMaxScaler, MaxAbsScaler, RandomizedPCA, Binarizer, and PolynomialFeatures.	VarianceThreshold, SelectKBest, SelectPercentile, SelectFwe, and Recursive Feature Elimination (RFE).
Classification operators store the classifier's predictions as a new feature as well as the classification for the pipeline.	Preprocessing operators modify the dataset in some way and return the modified dataset.	Feature selection operators reduce the number of features in the data set using some criteria and return the modified dataset.

Figure 3.18 – TPOT – a tree-based pipeline optimization tool for automating ML

Introducing Featuretools

Featuretools is an excellent Python framework that helps with automated feature engineering by using DFS. Feature engineering is a tough problem due to its very nuanced nature. However, this open source toolkit, with its robust timestamp handling and reusable feature primitives, provides a proper framework for us to build and extract combinations of features and their impact.

The toolkit is available on GitHub to be downloaded: `https://github.com/FeatureLabs/featuretools/`. The following steps will guide you through how to install Featuretools, as well as how to run an automated ML experiment using the library. Let's get started:

1. To start Featuretools in Colab, you will need to use pip to install the package. In this example, we will try to create features for the Boston Housing Prices dataset:

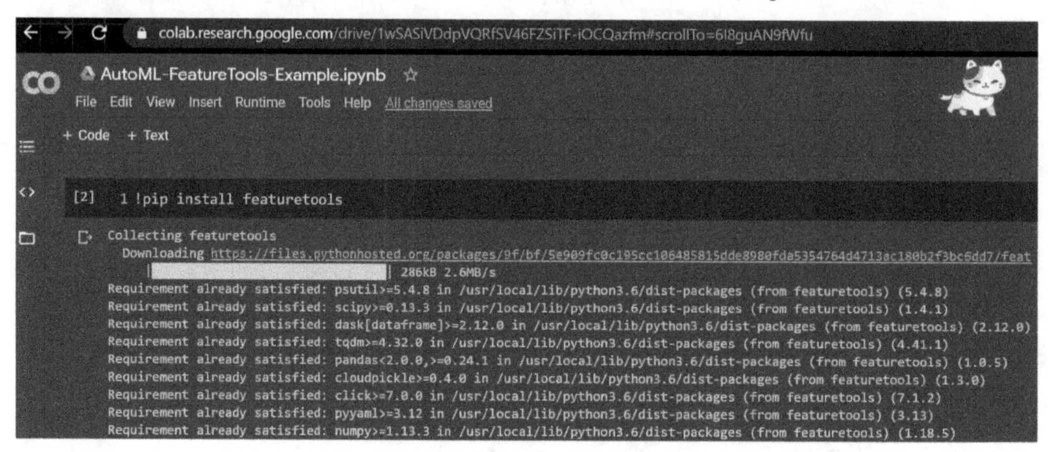

Figure 3.19 – AutoML with Featuretools – installing Featuretools

In this experiment, we will be using the Boston Housing Prices dataset, which is a well-known and widely used dataset in ML. The following is a brief description and the metadata of the dataset:

7.2.1. Boston house prices dataset

Data Set Characteristics:

Number of Instances:	506
Number of Attributes:	13 numeric/categorical predictive. Median Value (attribute 14) is usually the target.
Attribute Information (in order):	• CRIM per capita crime rate by town • ZN proportion of residential land zoned for lots over 25,000 sq.ft. • INDUS proportion of non-retail business acres per town • CHAS Charles River dummy variable (= 1 if tract bounds river; 0 otherwise) • NOX nitric oxides concentration (parts per 10 million) • RM average number of rooms per dwelling • AGE proportion of owner-occupied units built prior to 1940 • DIS weighted distances to five Boston employment centres • RAD index of accessibility to radial highways • TAX full-value property-tax rate per $10,000 • PTRATIO pupil-teacher ratio by town • B 1000(Bk - 0.63)^2 where Bk is the proportion of blacks by town • LSTAT % lower status of the population • MEDV Median value of owner-occupied homes in $1000's
Missing Attribute Values:	None
Creator:	Harrison, D. and Rubinfeld, D.L.

This is a copy of UCI ML housing dataset. https://archive.ics.uci.edu/ml/machine-learning-databases/housing/

Figure 3.20 – AutoML with Featuretools – Boston Housing Prices dataset

2. The Boston Housing Prices dataset is part of the `scikit-learn` dataset, which makes it very easy to import, as shown here:

```
from sklearn.datasets import load_boston
import pandas as pd
import featuretools as ft
```

```
# Load data and put into dataframe
boston = load_boston()
df = pd.DataFrame(boston.data, columns = boston.feature_names)
df['MEDV'] = boston.target
print (df.head(5))
```

```
      CRIM    ZN  INDUS  CHAS    NOX  ...    TAX  PTRATIO       B  LSTAT  MEDV
0  0.00632  18.0   2.31   0.0  0.538  ...  296.0     15.3  396.90   4.98  24.0
1  0.02731   0.0   7.07   0.0  0.469  ...  242.0     17.8  396.90   9.14  21.6
2  0.02729   0.0   7.07   0.0  0.469  ...  242.0     17.8  392.83   4.03  34.7
3  0.03237   0.0   2.18   0.0  0.458  ...  222.0     18.7  394.63   2.94  33.4
4  0.06905   0.0   2.18   0.0  0.458  ...  222.0     18.7  396.90   5.33  36.2

[5 rows x 14 columns]
```

Figure 3.21 – AutoML with Featuretools – installing Featuretools

3. Now, we will use **Featuretools** to build features. **Featuretools** helps us build new features by using the existing features and applying different operations to them. You can also link multiple tables and build relationships, but first, we would like to see it working on a single table. The following code shows how easily you can create an entity set (`boston`) using the `featuretools` **Deep Feature Synthesis** (**DFS**) API:

```
# Make an entityset and add the entity
es = ft.EntitySet(id = 'boston')
es.entity_from_dataframe(entity_id = 'data', dataframe = df,
                         make_index = True, index = 'index')

# Run deep feature synthesis with transformation primitives
feature_matrix, feature_defs = ft.dfs(entityset = es, target_entity = 'data',
                                      trans_primitives = ['add_numeric', 'multiply_numeric'])
```

Figure 3.22 – AutoML with Featuretools – loading the dataset as a pandas DataFrame

4. Let's create a feature tools entity set for the Boston table, and then define the target entries. In this case, we will just create some new features; that is, the products and the sum of existing features. Once **Featuretools** has run the DFS, you will have all the summation and product features:

```
                                      trans_primitives = ['add_numeric', 'multiply_numeric'])

feature_matrix.head()
```

index	CRIM	ZN	INDUS	CHAS	NOX	RM	AGE	DIS	RAD	TAX	PTRATIO	B	LSTAT	MEDV	AGE + B	AGE + CHAS	AGE + CRIM	AGE + DIS	AGE + INDUS	AGE + LSTAT	AGE + MEDV	AGE + NOX	AGE + PTRATIO	AGE + RAD	AGE + RM	AGE + TAX	AGE + ZN	B + CHAS	B + CRIM	B + DIS	B + INDUS
0	0.00632	18.0	2.31	0.0	0.538	6.575	65.2	4.0900	1.0	296.0	15.3	396.90	4.98	24.0	462.10	65.2	65.20632	69.2900	67.51	70.18	89.2	65.738	80.5	66.2	71.775	361.2	83.2	396.90	396.90632	400.9900	399.21
1	0.02731	0.0	7.07	0.0	0.469	6.421	78.9	4.9671	2.0	242.0	17.8	396.90	9.14	21.6	475.80	78.9	78.92731	83.8671	85.97	88.04	100.5	79.369	96.7	80.9	85.321	320.9	78.9	396.90	396.92731	401.8671	403.97
2	0.02729	0.0	7.07	0.0	0.469	7.185	61.1	4.9671	2.0	242.0	17.8	392.83	4.03	34.7	453.93	61.1	61.12729	66.0671	68.17	65.13	95.8	61.569	78.9	63.1	68.285	303.1	61.1	392.83	392.85729	397.7971	399.90
3	0.03237	0.0	2.18	0.0	0.458	6.998	45.8	6.0622	3.0	222.0	18.7	394.63	2.94	33.4	440.43	45.8	45.83237	51.8622	47.98	48.74	79.2	46.258	64.5	48.8	52.798	267.8	45.8	394.63	394.66237	400.6922	396.81
4	0.06905	0.0	2.18	0.0	0.458	7.147	54.2	6.0622	3.0	222.0	18.7	396.90	5.33	36.2	451.10	54.2	54.26905	60.2622	56.38	59.53	90.4	54.658	72.9	57.2	61.347	276.2	54.2	396.90	396.96905	402.9622	399.06

5 rows × 196 columns

Figure 3.23 – AutoML with Featuretools – results of DFS

The list of features continues:

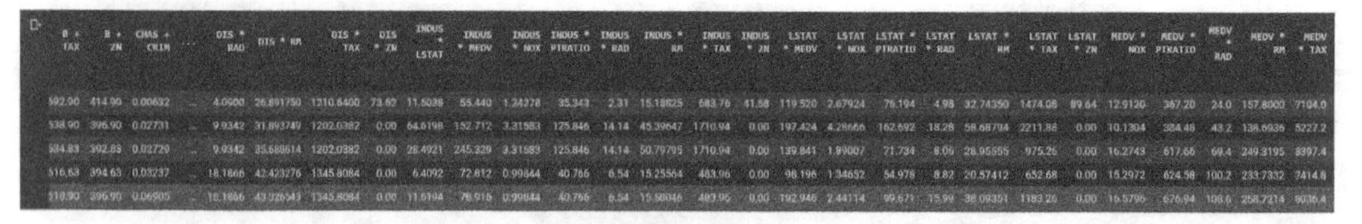

B + TAX	B + ZN	CHAS + CRIM	...	DIS * RAD	DIS * RM	DIS * TAX	DIS * ZN	INDUS * LSTAT	INDUS * MEDV	INDUS * NOX	INDUS * PTRATIO	INDUS * RAD	INDUS * RM	INDUS * TAX	INDUS * ZN	LSTAT * MEDV	LSTAT * NOX	LSTAT * PTRATIO	LSTAT * RAD	LSTAT * RM	LSTAT * TAX	LSTAT * ZN	MEDV * NOX	MEDV * PTRATIO	MEDV * RAD	MEDV * RM	MEDV * TAX
692.90	414.90	0.00632	...	4.0900	26.891750	1210.6400	73.62	11.5038	55.440	1.24278	35.343	2.31	15.18825	683.76	41.58	119.520	2.67924	76.194	4.98	32.74350	1474.08	89.64	12.9120	367.20	24.0	157.8000	7104.0
638.90	396.90	0.02731	...	9.9342	31.893749	1202.0382	0.00	64.6198	152.712	3.31583	125.846	14.14	45.39647	1710.94	0.00	197.424	4.28666	162.692	18.28	58.68794	2211.88	0.00	10.1304	384.48	43.2	138.6936	5227.2
634.83	392.83	0.02729	...	9.9342	35.688614	1202.0382	0.00	28.4921	245.329	3.31583	125.846	14.14	50.79795	1710.94	0.00	139.841	1.89007	71.734	8.06	28.95555	975.26	0.00	16.2743	617.66	69.4	249.3195	8397.4
616.63	394.63	0.03237	...	18.1866	42.423276	1345.8084	0.00	6.4092	72.812	0.99844	40.766	6.54	15.25564	483.96	0.00	98.196	1.34652	54.978	8.82	20.57412	652.68	0.00	15.2972	624.58	100.2	233.7332	7414.8
618.90	396.90	0.06905	...	18.1866	43.326543	1345.8084	0.00	11.6194	78.916	0.99844	40.766	6.54	15.58046	483.96	0.00	192.946	2.44114	99.673	15.99	38.09351	1183.26	0.00	16.5798	676.94	108.6	258.7214	8036.4

Figure 3.24 – AutoML with Featuretools – results of DFS – continued

At this point, you might be wondering, what is the point of doing DFS if it just contains the sums and products of existing features? I'm glad you asked. Think of these derived features as highlighting the latent relationships between multiple data points – and it's not related to sum and product. For example, you can link multiple tables with average order summation, and the algorithms will have additional pre-defined features to work with to find correlations. This is a very strong and significant quantitative value proposition that's provided by DFS, and it is typically used in machine learning algorithmic competitions:

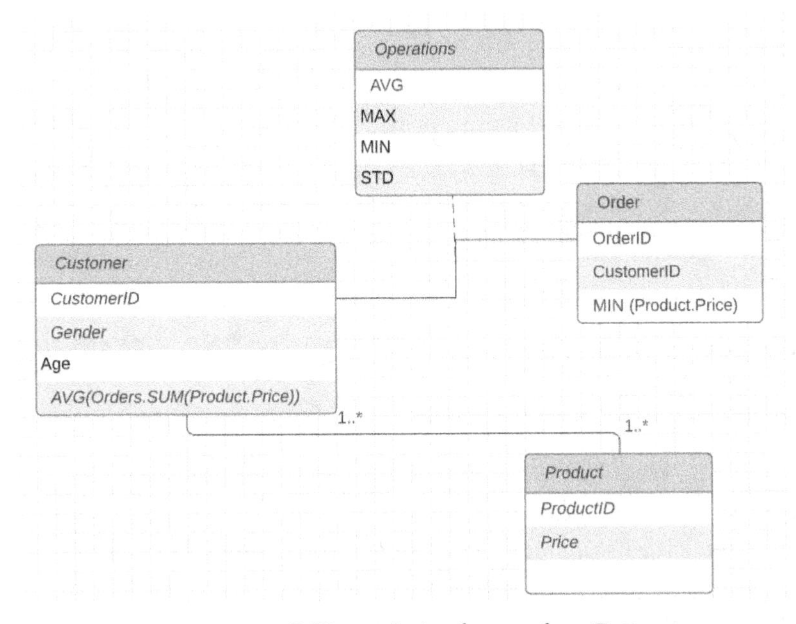

Figure 3.25 – DFS – analyzing features from Entity

The **Featuretools** website contains an excellent set of demos for predicting next purchases, remaining useful life, appointment no-shows, loan repayment likelihood, customer churn, household poverty, and malicious internet traffic, among many other use cases: `https://www.featuretools.com/demos/`.

Introducing Microsoft NNI

Microsoft Neural Network Intelligence (**NNI**) is an open source platform that addresses the three key areas of any automated ML life cycle – automated feature engineering, architectural search (also referred to as **neural architectural search** or **NAS**), and **hyperparameter tuning** (**HPI**). The toolkit also offers model compression features and operationalization. NNI comes with many hyperparameter tuning algorithms already built in.

A high-level architecture diagram of NNI is as follows:

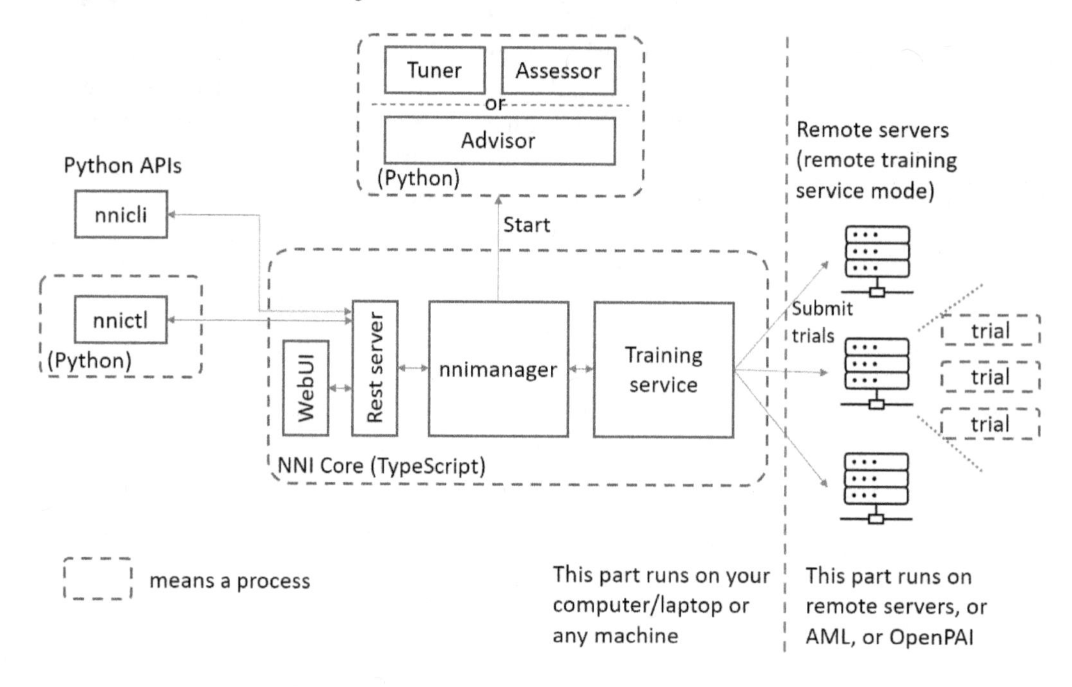

Figure 3.26 – Microsoft NNI high-level architecture

NNI has several state-of-the-art **hyperparameter** optimization algorithms built in, and they are called **tuners**. The list includes **TPE**, **Random Search**, **Anneal**, **Naive Evolution**, **SMAC**, **Metis Tuner**, **Batch Tuner**, **Grid Search**, **GP Tuner**, **Network Morphism**, **Hyperband**, **BOHB**, **PPO Tuner**, and **PBT Tuner**.

The toolkit is available on GitHub to be downloaded: `https://github.com/microsoft/nni`. More information about its built-in tuners can be found here: `https://nni.readthedocs.io/en/latest/builtin_tuner.html`.

Now, let's learn how to install Microsoft NNI and how to run an automated ML experiment using this library.

Let's go ahead and install the NNI on our machine using `pip`:

```
Anaconda Prompt (Anaconda3)
(base) C:\Users\u53704\Desktop\Adnan Masood\My Books\automl-book\src\nni>python -m pip install --upgrade nni
Collecting nni
  Downloading nni-1.8-py3-none-win_amd64.whl (32.9 MB)
     |████████████████████████████████| 32.9 MB 3.2 MB/s
Requirement already satisfied, skipping upgrade: scipy in d:\anaconda3\lib\site-packages (from nni) (1.4.1)
Requirement already satisfied, skipping upgrade: numpy in d:\anaconda3\lib\site-packages (from nni) (1.18.1)
Collecting astor
  Downloading astor-0.8.1-py2.py3-none-any.whl (27 kB)
Collecting hyperopt==0.1.2
  Downloading hyperopt-0.1.2-py3-none-any.whl (115 kB)
     |████████████████████████████████| 115 kB 3.3 MB/s
Requirement already satisfied, skipping upgrade: requests in d:\anaconda3\lib\site-packages (from nni) (2.22.0)
Requirement already satisfied, skipping upgrade: psutil in d:\anaconda3\lib\site-packages (from nni) (5.6.7)
Collecting coverage
  Downloading coverage-5.3-cp37-cp37m-win_amd64.whl (208 kB)
     |████████████████████████████████| 208 kB 6.4 MB/s
Requirement already satisfied, skipping upgrade: pkginfo in d:\anaconda3\lib\site-packages (from nni) (1.5.0.1)
Collecting websockets
  Downloading websockets-8.1-cp37-cp37m-win_amd64.whl (66 kB)
     |████████████████████████████████| 66 kB 4.5 MB/s
Requirement already satisfied, skipping upgrade: colorama in d:\anaconda3\lib\site-packages (from nni) (0.4.3)
Collecting netifaces
  Downloading netifaces-0.10.9-cp37-cp37m-win_amd64.whl (16 kB)
Collecting schema
  Downloading schema-0.7.2-py2.py3-none-any.whl (16 kB)
Collecting PythonWebHDFS
  Downloading PythonWebHDFS-0.2.3-py3-none-any.whl (10 kB)
Collecting scikit-learn>=0.23.2
  Downloading scikit_learn-0.23.2-cp37-cp37m-win_amd64.whl (6.8 MB)
```

Figure 3.27 – AutoML with Microsoft NNI – installation via Anaconda

One of the best features offered by NNI is that it has both a **command-line interface** (**CLI**) and a **web UI** so that we can view the trials and experiments. NNICtl is the command line that's used to manage the NNI application. You can see the options for experiments in the following screenshot:

```
Anaconda Prompt (Anaconda3)
(base) C:\Users\u53704\Desktop\Adnan Masood\My Books\automl-book\src\nni\keras-mnist>nnictl --help
usage: nnictl [-h] [--version]
              {ss_gen,create,resume,view,update,stop,trial,experiment,platform,webui,config,log,package,tensorboard
,top}
              ...

use nnictl command to control nni experiments

positional arguments:
  {ss_gen,create,resume,view,update,stop,trial,experiment,platform,webui,config,log,package,tensorboard,top}
    ss_gen              automatically generate search space file from trial
                        code
    create              create a new experiment
    resume              resume a new experiment
    view                view a stopped experiment
    update              update the experiment
    stop                stop the experiment
    trial               get trial information
    experiment          get experiment information
    platform            get platform information
    webui               get web ui information
    config              get config information
    log                 get log information
    package             control nni tuner and assessor packages
    tensorboard         manage tensorboard
    top                 monitor the experiment

optional arguments:
  -h, --help            show this help message and exit
  --version, -v

(base) C:\Users\u53704\Desktop\Adnan Masood\My Books\automl-book\src\nni\keras-mnist>_
```

Figure 3.28 – AutoML with Microsoft NNI – the nnictl command

NNI can have a learning curve if you do not understand how it works. You need to become familiar with the three primary NNI elements for it to work. First, you must define the search space, which you can find in the `search_space.json` file. You also need to update the model code (`main.py`) so that it incorporates **hyperparameter** tuning. Finally, you must define the experiment (`config.yml`) so that you can define the tuner and trial (execution model code) information:

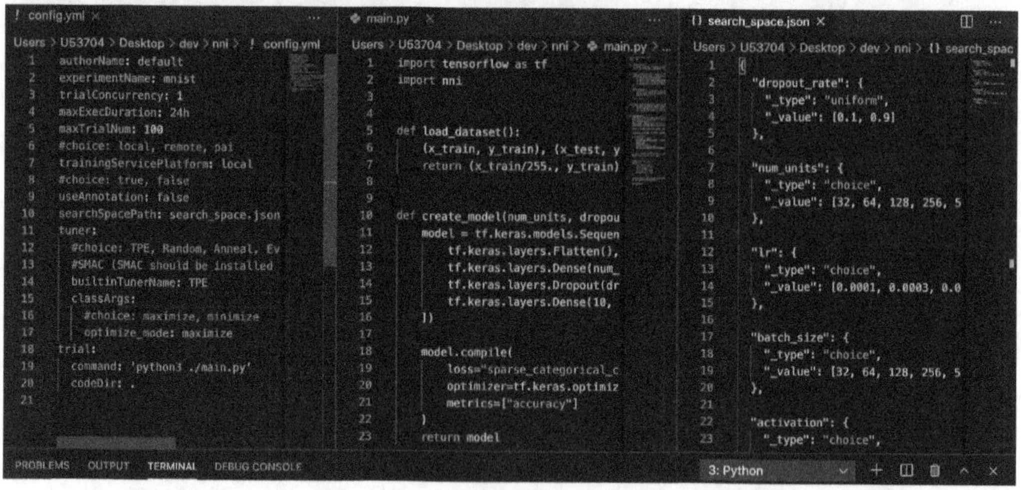

Figure 3.29 – AutoML with Microsoft NNI – the configuration and execution files

As a reminder, the search space describes the value range of each hyperparameter and for each trial, various hyperparameter values from this space are picked. While creating a configuration for a hyperparamter tuning experiment, we can limit the maximum number of trials. Also, while creating a hyperparameter search space, we can list the values that we want to try out in the tuning experiment when using the **choice** type hyperparameter.

In this case, we have taken a simple **Keras MNIST** model and retrofitted it to use NNI for tuning the parameters. Now that the code files are ready, we can run the experiment using the `nnictl create` command:

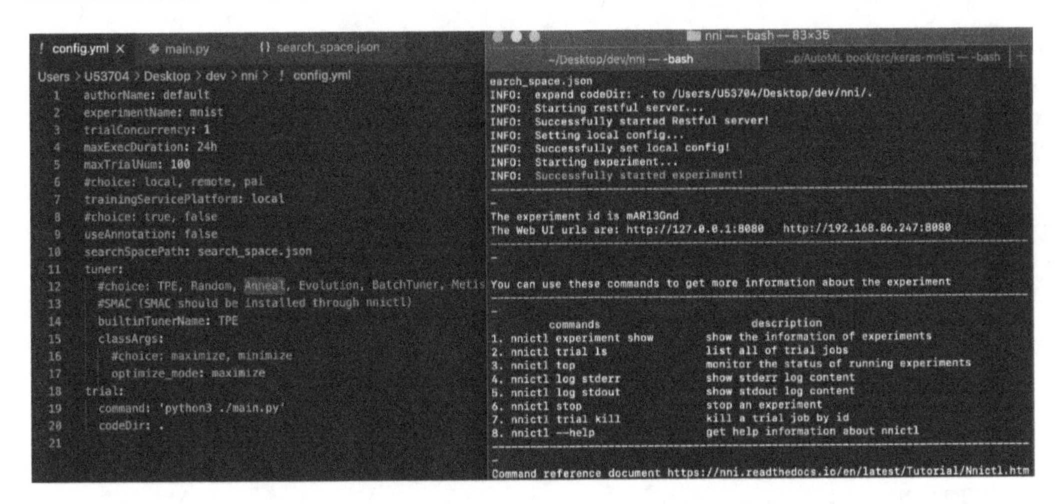

Figure 3.30 – AutoML with Microsoft NNI – running the experiment

You can use the following commands to find out more about the experiment:

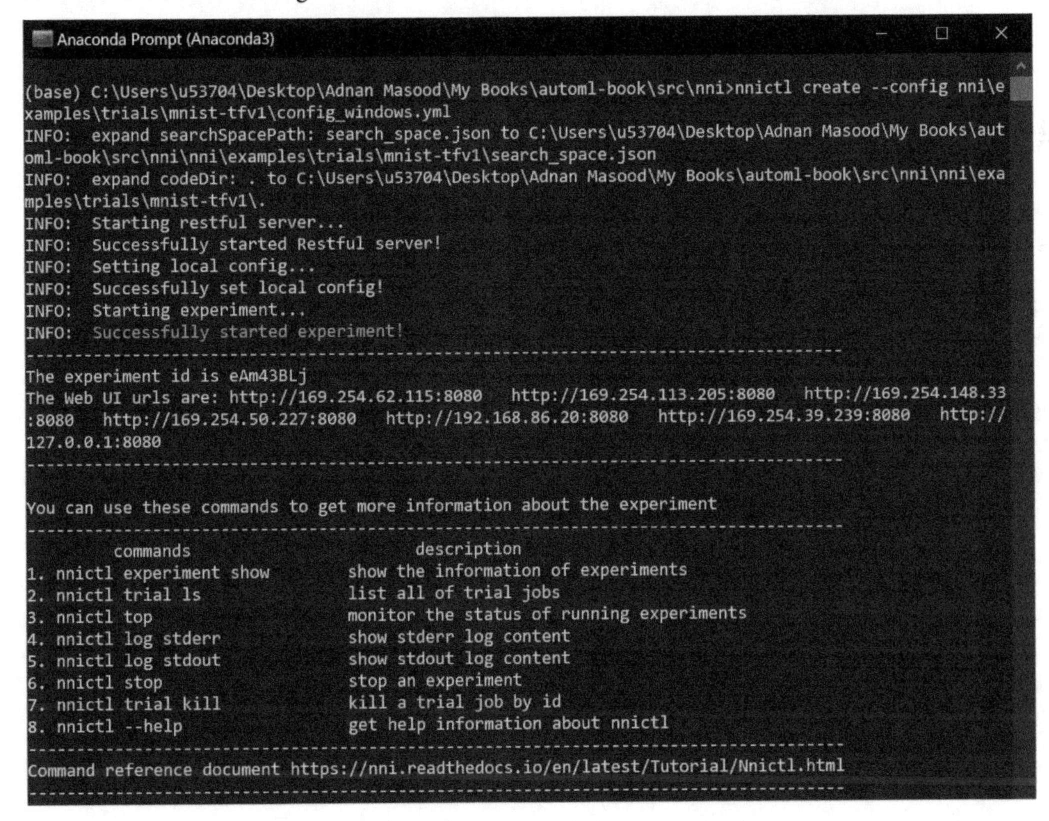

Figure 3.31 – AutoML with Microsoft NNI – the nnictrl parameters

Now, let's look at NNI's secret weapon – its UI. The NNI UI can be accessed via the web UI URL shown in the output console shown in *Figure 3.29*. Here, you can see the experiment running, its parameters, and its details. For instance, in this case, we only ran 19 trials, so it ran through these quickly. However, there were no meaningful results, such as us finding out what the best metric is (**N/A**), as shown in the following screenshot:

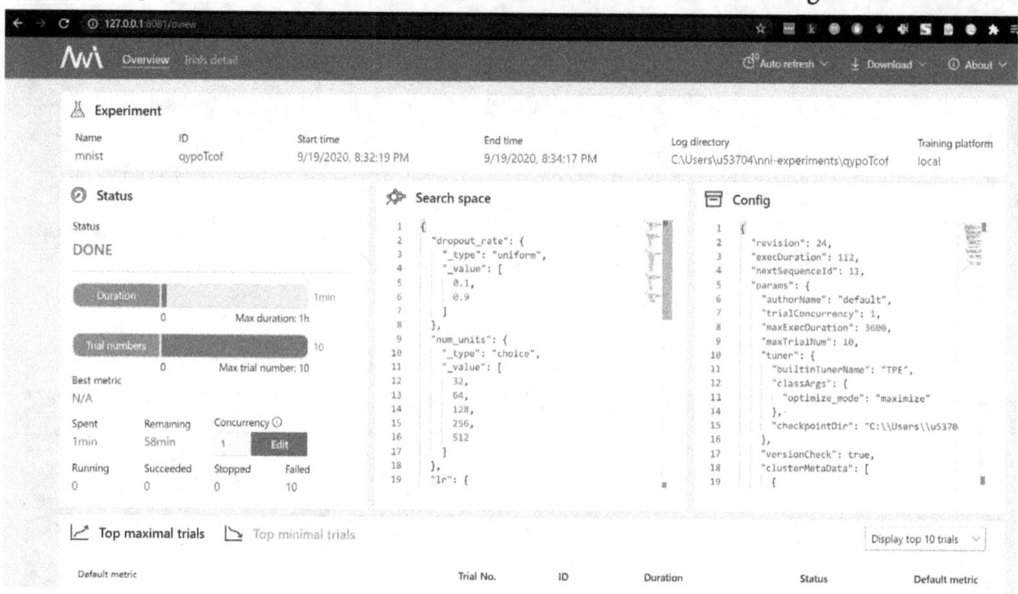

Figure 3.32 – AutoML with the Microsoft NNI UI

Increasing the number of trials to 30 takes longer, but it also gives you better accuracy in the results. Microsoft NNI helps you report intermediate results (results during a trial or training process before the training is complete). For instance, if the value of the metric being reported is stored in a variable, "x", you can do intermediate reporting using NNI like this:

```
nni.report_intermediate_result(x)
```

The following will be displayed on your screen:

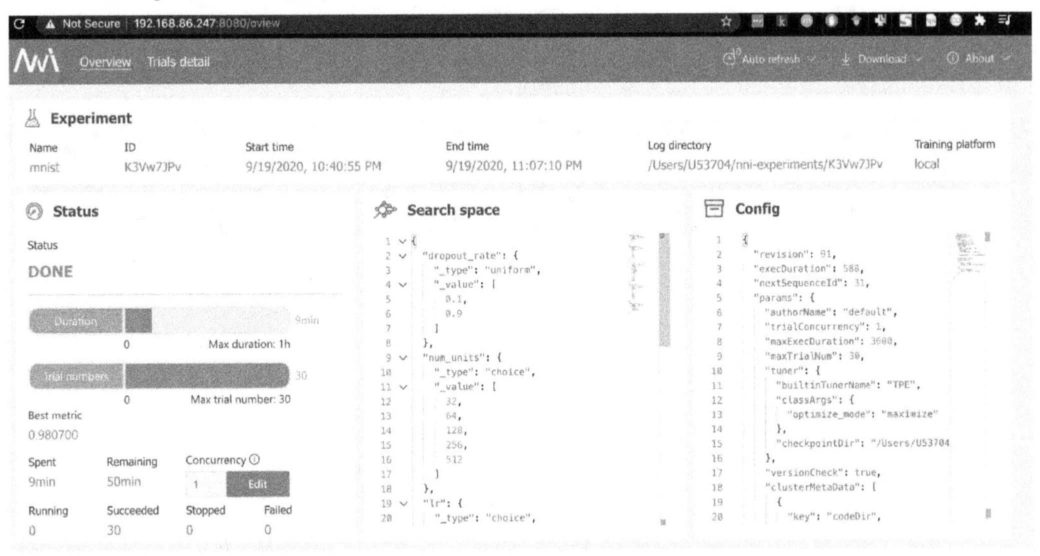

Figure 3.33 – AutoML with Microsoft NNI – the UI after completing an experiment

The NNI UI also provides you with views of the **default metrics, hyperparameters, duration**, and **intermediate results** of each trial. The **hyperparameter** view is especially amazing because you can visualize how each **hyperparameter** was selected. For example, in this case, it seems like RELU with a batch size of 1,024 provided significantly good results. This gives you an idea of what underlying algorithm can be used for model selection, as shown in the following screenshot:

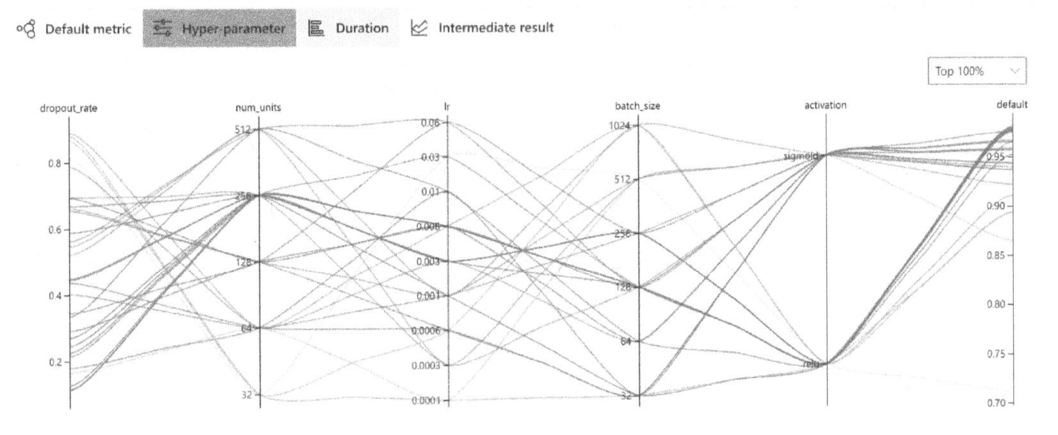

Figure 3.34 – AutoML with Microsoft NNI – the hyperparameters in the experiment

As we learned earlier regarding diminishing returns, increasing the number of trials doesn't increase the accuracy of the model significantly. In this case, the experiment spent 40 minutes completing 100 trials and provided a best metric of **0.981** compared to **0.980** from earlier, as seen in the following screenshot:

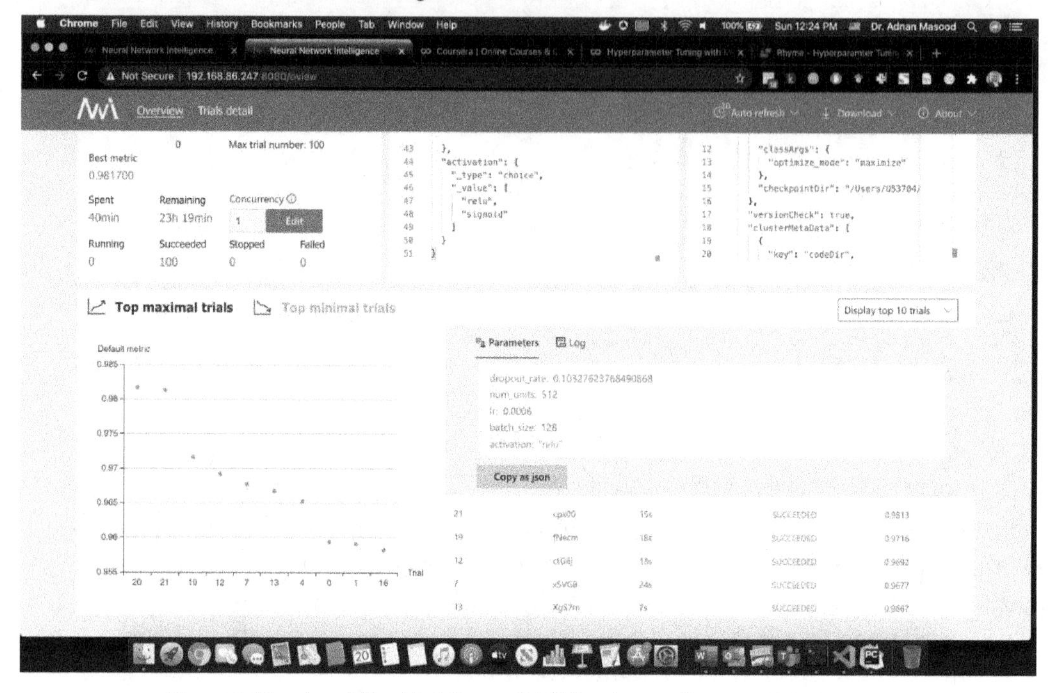

Figure 3.35 – AutoML with Microsoft NNI – the configuration parameters

You can also select a different top percentage of results for the **hyperparameters** to see what **hyperparameters** we used to get the best performing results:

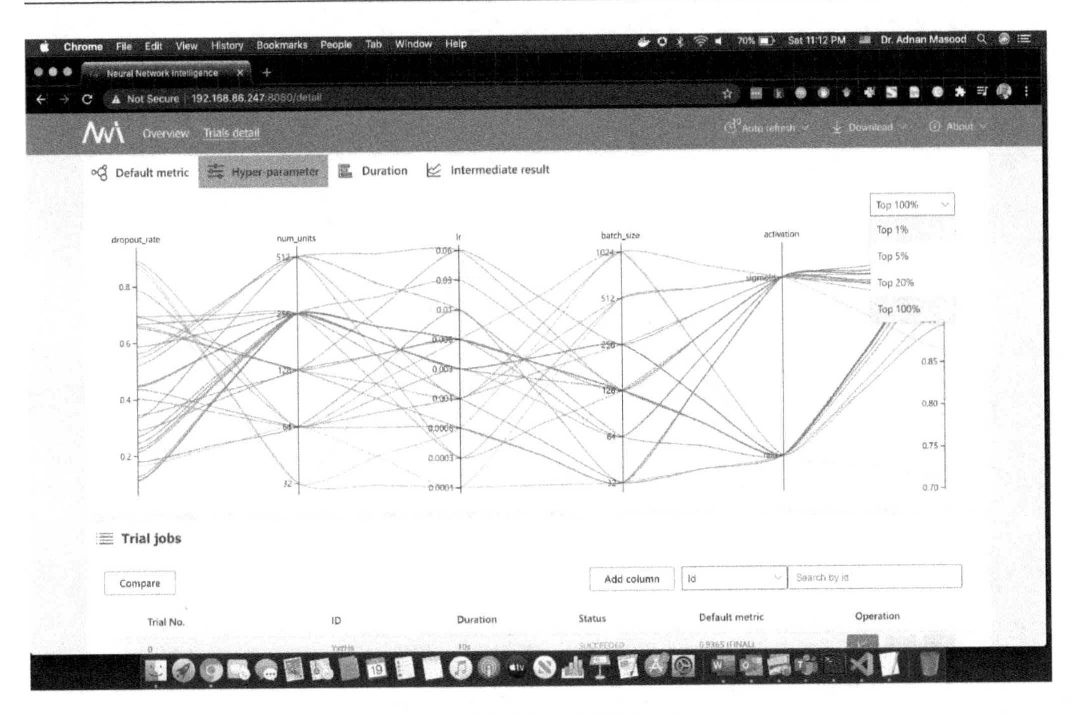

Figure 3.36 – AutoML with Microsoft NNI – the hyperparameters

Alternatively, you can just look at the top 5% of the results by selecting **Top 5%** from the dropdown on the top right of the graph:

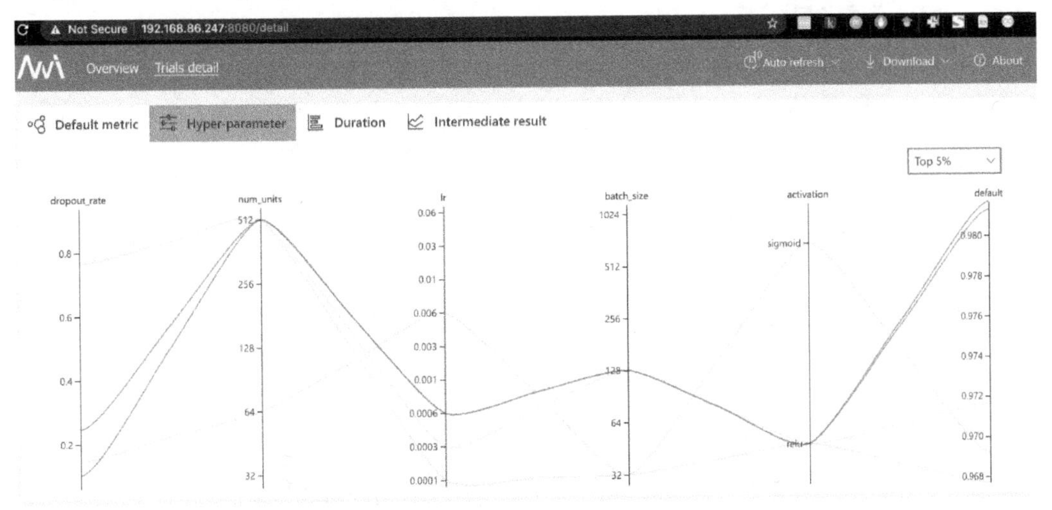

Figure 3.37 – AutoML with Microsoft NNI – the hyperparameters for the top 5%

NNI also allows you to drill down into each trial visually. You can see all the trial jobs in the following screenshot:

Trial jobs

Compare | Add column | Id | Search by id

Trial No.	ID	Duration	Status	Default metric ↓	Operation
28	IS2fj	11s	SUCCEEDED	0.9807 (FINAL)	
27	mxn9S	8s	SUCCEEDED	0.9797 (FINAL)	
26	tMNvv	9s	SUCCEEDED	0.9794 (FINAL)	
21	IJJC1	8s	SUCCEEDED	0.9781 (FINAL)	
20	Rdm9Z	9s	SUCCEEDED	0.978 (FINAL)	
29	Ztsfn	9s	SUCCEEDED	0.9778 (FINAL)	
24	GVCUo	9s	SUCCEEDED	0.9777 (FINAL)	
14	z5bYY	16s	SUCCEEDED	0.9769 (FINAL)	
22	BHaar	8s	SUCCEEDED	0.9766 (FINAL)	
17	hHsqt	8s	SUCCEEDED	0.9757 (FINAL)	
23	pRRcL	8s	SUCCEEDED	0.9754 (FINAL)	
3	bbNIr	11s	SUCCEEDED	0.9663 (FINAL)	
18	sNSbm	14s	SUCCEEDED	0.9648 (FINAL)	
8	pof1Y	16s	SUCCEEDED	0.9648 (FINAL)	

Figure3.38 – AutoML with Microsoft NNI – the experiments list

Alternatively, you can drill down into individual jobs and view various hyperparameters, including `dropout_rate`, `num_units`, learning rate, `batch_size`, and `activation` function:

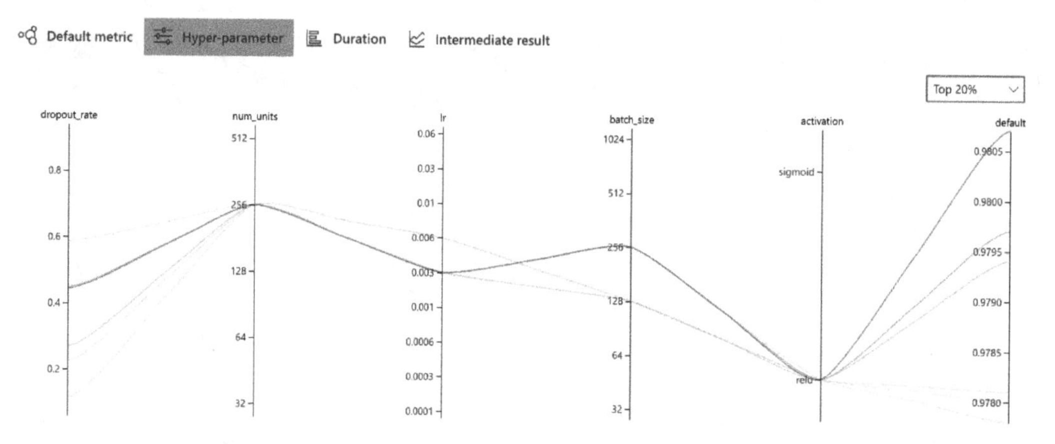

Figure 3.39 – AutoML with Microsoft NNI – the path for the top 20% of hyperparameters

Being able to see this level of detail about experiments and hyperparameters is phenomenal, and makes NNI one of our top open source tools for automated ML.

Before we move on, it is important to note that, like AutoGluon is part of AWS's automated ML offering, NNI is part of Microsoft Azure's automated ML toolset, which makes it much more powerful and verstaile when it comes to reusing it.

Introducing auto-sklearn

scikit-learn (also known as **sklearn**) is a very popular ML library for Python development – so popular that it has its own memes:

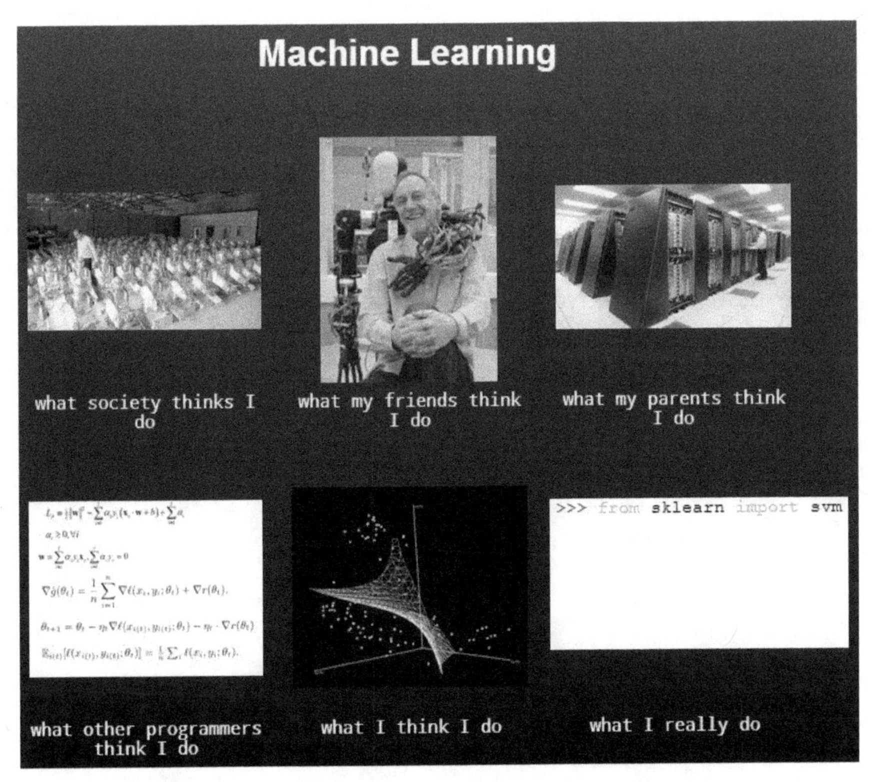

Figure 3.40 – An ML meme

As part of this ecosystem and based on *Efficient and Robust Automated Machine Learning* by *Feurer et al.*, **auto-sklearn** is an automated ML toolkit that performs algorithm selection and **hyperparameter tuning** using **Bayesian optimization**, **meta-learning**, and **ensemble construction**.

The toolkit is available on GitHub to be downloaded: `github.com/automl/auto-sklearn`.

`auto-sklearn` touts its ease of use for performing automated ML since it's a four-line automated ML solution:

```
>>> import autosklearn.classification
>>> cls = autosklearn.classification.AutoSklearnClassifier()
>>> cls.fit(X_train, y_train)
>>> predictions = cls.predict(X_test)
```

Figure 3.41 – AutoML with auto-sklearn – getting started

If the preceding syntax looks familiar, then it's because this is how `scikit-learn` does predictions and therefore makes `auto-sklearn` one of the easiest libraries to use. `auto-sklearn` uses `scikit-learn` as its backend framework and supports **Bayesian optimization** with automated **ensembled construction.**

Based on **Combined Algorithm Selection and Hyperparameter optimization** (**CASH**), as discussed earlier in this book, `auto-sklearn` addresses the problem of finding the best model and its hyperparameters at the same time. The following diagram shows how `auto-sklearn` describes its internal pipeline:

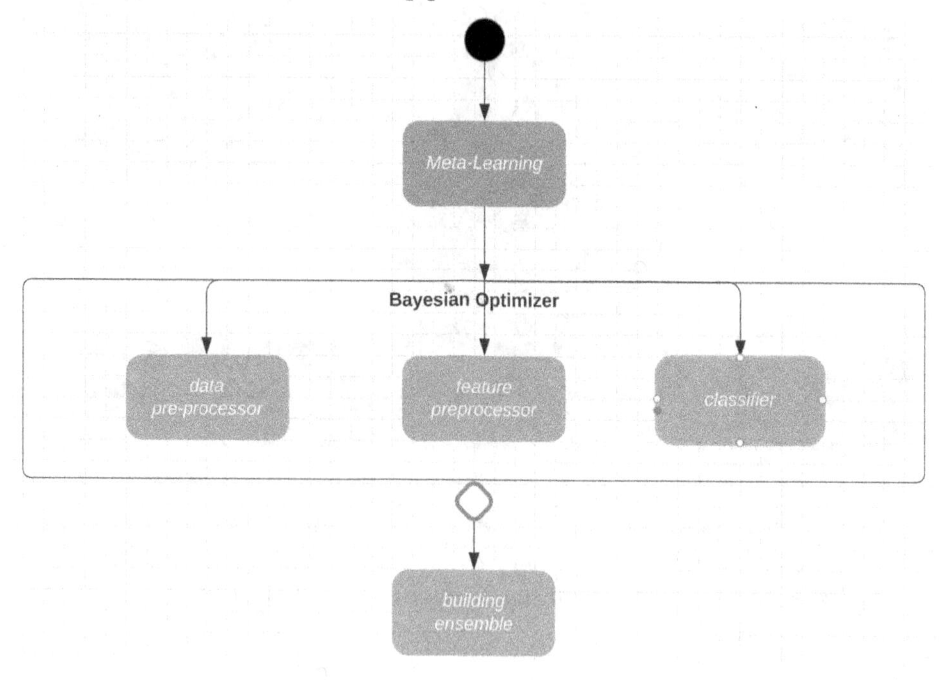

Figure 3.42 – An auto-sklearn automated ML pipeline

The underlying automated ML "engine" uses **Information Retrieval** (**IR**) and statistical meta feature approaches to select a variety of configurations, all of which are used as part of Bayesian optimization input. This process is iterative, and auto-sklearn keeps the models to create an ensemble, thus iteratively building a model to maximize performance. Setting up
auto-sklearn on Colab can be tricky as you will need to install the following packages to get started:

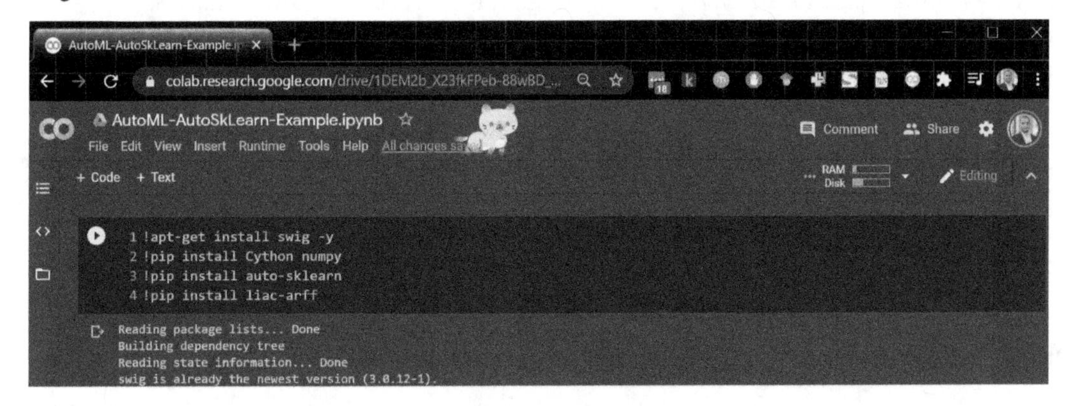

Figure 3.43 – AutoML with auto-sklearn – installing the necessary libraries

You may have to restart the runtime in Colab upon installation. You can also set up auto-sklearn on your local machine by following the instructions specified here: `https://automl.github.io/auto-sklearn/master/installation.html`.

Once you have completed the installation, you can run the **auto-sklearn classifier** and get great accuracy and **hyperparameters** via the magic of automated ML:

```
import autosklearn.classification
import sklearn.model_selection as cv
import sklearn.datasets
import sklearn.metrics
#from autosklearn.experimental.askl2 import AutoSklearn2Classifier


X, y = sklearn.datasets.load_digits(return_X_y=True)
X_train, X_test, y_train, y_test = \
        sklearn.model_selection.train_test_split(X, y, random_state=1)
automl = autosklearn.classification.AutoSklearnClassifier()
automl.fit(X_train, y_train)
y_hat = automl.predict(X_test)
print("Accuracy score", sklearn.metrics.accuracy_score(y_test, y_hat))

Accuracy score 0.9888888888888889
```

Figure 3.44 – AutoML with AutoSkLearn – running a simple experiment for the auto-sklearn classifier

We should point out that **auto-sklearn 2**, an experimental version of auto-sklearn, is also out and includes the latest work that's been done on automatic configuration and performance improvements. You can import **auto-sklearn 2** like so:

```
from auto-sklearn.experimental.askl2 import Auto-
sklearn2Classifier
```

Examples of basic classification, regression, and multi-label classification datasets, as well as advanced examples of customizing auto-sklearn, are available here: `https://automl.github.io/auto-sklearn/master/examples/`.

If you wish, you can try out the advanced use cases of changing the optimization metrics, the train-validation split, providing different feature types, using pandas DataFrames, and inspecting search procedures. These advanced examples also demonstrate how auto-sklearn can be used to **extend regression, classification**, and **preprocessor components**, as well as how a number of **hyperparameters** can be restricted.

AutoKeras

Keras is one of the most widely used deep learning frameworks and is an integral part of the TensorFlow 2.0 ecosystem. **Auto-Keras** is based on the paper by Jin et al., (`https://arxiv.org/abs/1806.10282`) which proposed "*a novel method for efficient neural architecture search with network morphism, enabling Bayesian optimization*". **AutoKeras** is built on the concept that since existing neural architecture search algorithms such as **NASNet** and **PNAS** are computationally quite expensive, using **Bayesian optimization** to guide the network's morphism is an efficient approach to explore the search space.

The toolkit is available on GitHub to be downloaded: `github.com/jhfjhfj1/autokeras`.

The following steps will guide you through how to install AutoKeras and how to run an automated ML experiment using the library. Let's get started:

1. To get started with Auto-Keras, run the following `install` commands in Colab or in a Jupyter Notebook. Doing this will install **AutoKeras** and the **Keras** tuner. **AutoKeras** needs its tuner to be greater than version 1.0.1, and the release candidate version can be found in `git uri`:

```
!pip install autokeras
!pip install git+https://github.com/keras-team/keras-tuner.git@1.0.2rc1
!pip install tensorflow
```

Figure 3.45 – AutoML with AutoKeras – installation

2. Once you have met the dependencies, you can load the **MNIST dataset**:

```
import tensorflow as tf
from tensorflow.keras.datasets import mnist

(x_train, y_train), (x_test, y_test) = mnist.load_data()
print(x_train.shape)
print(y_train.shape)
print(y_train[:3])
```

```
Downloading data from https://storage.googleapis.com/tensorflow/tf-keras-datasets/mnist.npz
11493376/11490434 [==============================] - 0s 0us/step
(60000, 28, 28)
(60000,)
[5 0 4]
```

Figure 3.46 – AutoML with AutoKeras – loading the training data

3. Now, you can get **AutoKeras** and go through the code for a classifier – in this case, an image classifier. **AutoKeras** shows the accuracy of data as it calculates the classification metrics:

```
import autokeras as ak

# Initialize the image classifier.
clf = ak.ImageClassifier(
    overwrite=True,
    max_trials=1)
# Feed the image classifier with training data.
clf.fit(x_train, y_train, epochs=10)
```

```
Search: Running Trial #1

Hyperparameter      |Value      |Best Value So Far
image_block_1/block_type|vanilla      |?
image_block_1/normalize|True      |?
image_block_1/augment|False      |?
image_block_1/conv_block_1/kernel_size|3      |?
image_block_1/conv_block_1/num_blocks|1      |?
image_block_1/conv_block_1/num_layers|2      |?
image_block_1/conv_block_1/max_pooling|True      |?
image_block_1/conv_block_1/separable|False      |?
image_block_1/conv_block_1/dropout|0.25      |?
image_block_1/conv_block_1/filters_0_0|32      |?
image_block_1/conv_block_1/filters_0_1|64      |?
classification_head_1/spatial_reduction_1/reduction_type|flatten      |?
classification_head_1/dropout|0.5      |?
optimizer      |adam      |?
learning_rate      |0.001      |?

Epoch 1/10
  90/1500 [>.............................] - ETA: 1:42 - loss: 0.7316 - accuracy: 0.7750
```

Figure 3.47 – AutoML with AutoKeras – running the epochs

4. Fast forwarding through the fit procedure, now that you have discovered the **hyperparameters** and model, you can predict the results for the test features:

```
# Predict with the best model.
print (x_test)
predicted_y = clf.predict(x_test)
print(predicted_y)
```

```
[[[0 0 0 ... 0 0 0]
  [0 0 0 ... 0 0 0]
  [0 0 0 ... 0 0 0]
  ...
  [0 0 0 ... 0 0 0]
  [0 0 0 ... 0 0 0]
  [0 0 0 ... 0 0 0]]

 [[0 0 0 ... 0 0 0]
  [0 0 0 ... 0 0 0]
  [0 0 0 ... 0 0 0]
```

Figure 3.48 – AutoML with AutoKeras – predicting the best model using the predict command

You will get the following results:

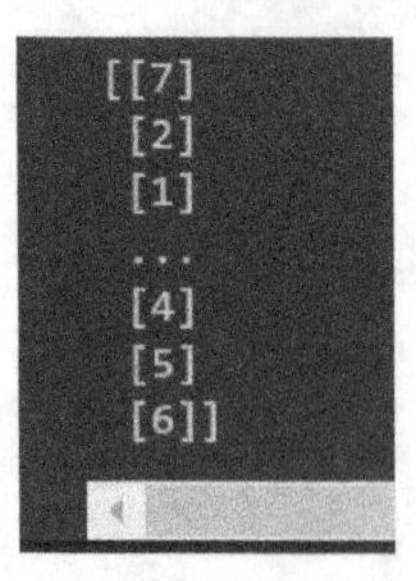

```
[[7]
 [2]
 [1]
 ...
 [4]
 [5]
 [6]]
```

Figure 3.49 – AutoML with AutoKeras – results from the best model

With these results, you can evaluate the accuracy metrics on the training and test datasets, respectively:

```
# Evaluate the best model with testing data.
print(clf.evaluate(x_test, y_test))
```

```
WARNING:tensorflow:Unresolved object in checkpoint: (root).optimizer.iter
WARNING:tensorflow:Unresolved object in checkpoint: (root).optimizer.beta_1
WARNING:tensorflow:Unresolved object in checkpoint: (root).optimizer.beta_2
WARNING:tensorflow:Unresolved object in checkpoint: (root).optimizer.decay
WARNING:tensorflow:Unresolved object in checkpoint: (root).optimizer.learning_rate
WARNING:tensorflow:A checkpoint was restored (e.g. tf.train.Checkpoint.restore or tf.keras.Mo
313/313 [==============================] - 5s 16ms/step - loss: 0.0332 - accuracy: 0.9893
[0.03324095159769058, 0.989300012588501]
```

Figure 3.50 – AutoML with AutoKeras – evaluating for the best model with the test data

Like TPOT, you can easily export the model using the `model.save` method and use it for `eval` at a later date. You can see the model stored in the `model_autokeras` folder on the left-hand pane of your Colab notebook, as shown in the following screenshot:

```
▸ image_classifier
▾ model_autokeras
  ▸ assets
  ▸ variables
    saved_model.pb
▸ sample_data
```

```
# Export as a Keras Model.
model = clf.export_model()
print(type(model))  # <class 'tensorflow.python.keras.engine.training.Model'>

try:
    model.save("model_autokeras", save_format="tf")
except:
    model.save("model_autokeras.h5")
```

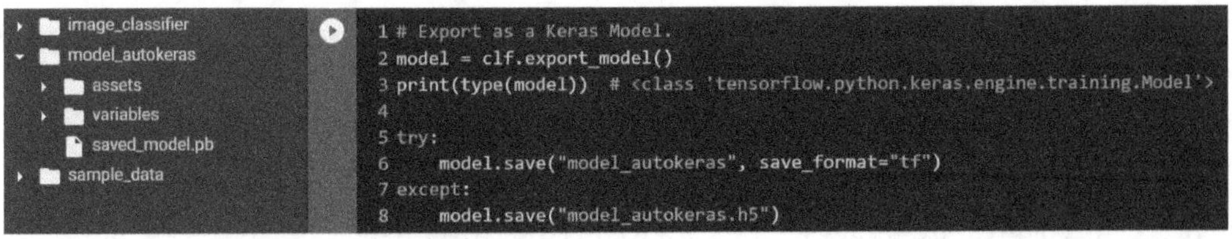

Figure 3.51 – AutoML with AutoKeras – exporting as a Keras model

5. Once the model has been saved, it can be used to retrieve data using `load_model` and make predictions against it, as shown in the following screenshot:

```
from tensorflow.keras.models import load_model

loaded_model = load_model("model_autokeras", custom_objects=ak.CUSTOM_OBJECTS)

predicted_y = loaded_model.predict(x_test)
print(predicted_y)
```

```
[[1.04031775e-11 9.33228213e-13 3.23597726e-09 ... 1.00000000e+00
  1.17060192e-12 1.85679241e-08]
 [4.87752505e-10 1.90604410e-07 9.99998689e-01 ... 7.58147953e-14
  1.23664350e-08 2.22702485e-13]
 [6.41350306e-10 9.99989390e-01 3.39003947e-07 ... 2.34114125e-07
  9.20917387e-08 2.21865425e-11]
 ...
 [4.93693844e-13 1.76908531e-12 3.52099534e-14 ... 6.91528346e-09
  1.48345379e-07 4.70826649e-08]
 [1.18143204e-10 1.15603967e-15 7.66210359e-12 ... 1.39525295e-12
  4.68984922e-07 8.58040028e-11]
 [1.64026692e-09 3.16855014e-16 2.26211161e-09 ... 1.00495569e-16
  8.35135960e-09 3.97002526e-12]]
```

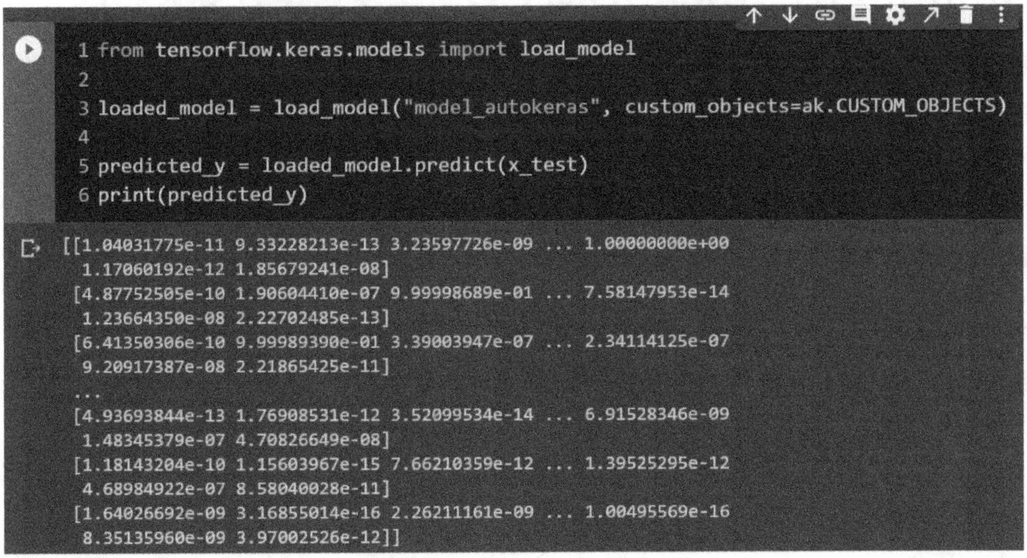

Figure 3.52 – AutoML with AutoKeras – predicting the values

AutoKeras uses **Efficient Neural Architecture Search** (**ENAS**), an approach similar to transfer learning. Like ensembles, the **hyperparameters** that are learned during the search are reused for other models, which helps us avoid having to retrain and provide improved performance.

As we conclude our overview of open source libraries, some honorable mentions go to two excellent and easy-to-use AutoML frameworks: **Ludwig** and **AutoGluon**.

Ludwig – a code-free AutoML toolbox

Uber's automated ML tool, Ludwig, is an open source deep learning toolbox used for experimenting with, testing, and training ML models. Built on top of TensorFlow, Ludwig enables users to create model baselines and perform automated ML-style experimentation with different network architectures and models. In its latest release, Ludwig now integrates with CometML and supports BERT text encoders.

The toolkit is available on GitHub to be downloaded: `https://github.com/uber/ludwig`.

There are tons of great examples with regard to this topic over here: `https://ludwig-ai.github.io/ludwig-docs/examples/#image-classification-mnist`.

AutoGluon – the AutoML toolkit for deep learning

From AWS Labs, with the goal of democratization of ML in mind, AutoGluon is described as being developed to enable "*easy-to-use and easy-to-extend AutoML with a focus on deep learning and real-world applications spanning image, text, or tabular data*". AutoGluon, an integral part of AWS's automated ML strategy, enables both junior and seasoned data scientists to build deep learning models and end-to-end solutions with ease. Like other automated ML toolkits, AutoGluon offers network architecture search, model selection, and the ability for you to improve custom models.

The toolkit is available on GitHub to be downloaded: `https://github.com/awslabs/autogluon`.

Summary

In this chapter, you reviewed some major open source tools that are used for AutoML, including **TPOT**, **AutoKeras**, **auto-sklearn**, **Featuretools**, and **Microsoft NNI**. These tools have been provided to help you understand the concepts we discussed in *Chapter 2, Automated Machine Learning, Algorithms, and Techniques*, and the underlying approaches that are used in each of these libraries.

In the next chapter, we will do an in-depth review of commercial automated ML offerings, starting with the Microsoft Azure platform.

Further reading

For more information on the topics that were covered in this chapter, please refer to the resources and links:

- TPOT for Automated ML in Python:
`https://machinelearningmastery.com/tpot-for-automated-machine-learning-in-python/`
- Featuretools Demos:
`https://www.featuretools.com/demos/`
- Boston Dataset:
`https://scikit-learn.org/stable/modules/generated/sklearn.datasets.load_boston.html`
- How to Automate ML:
`https://www.knime.com/blog/how-to-automate-machine-learning`
- *Data-driven advice for applying ML to bioinformatics problems*, by Randal S. Olson:
`https://www.ncbi.nlm.nih.gov/pmc/articles/PMC5890912/`
- TPOT Automated ML in Python:
`https://towardsdatascience.com/tpot-automated-machine-learning-in-python-4c063b3e5de9`
- Microsoft NNI:
`https://github.com/microsoft/nni`
- auto-sklearn:
`https://automl.github.io/auto-sklearn/master/examples/20_basic/example_regression.html#sphx-glr-examples-20-basic-example-regression-py`
- TPOT Demos:
`https://github.com/EpistasisLab/tpot/blob/master/tutorials/Digits.ipynb`

Section 2: AutoML with Cloud Platforms

This part provides a detailed overview of building automated machine learning solutions using Microsoft Azure, Amazon Web Services, and Google Cloud Platform.

This section comprises the following chapters:

- *Chapter 4, Getting Started with Azure Machine Learning*
- *Chapter 5, Automated Machine Learning with Microsoft Azure*
- *Chapter 6, Machine Learning with Amazon Web Services*
- *Chapter 7, Doing Automated Machine Learning with Amazon SageMaker Autopilot*
- *Chapter 8, Machine Learning with Google Cloud Platform*
- *Chapter 9, Automated Machine Learning with GCP Cloud AutoML*

4
Getting Started with Azure Machine Learning

"As a technologist, I see how AI and the fourth industrial revolution will impact every aspect of people's lives."

– Fei-Fei Li, Professor of Computer Science at Stanford University

In the previous chapter, you were introduced to the major AutoML **Open Source Software (OSS)** tools and libraries. We did a tour of the major OSS offerings, including TPOT, AutoKeras, auto-sklearn, Featuretools, and Microsoft NNI, which will have helped you, the reader, understand the differential value propositions and approaches used in each of these libraries.

In this chapter, we will start exploring the first of many commercial offerings, namely Microsoft's Azure capabilities in automated **Machine Learning** (**ML**). Azure Machine Learning is part of the Microsoft AI ecosystem, which helps accelerate the end-to-end ML life cycle using the power of the Windows Azure platform and services. We will start with an enterprise-grade ML service for building and deploying models that empowers developers and data scientists to build, train, and deploy ML models faster. With examples and walk-throughs, you will learn the fundamentals to build and deploy automated ML solutions using Azure.

In this chapter, we will cover the following topics:

- Getting started with Azure Machine Learning
- The Azure Machine Learning stack
- Getting started with the Azure Machine Learning service
- Modeling with Azure Machine Learning
- Deploying and testing models with Azure Machine Learning

Getting started with Azure Machine Learning

Not so long ago, if you wanted to use ML in a production environment on the Azure platform, you needed to bring together a bunch of different services to support the full ML life cycle.

For example, to use the datasets, you would need storage repositories such as Azure Blob storage or Azure Data Lake storage. For compute, you would either need individual virtual machines, Spark clusters using HDInsight, or Azure Databricks to actually run your model code. To protect your data for enterprise readiness, you'd need to bring in your virtual networks or configure your compute and data inside the same virtual network, along with Azure Key Vault to manage and secure your credentials. In order to provide repeatability for your experiments by using a consistent set of ML libraries, and the different versions thereof, you'd create Docker containers and use Azure Container Registry to store those Docker containers. You would need to put that inside your virtual network and then use Azure Kubernetes Service. Woah! This all sounds like a whole lot of stuff to piece together to get the ML and all the models and everything to work.

But things have progressed since, and you are in luck. With the Azure Machine Learning service, Microsoft removed that complexity. As a managed service, Azure Machine Learning comes with its own compute, hosted notebooks, and capabilities for model management, version control, and model reproducibility built right in. You can layer that on top of your existing Azure services. For example, you can plug in the compute and storage that you already have as well as your other infrastructure services. Azure Machine Learning connects and orchestrates them within a single environment so that you have one end-to-end modular platform for your entire ML life cycle, as you prepare data, build, train, package, and deploy your ML models.

Microsoft's Azure Machine Learning development group offers a great cheat sheet (`aka.ms/mlcheatsheet`) that helps you choose the best ML algorithm for the predictive analytics task at hand. Whether you want to predict between categories, discover a structure, or extract information from text, in the following flow you can find out the best algorithm based on your needs:

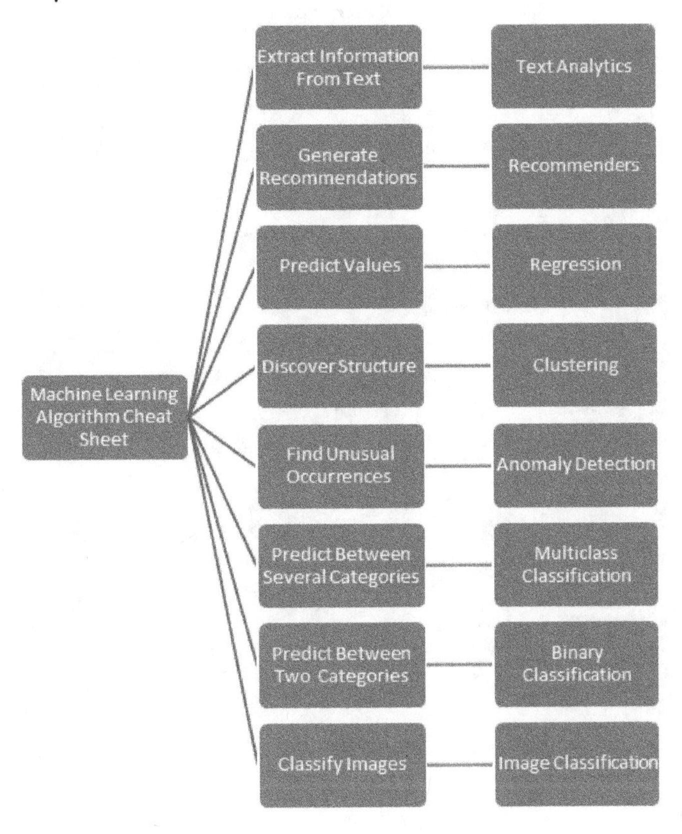

Figure 4.1 – The Machine Learning Algorithm Cheat Sheet overview

The Azure Machine Learning service makes for an ideal working environment for citizen data scientists. They get the services, tools, and virtually everything configured for them and the luxury of not having to compile all these individual services helps them do what matters most – solve the business problem. These subject matter experts also don't need to learn how to use new tools. You can use the Jupyter Notebooks, your favorite Python editor – including VS Code and PyCharm, and Azure Machine Learning works with any Python ML framework library, such as TensorFlow, PyTorch, scikit-learn, and so on. This is amazing in terms of making the life cycle a lot shorter, in terms of getting everything up and running. You will see some of these examples in the rest of this chapter as we walk through building a classification model using the Azure Machine Learning service.

The Azure Machine Learning stack

The Microsoft Azure ecosystem is quite broad; in this chapter, we will focus on its AI and ML related cloud offerings, especially the Azure Machine Learning service.

The following figure shows the offerings available for ML in the Azure cloud:

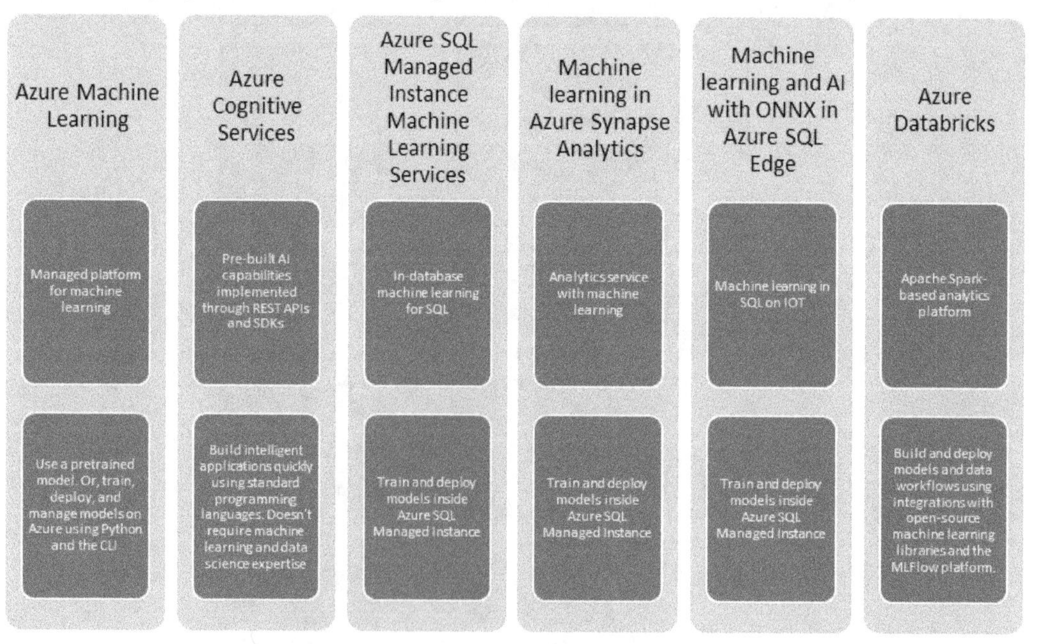

Figure 4.2 – Azure cloud ML offerings

You can visit the following link for more information about the offerings in the preceding table: `https://docs.microsoft.com/en-us/azure/architecture/data-guide/technology-choices/data-science-and-machine-learning`.

It can be confusing to know which Azure Machine Learning offering should be chosen among the many described in the preceding table. The following diagram helps with choosing the right offering based on the given business and technology scenario:

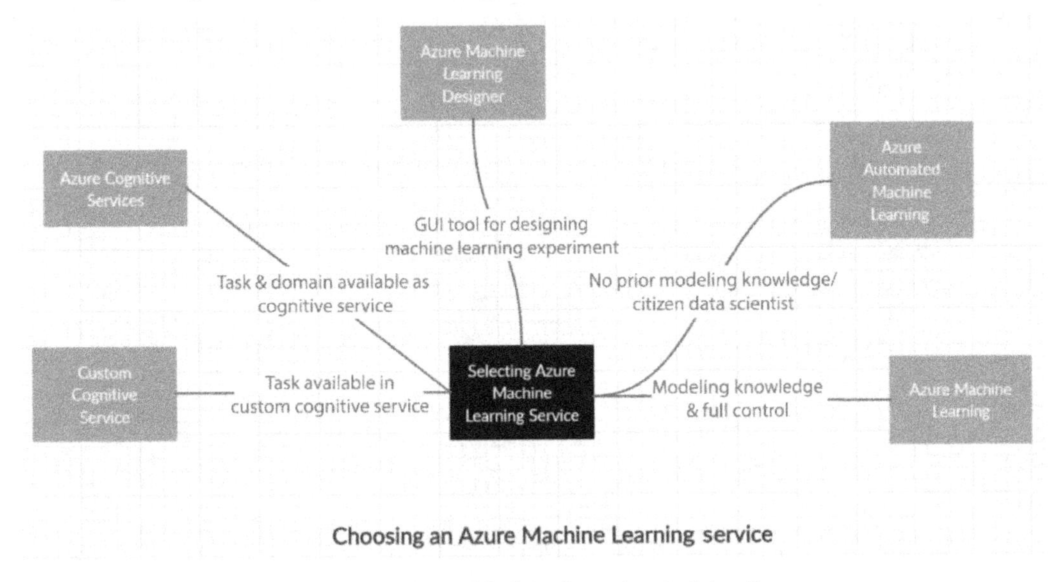

Figure 4.3 – Azure Machine Learning decision flow

Automated ML is a part of the Azure Machine Learning service capabilities. Other capabilities include collaborative notebooks, data labeling, ML operations, a drag-and-drop designer studio, autoscaling capabilities, and many other tools described in the following table.

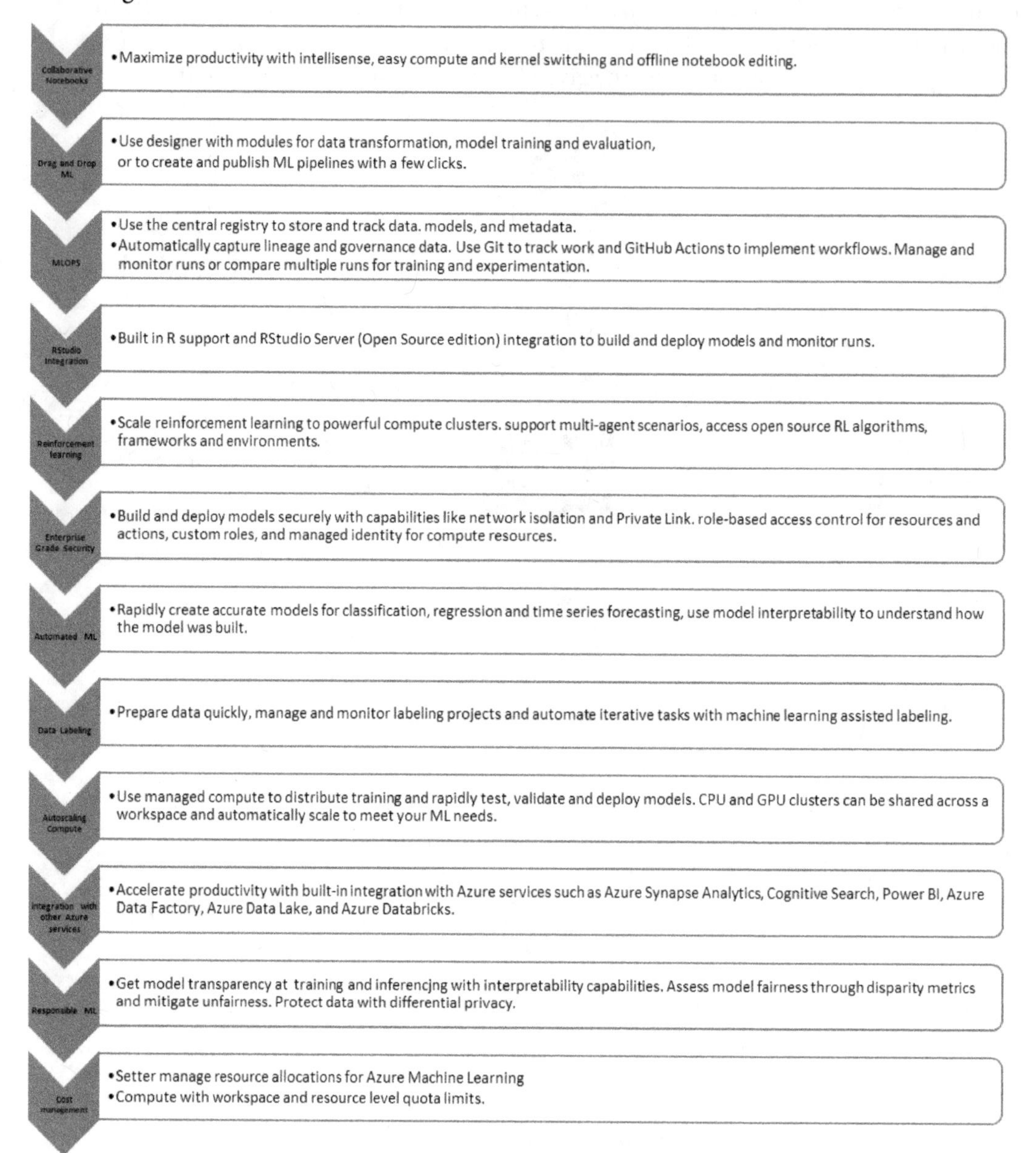

Figure 4.4 – Azure Machine Learning key service capabilities

The AI and ML capabilities offered in Azure are quite comprehensive and the best place to start looking into these in detail is the Azure AI Fundamentals Learning Path on Microsoft Learn: `https://aka.ms/AzureLearn_AIFundamentals`.

The Azure Machine Learning service arguably provides virtually everything a data scientist needs. It includes environments, experiments, pipelines, datasets, models, endpoints, and workspaces, which helps enabling other Azure resources such as the following:

- **Azure Container Registry** (**ACR**): This registry stores information about the Docker containers used during training and deployment.
- **Azure storage account**: The default datastore for the ML workspace; it also stores the associated Jupyter notebooks.
- **Azure Application Insights**: This stores model monitoring information.
- **Azure Key Vault**: Credentials and other secrets for compute and workspace needs.

To train any ML algorithm, you need processing capabilities, that is, compute resources. Azure supports a variety of different compute resources, ranging from your local machine to remote VMs. The following table outlines different training targets for workloads. As shown in the table, you can use a local computer, a compute cluster, a remote virtual machine, and a variety of other training targets for automated ML workloads:

Training Targets	Automated Machine Learning	Machine Learning Pipelines
Local Computer	Supported	
Azure Machine Learning Compute Cluster	Supported with Hyperparameter Tuning	Supported
Azure Machine Learning Compute Instance	Supported with Hyperparameter Tuning	Supported
Remote VM	Supported with Hyperparameter Tuning	Supported
Azure Databricks	Supported (SDK Local Mode Only)	Supported
Azure Data Lake Analytics		Supported
Azure HDInsight		Supported
Azure Batch		Supported

Figure 4.5 – Azure Machine Learning training targets

In the step-by-step walk-through, you will see how to create and select a compute target. Lastly, all ML algorithms need operationalization and deployment. The infamous adage *it works on my machine* only goes so far; therefore, you need a deployment compute resource to be able to host your model and provide an endpoint.

This is where your service resides and will be consumed from, also known as the compute target used for deployment. The following table shows different types of deployment targets:

Compute Target	Usage	GPU / FPGA Support	Description
Local web service	Testing/debugging		Use for limited testing and troubleshooting. Hardware acceleration depends on use of libraries in the local system.
Azure Machine Learning compute instance web service	Testing/debugging		Use for limited testing and troubleshooting.
Azure Kubernetes Service (AKS)	Real-time inference	GPU supported with web service deployment. FPGA supported.	Use for high-scale production deployments. Provides fast response time and autoscaling of the deployed service. Cluster autoscaling isn't supported through the Azure Machine Learning SDK. To change the nodes in the AKS cluster, use the Ul for your AKS cluster in the Azure portal. AKS is the only option available for the designer.
Azure Container Instances	Testing or development		Use for low-scale CPU-based workloads that require less than 48 GB of RAM.
Azure Machine Learning compute clusters	Batch inference	GPU supported via machine learning pipeline.	Run batch scoring on serverless compute. Supports normal and low-priority VMs.
Azure Functions	(Preview) Real-time inference		
Azure IOT Edge	(Preview) IOT module		Deploy and serve ML models on IOT devices.
Azure Data Box Edge	Via IOT Edge	FPGA support	Deploy and serve ML models on IOT devices.

Figure 4.6 – Azure machine compute target

With this introduction to Azure Machine Learning capabilities, let's explore, step by step, how to build a classification model using the Azure Machine Learning service.

Getting started with the Azure Machine Learning service

In this section, we will explore a step-by-step walk-through of creating a classification model using Azure Machine Learning:

1. Sign up for a Microsoft account, unless if you already have one, then log into the Azure Machine Learning portal at `ml.azure.com`. Here, you will see the ML studio as shown in the following figure. An Azure subscription is essentially the way you pay for services. You can either use your existing subscription if you have one or sign up for a new one. For a brand-new user, the nice folks at Azure offer a $200 credit to get you acquainted. Make sure to turn off the resources when you are not using them; don't leave the data center lights on:

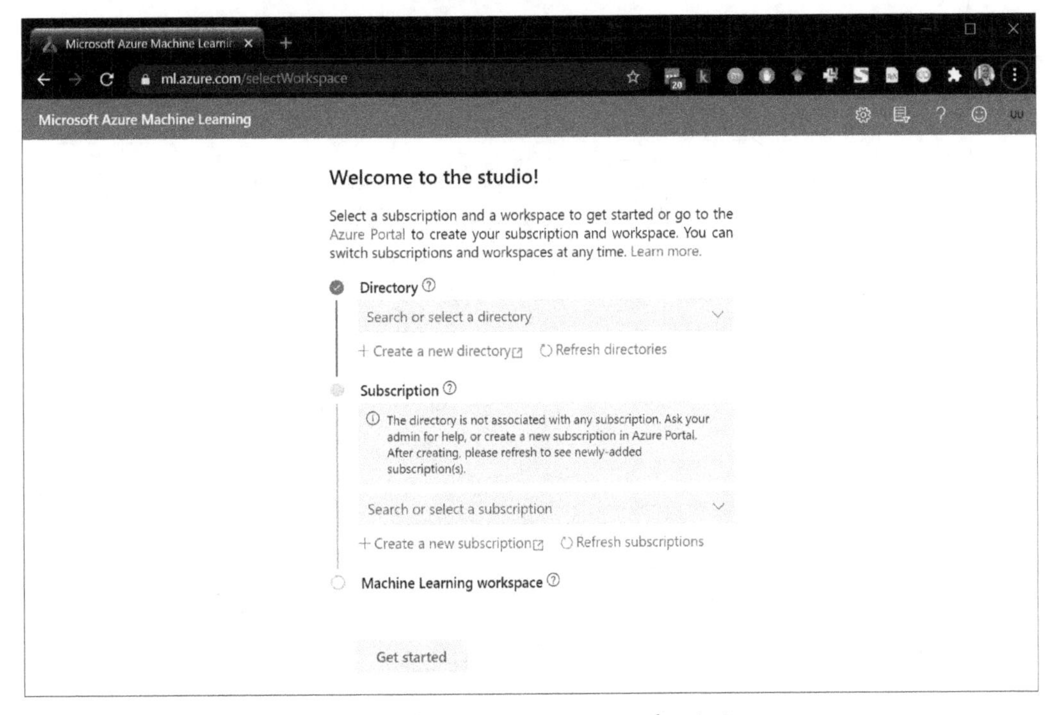

Figure 4.7 – Azure Machine Learning service subscription startup page

2. In the following figure, you can see we have now been asked to select a subscription. In this case, we'll choose **Free Trial** to explore the services. You can also choose **Pay-As-You-Go**, in which case, your account will be charged based on your compute, storage, and other service usage:

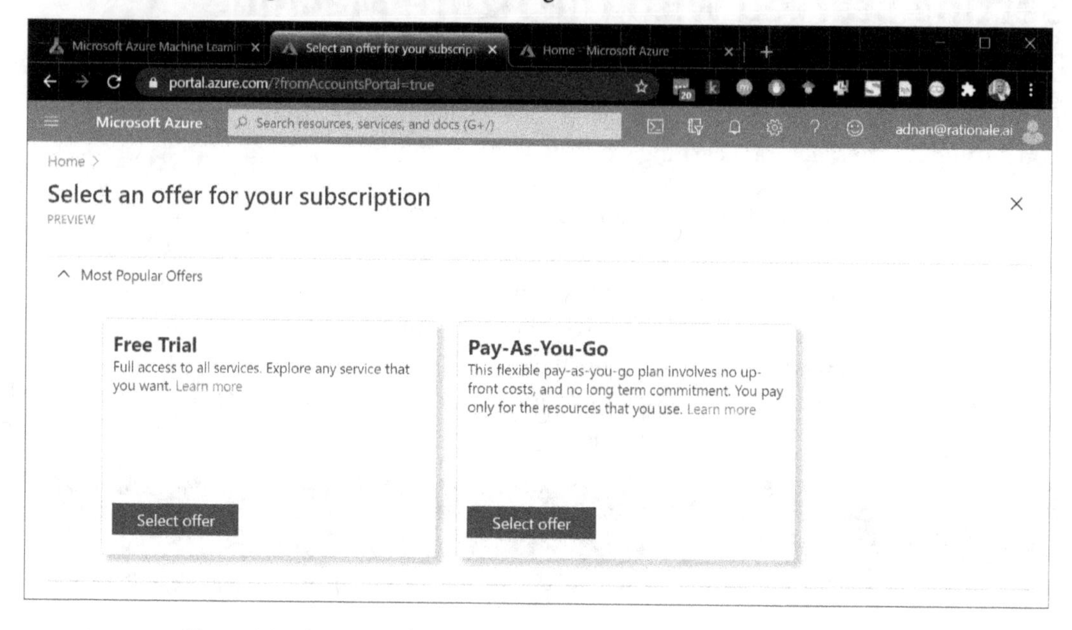

Figure 4.8 – Azure Machine Learning service subscription selection page

3. Now that you have chosen your subscription, you will be taken to the Azure portal, where you can do an overwhelming number of things. In this case, click on **Create a resource** and select the **Machine Learning** service:

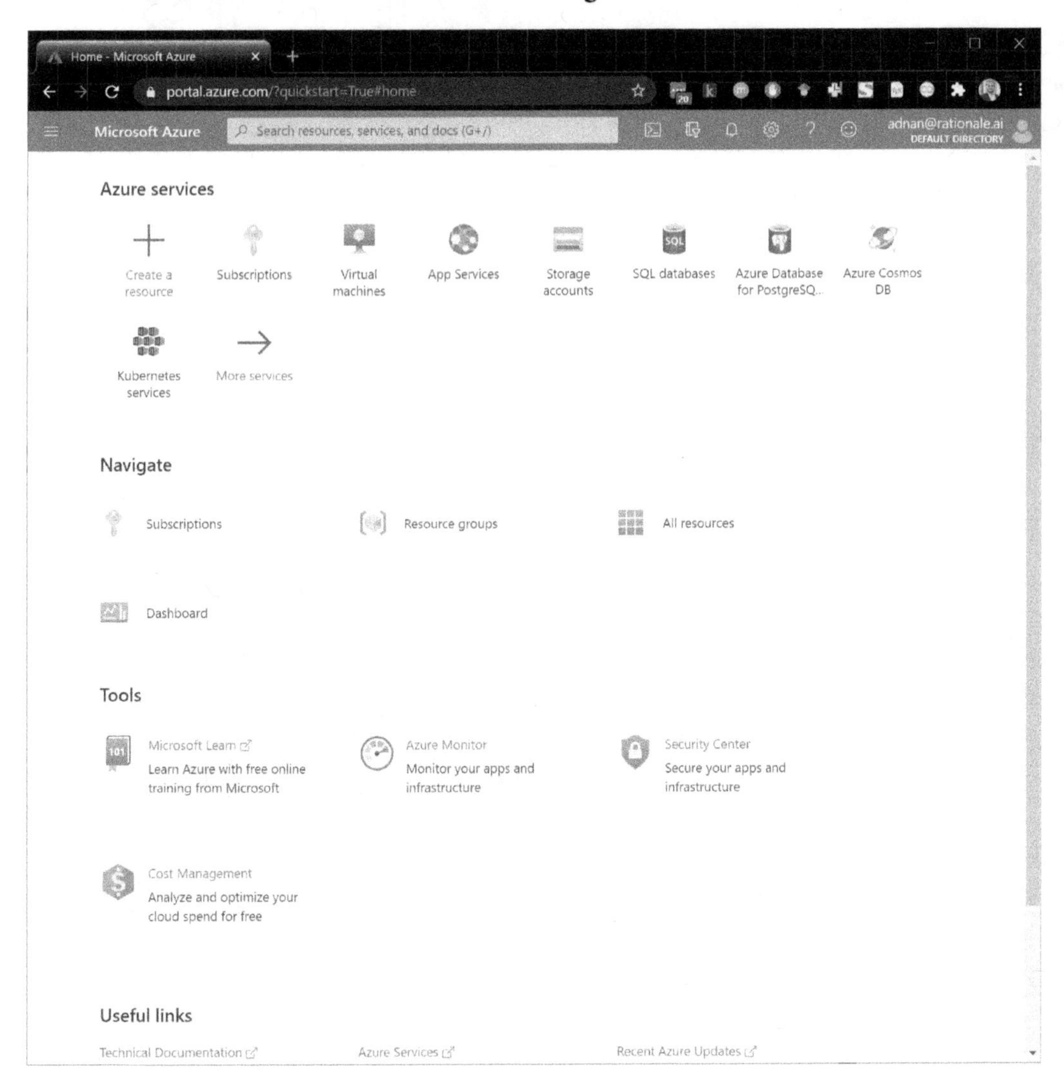

Figure 4.9 – Azure portal home page

Once you have selected the **Machine Learning** service, you will see the following screen to create an Azure Machine Learning service. This is where you can create one or more ML workspaces. Now proceed by clicking on the **Create Azure Machine Learning** button:

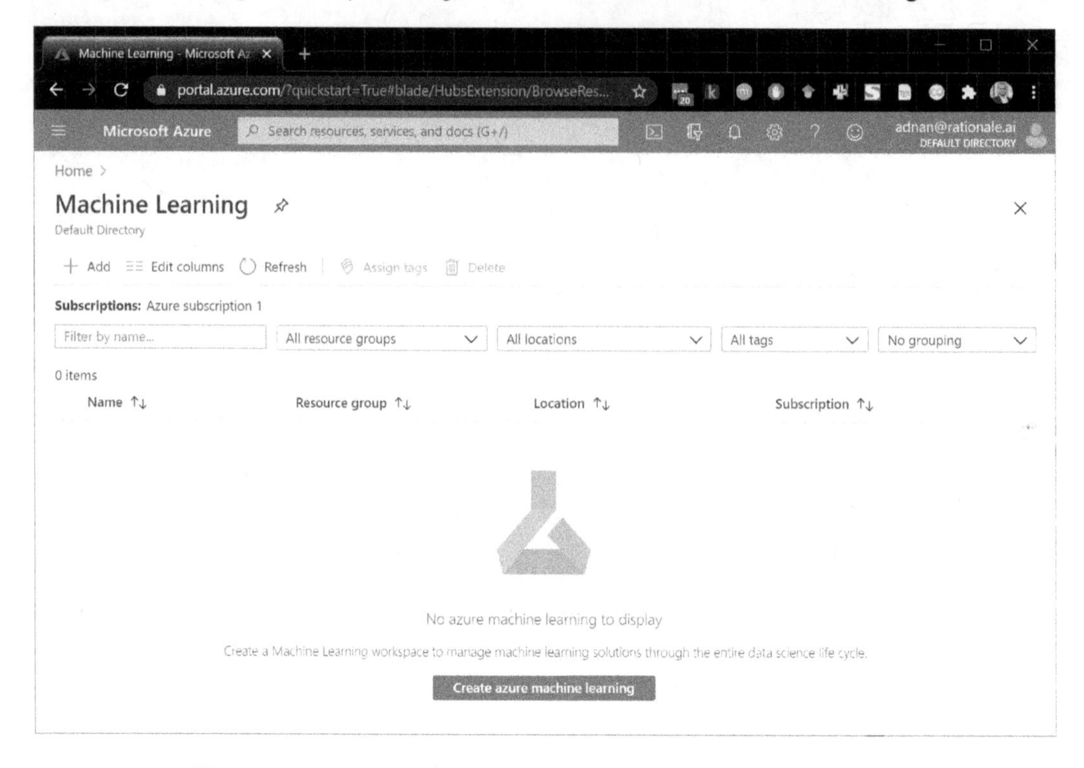

Figure 4.10 – Azure Machine Learning service startup page in the portal

4. Upon clicking **Create Azure Machine Learning**, you will be brought to the following page to create an ML workspace. This is where you will select your subscription, create a resource group, and select a geographical region:

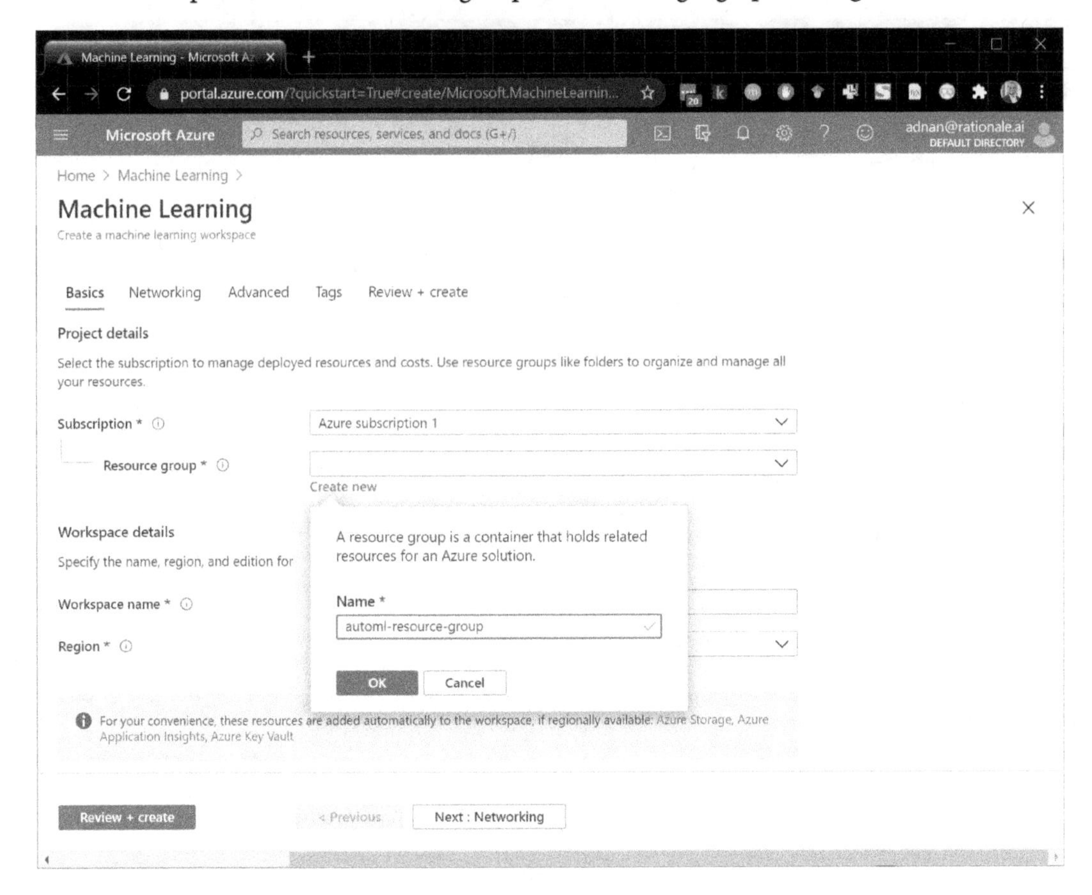

Figure 4.11 – Azure Machine Learning service – creating a workspace

5. The following figure shows the filled-out form for creating an ML workspace I called `auto-ml-workspace`. A **resource group** acts as a container for a collection of different resources, and hence associated assets (compute, storage, and so on):

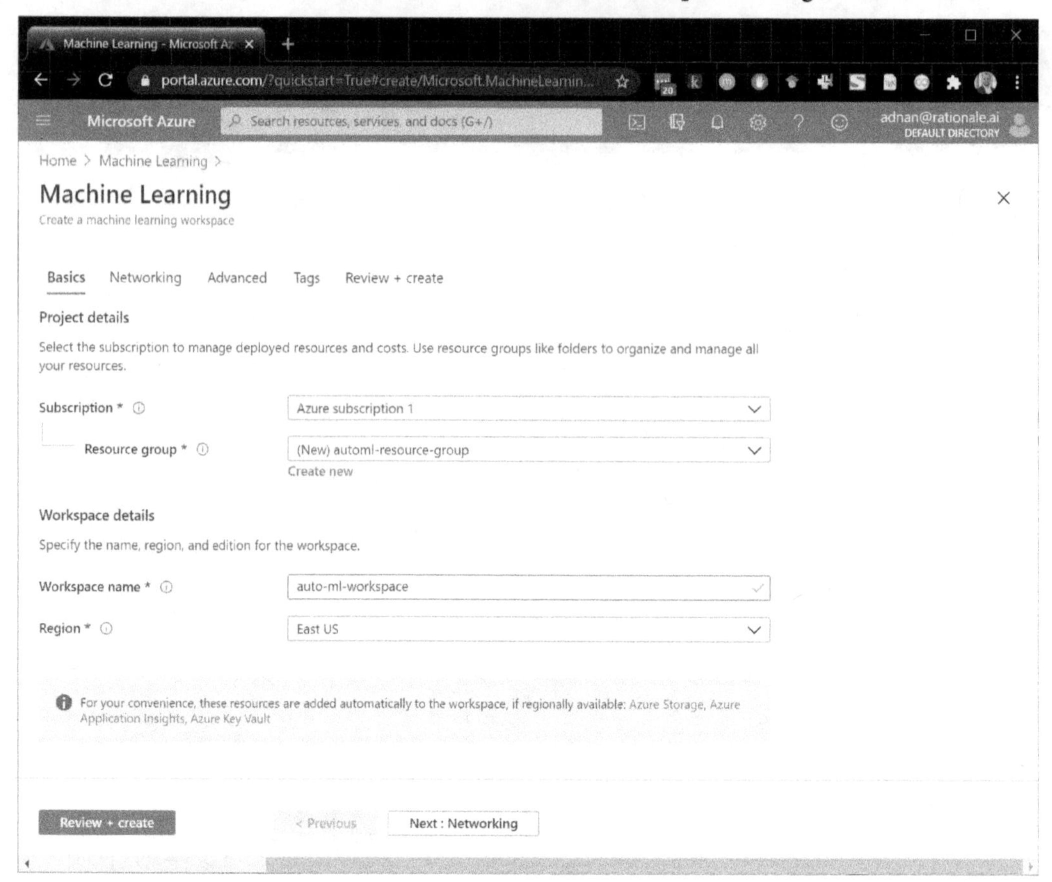

Figure 4.12 – Azure Machine Learning service page

A workspace can contain compute instances, experiments, datasets, models, deployment endpoints, and so on, and is the top-level resource in Azure Machine Learning. You can refer to the breakdown of Azure workspaces shown in the following figure:

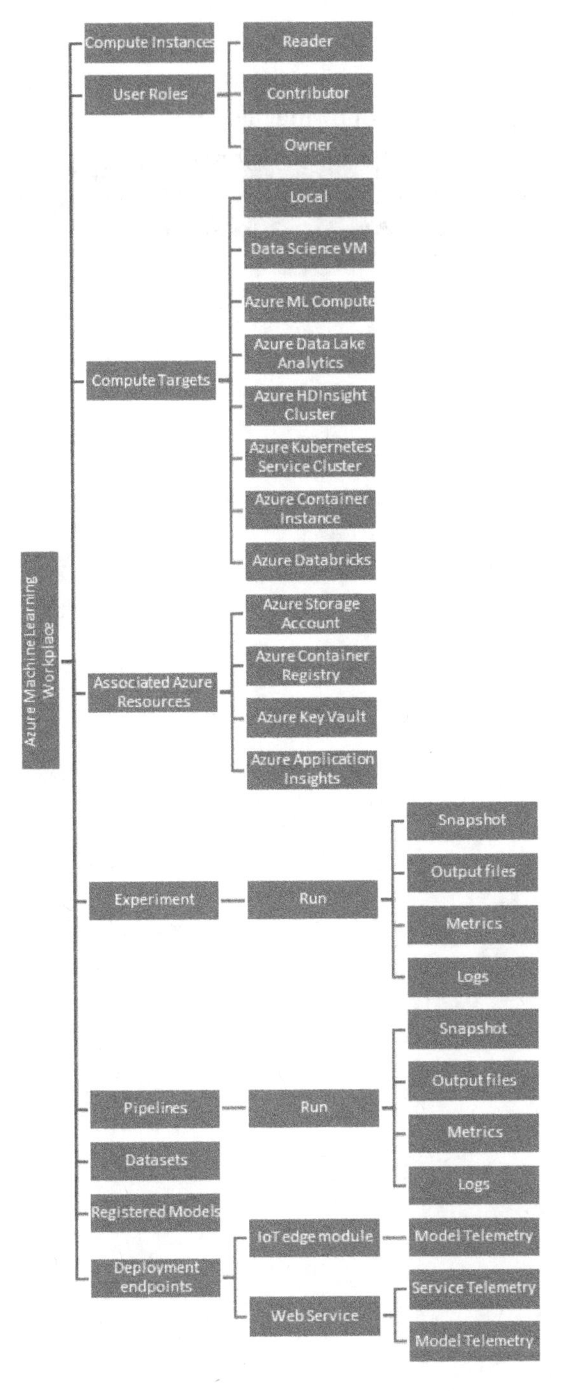

Figure 4.13 – What is an Azure Machine Learning workspace?

Upon clicking the **Create** button under **Project details**, your ML service will be deployed, and the associated dependencies will be created. It can take a few minutes. You will see something like the following figure:

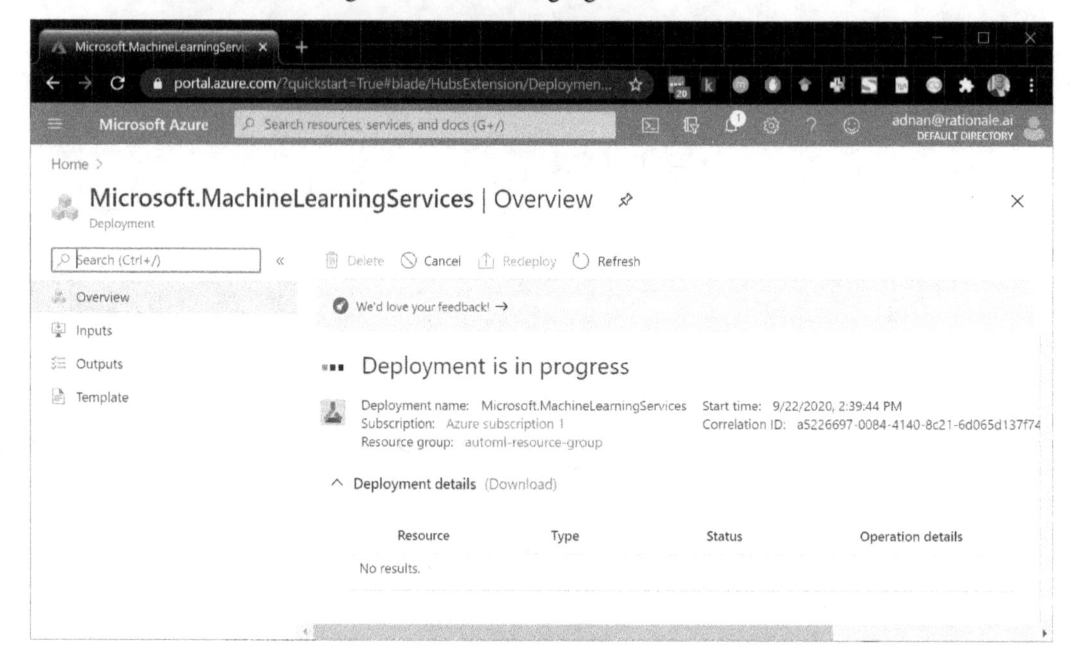

Figure 4.14 – Azure Machine Learning service deployment

6. Once the deployment is complete, you will see the resource group and its associated resources, as shown in the following figure. You can now click on the **Go to resource** button, which will take you to the ML workspace:

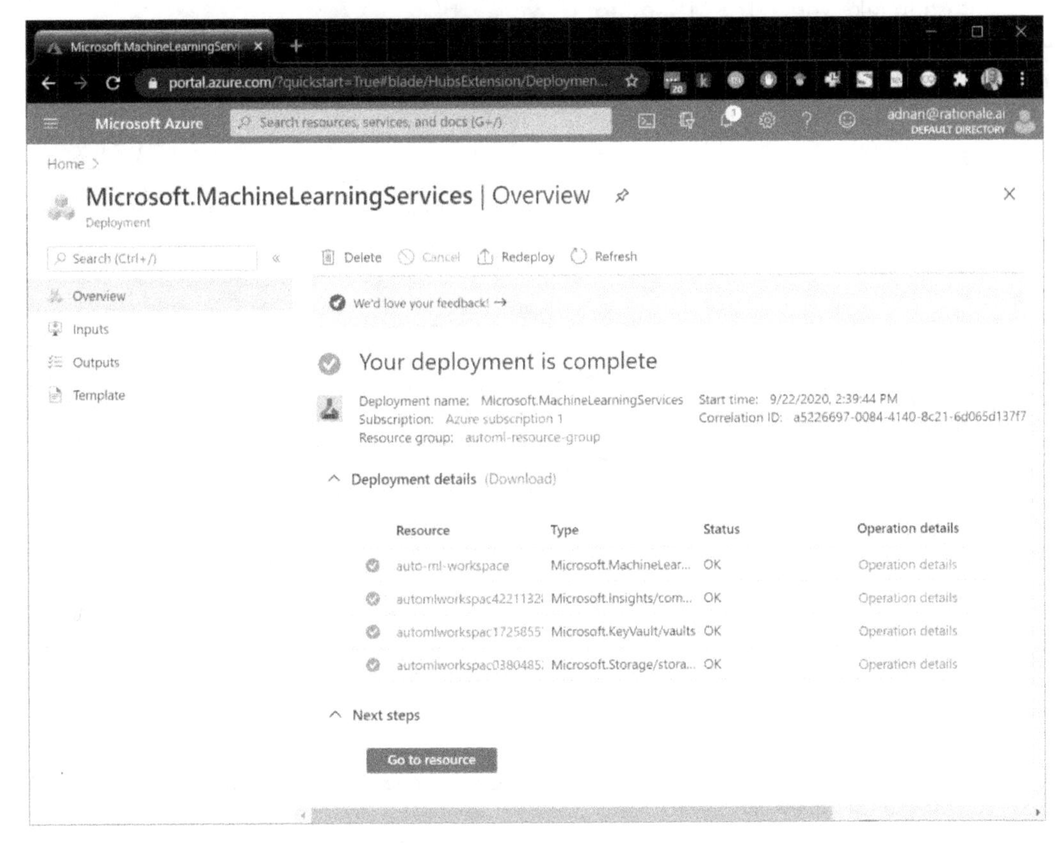

Figure 4.15 – Azure Machine Learning deployment completion screen

7. This is your ML workspace. Here, you can see your **Resource group**, **Subscription**, **Key Vault**, and all the important high-level details, but you are not in the ML arena, yet. There is one more click to go. So now, go ahead and click on the **Launch studio** button to take you to the ML studio:

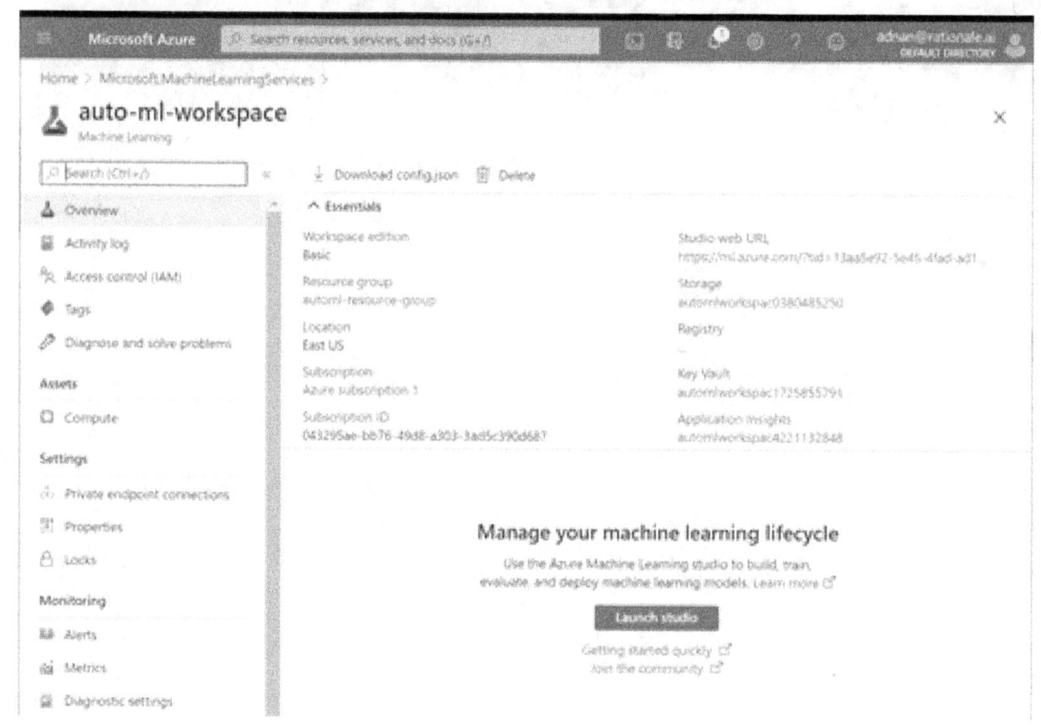

Figure 4.16 – Azure Machine Learning workspace console to launch studio

8. Now, the following is the screen you have been patiently waiting for all this time: the ML studio! Go ahead, take the tour by clicking on the **Start the tour** button – we will wait:

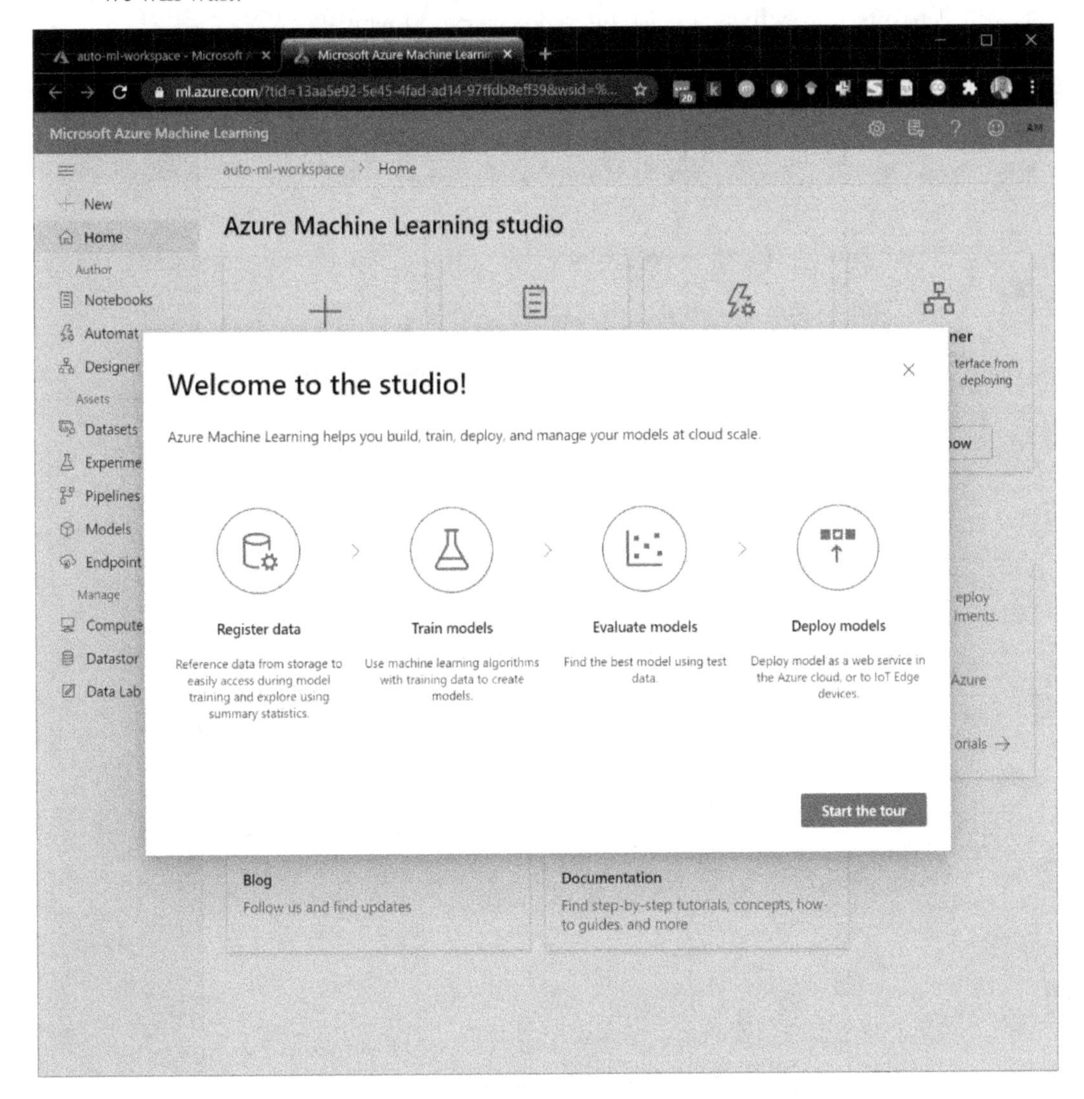

Figure 4.17 – Azure Machine Learning studio home page

The following figure shows the Azure Machine Learning studio. There used to be a "classic" version of the ML studio but it's no longer active so we will just focus on this new shiny web-based portal, which helps you do all things ML. On the left pane, you can see all the different offerings, which are also offered via a dropdown:

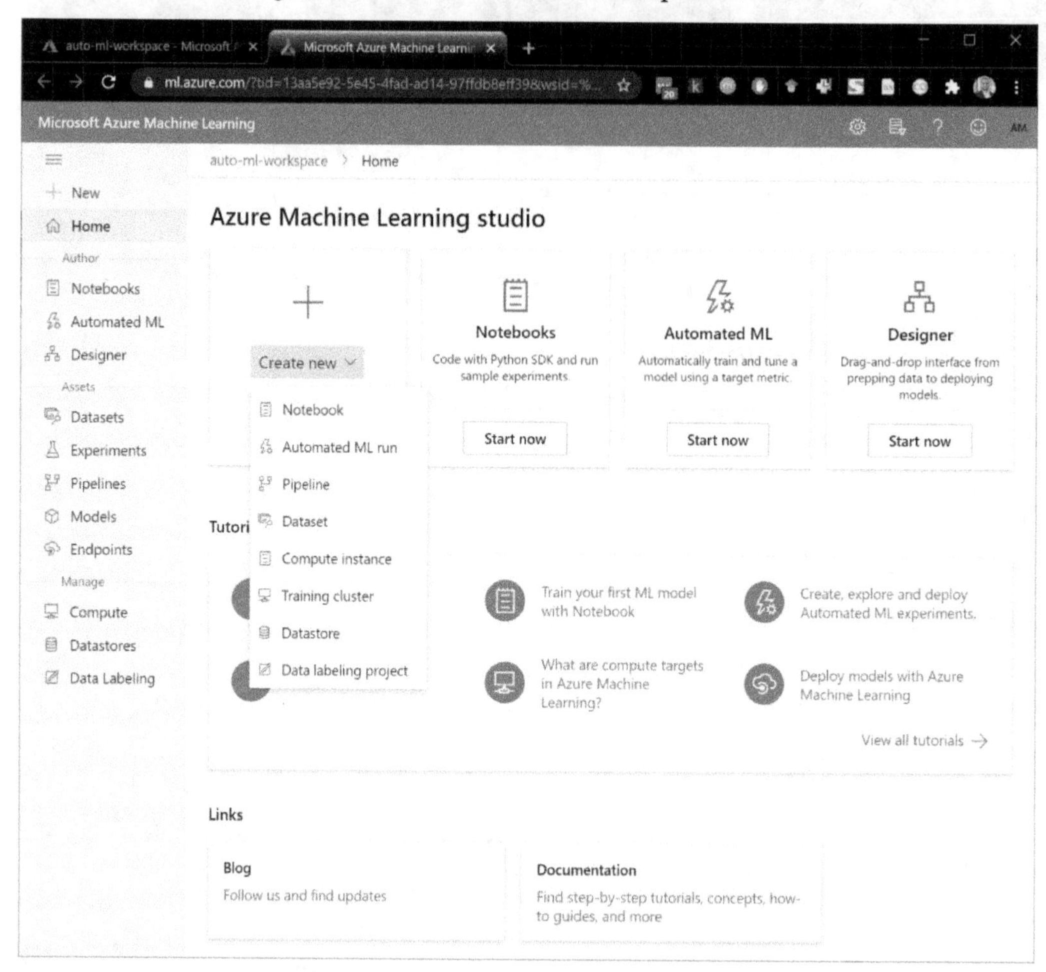

Figure 4.18 – Azure Machine Learning studio home page

The UI is modern, clean, and efficient. You can start by creating a new **Notebook**, an automated ML experiment, or a designer. Each of these offerings has a different purpose, yet they have a similar underlying infrastructure. Notebooks are great tools for hands-on data scientists, while automated ML experiments are targeted at AI democratization. The designer interface provides drag-and-drop capabilities to prepare data and deploy models.

Modeling with Azure Machine Learning

Before we create an automated ML workflow, let's start with a simple Azure notebook:

1. Azure notebooks are an integrated part of the Azure Machine Learning service, and you can either create or use a sample notebook to get started:

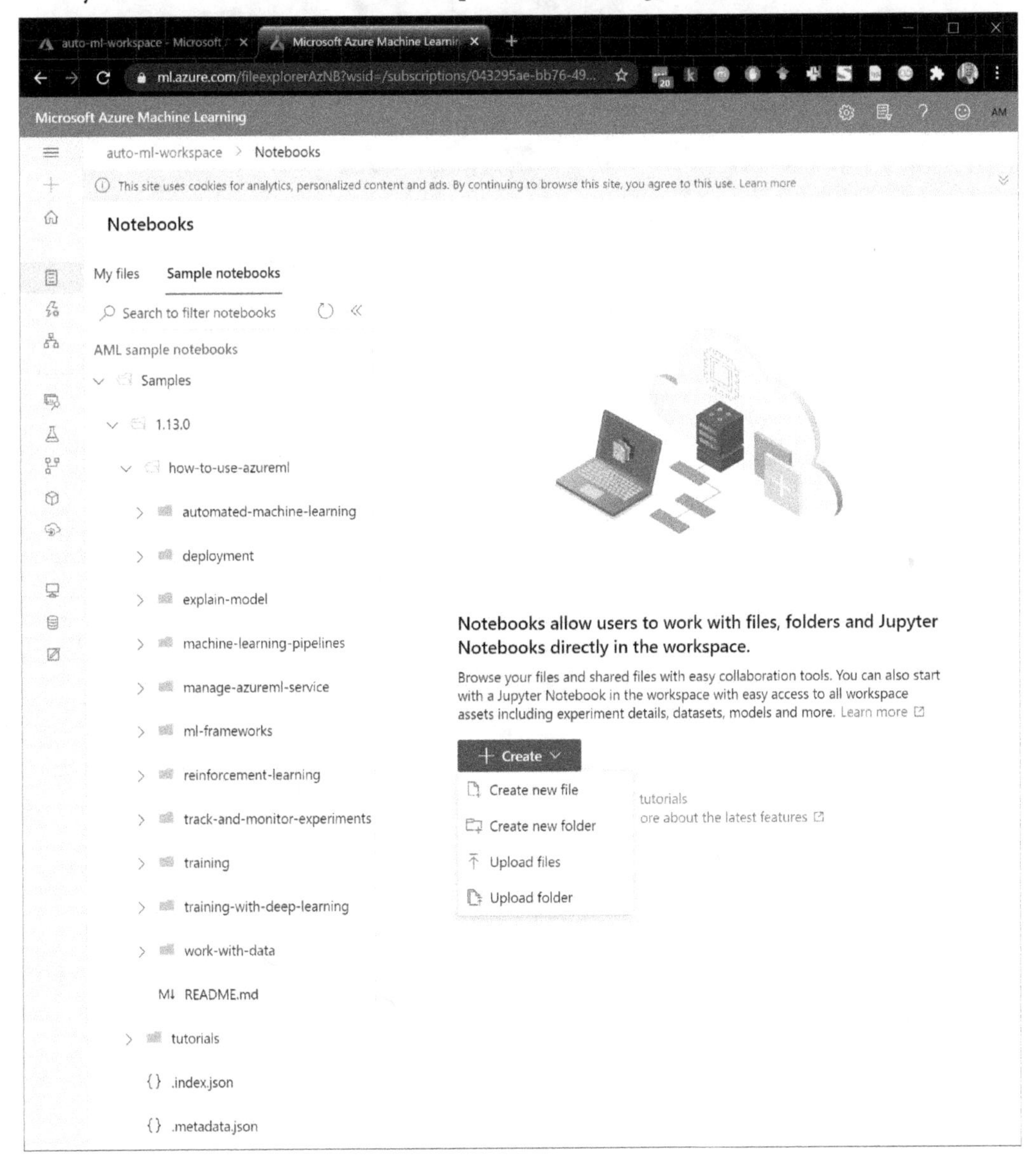

Figure 4.19 – Azure Machine Learning sample notebooks

2. In the **Search to filter notebooks** box in the left pane, as shown in the following figure, search for `MNIST` and it will filter to show you the notebooks. Select the `image-classification-part1-training.ipynb` file to see the notebook in the right pane, and click on **Clone this notebook** to create your own copy:

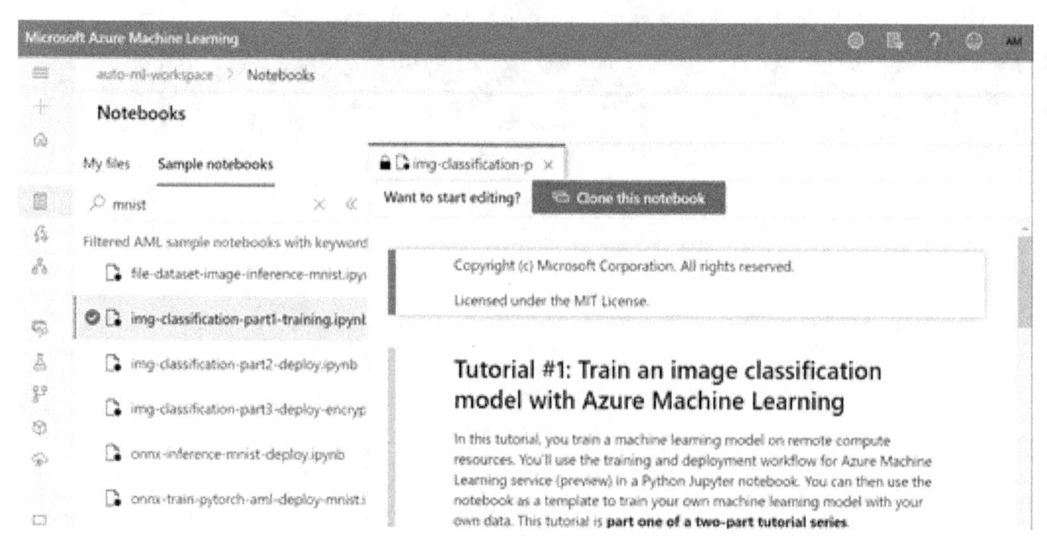

Figure 4.20 – MNIST image classification notebook

3. Click on the **Clone this notebook** button to clone the notebook. Cloning the notebook copies the notebook and associated configurations into your user folder as shown in the following figure. This step copies the notebooks and `yml` configuration files to the `user` directory:

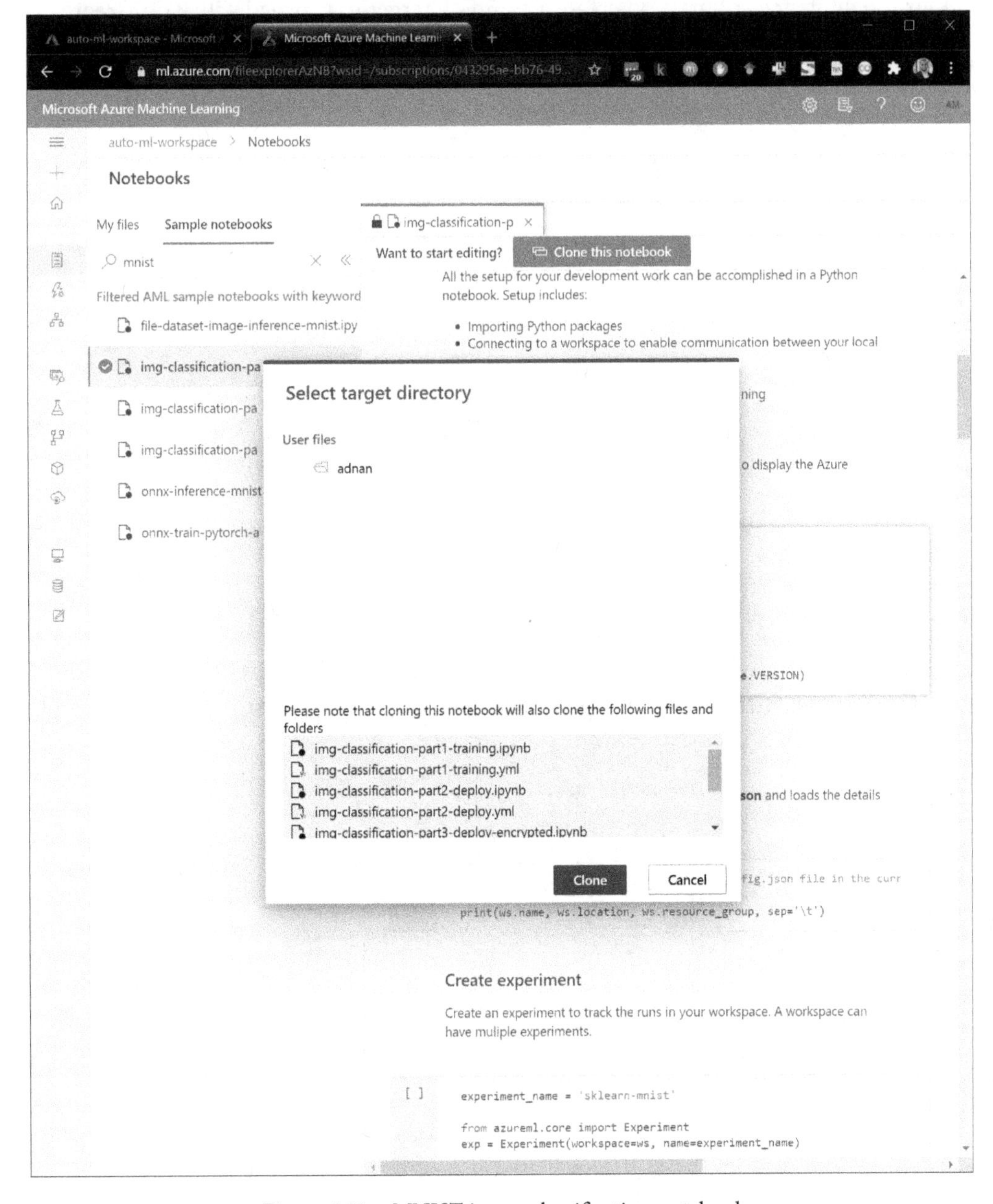

Figure 4.21 – MNIST image classification notebook

4. Now that you have cloned the assets, it's time to have a compute target. You cannot run a notebook without, well, actually having a machine to run it on. As you saw earlier, we were able to run code either on Google Colab or locally. In this case, we are running this workload on Azure, which requires us to be more explicit about our intent, hence you have to create a compute target to proceed. Click on **Create compute**:

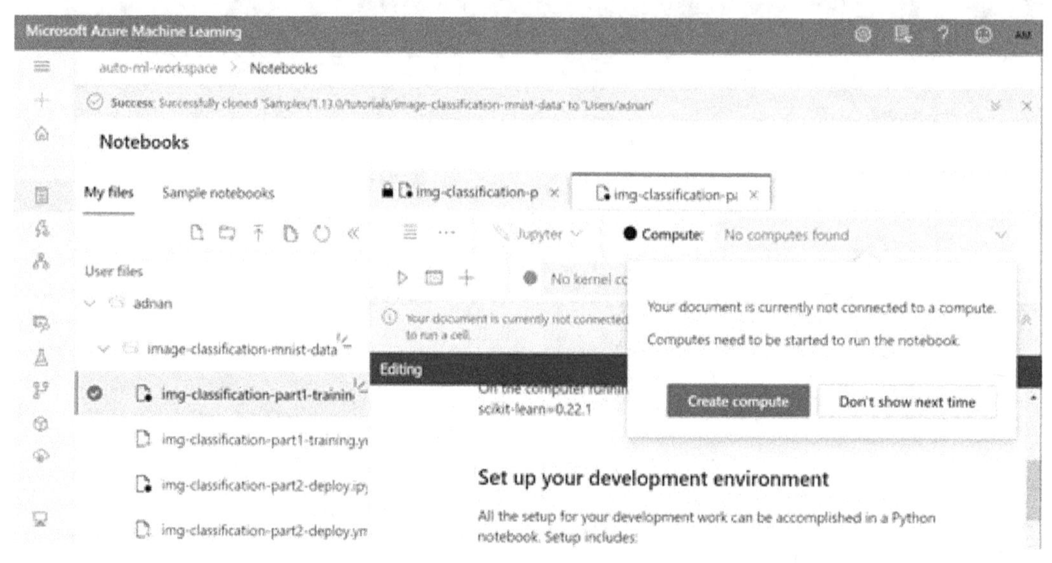

Figure 4.22 – MNIST image classification notebook

5. Once you've clicked on **Create compute**, you will be shown what kind of compute options are on offer, along with their associated cost. It's no surprise that bigger and better hardware will cost you more:

> **Note**
>
> You can read about different types of VMs and their associated costs at `https://azure.microsoft.com/en-us/pricing/details/virtual-machines/series/`.

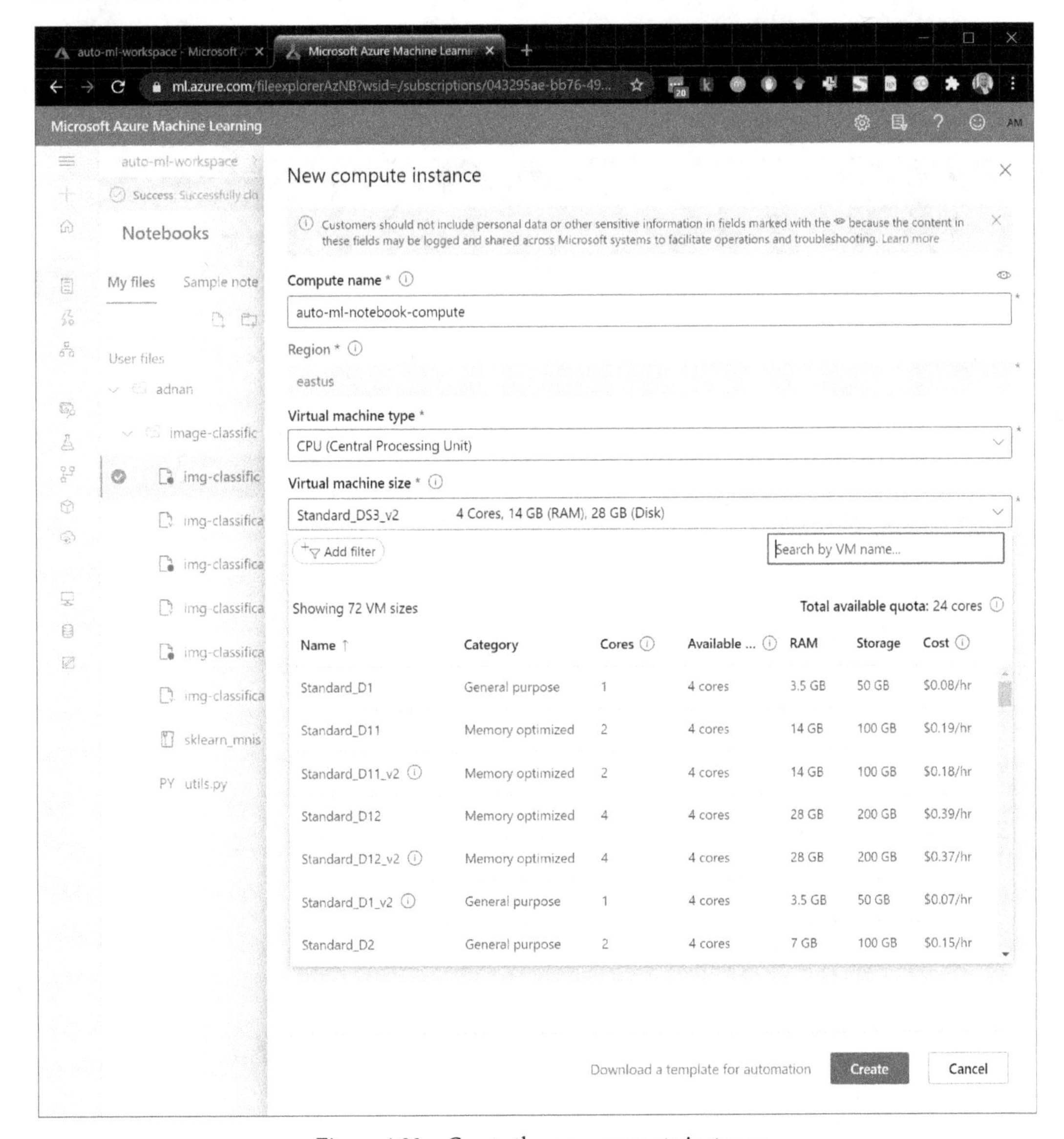

Figure 4.23 – Create the new compute instance

6. For the purpose of this AutoML notebook, we will select a standard small VM, as seen in the following figure, and create the compute instance. You can create multiple compute instances and assign them to different experiments as needed:

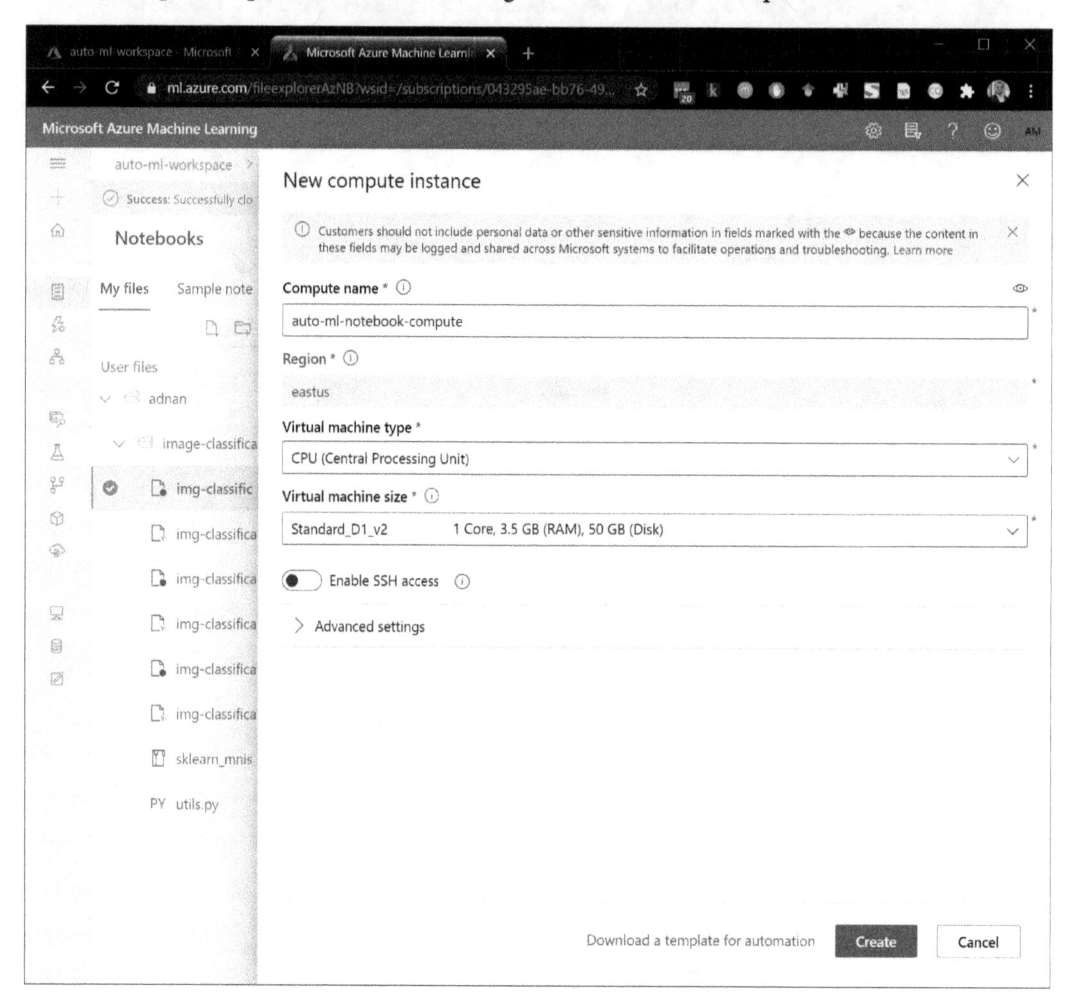

Figure 4.24 – Creating a new compute instance

7. Click on **Create** to create a new compute instance – as you create a compute instance, which can take some time as you stare at the revolving circle of patience, you can explore other parts of the Azure Machine Learning portal:

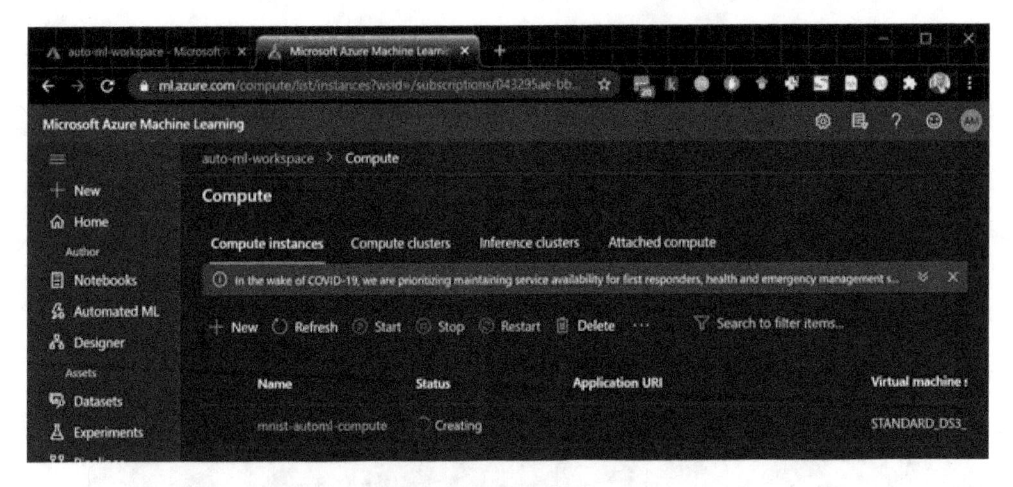

Figure 4.25 – MNIST automated ML compute instance

Now, the compute is ready to be used, as seen in the following figure. First, you get the status **Starting**, seen in the following figure:

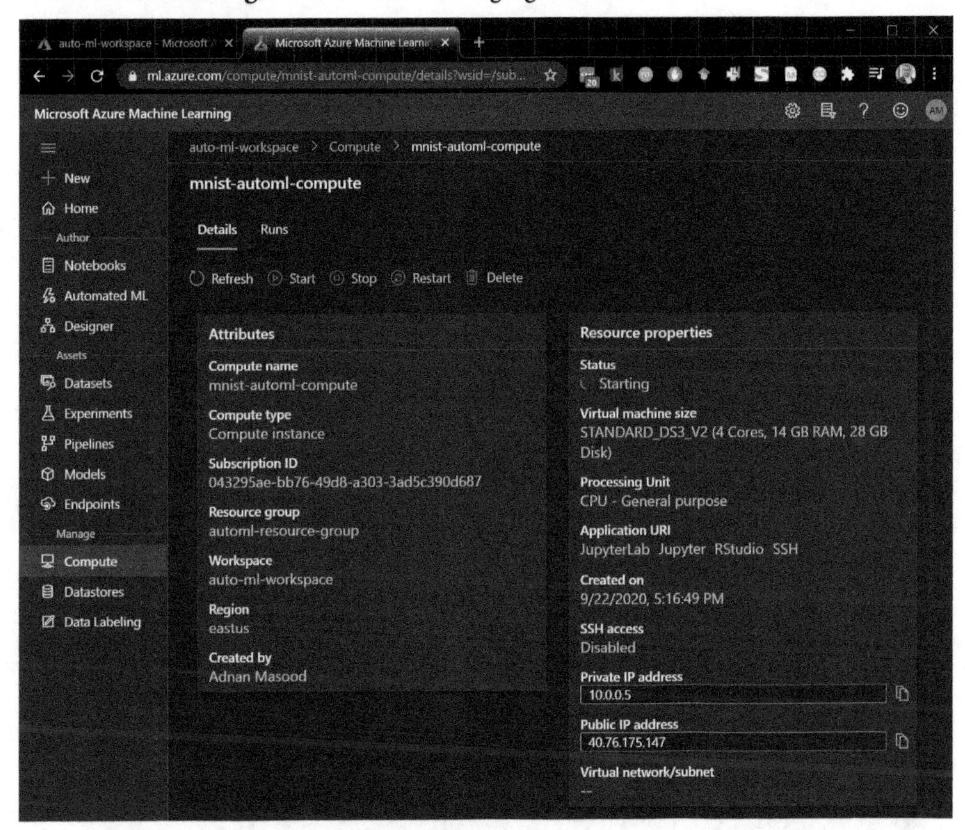

Figure 4.26 – Compute instance attributes

The compute instance status will change to **Running** when the compute instance is ready. You can choose to **Stop**, **Restart**, or **Delete** a compute resource as you please. Just remember that it will jeopardize its dependencies (things using or planning to use this compute resource):

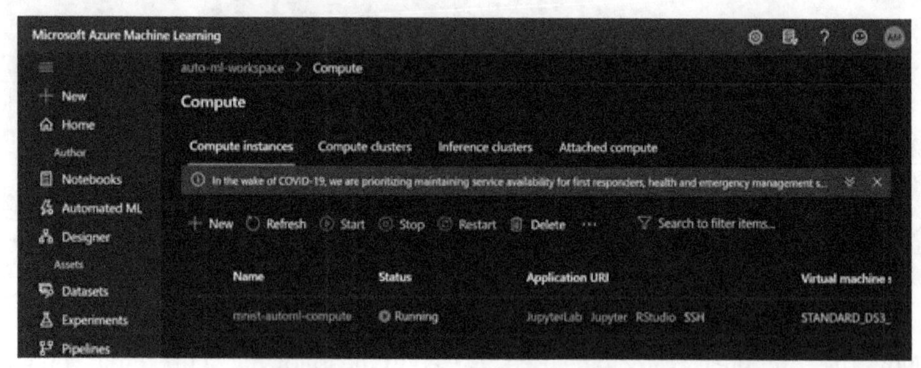

Figure 4.27 – Compute instance running

8. Navigate back to the notebook by clicking on the **Notebooks** link in the left pane. Now that we have the `image-classification-mnist-data` notebook open, we can run the code to make sure that it works. And you can see the Azure Machine Learning SDK version gets printed in the following figure:

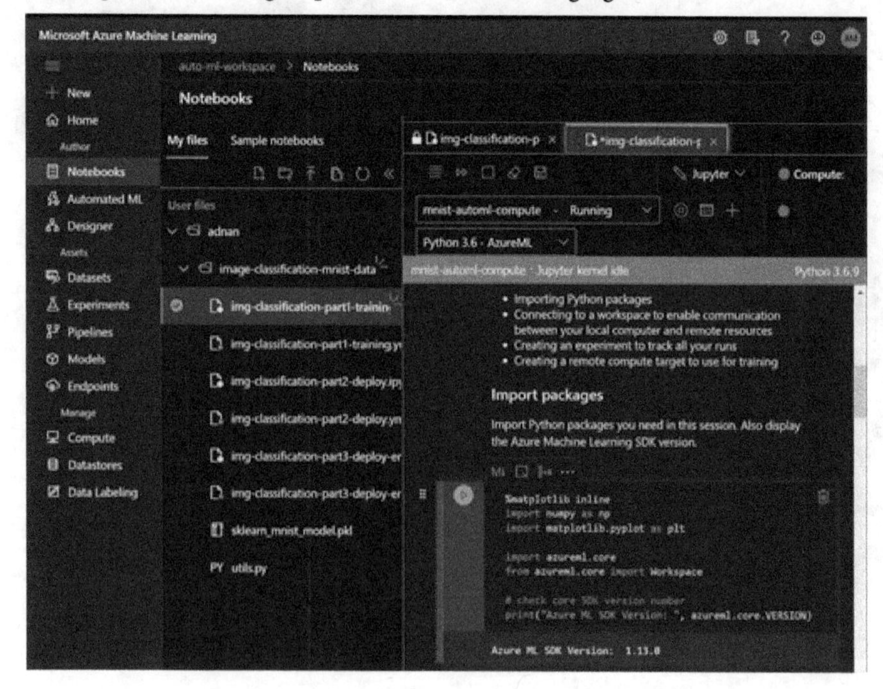

Figure 4.28 – MNIST image classification notebook

9. One configuration step still remains – you have to authenticate to use the workspace to be able to utilize the resources. In order to do that, there is an interactive authentication system built in where you click the link (`devicelogin`) shown in the following figure and enter the code (also in the following figure) to authenticate. I am sure you won't be using the code in the following figure – it won't work, but nice try!

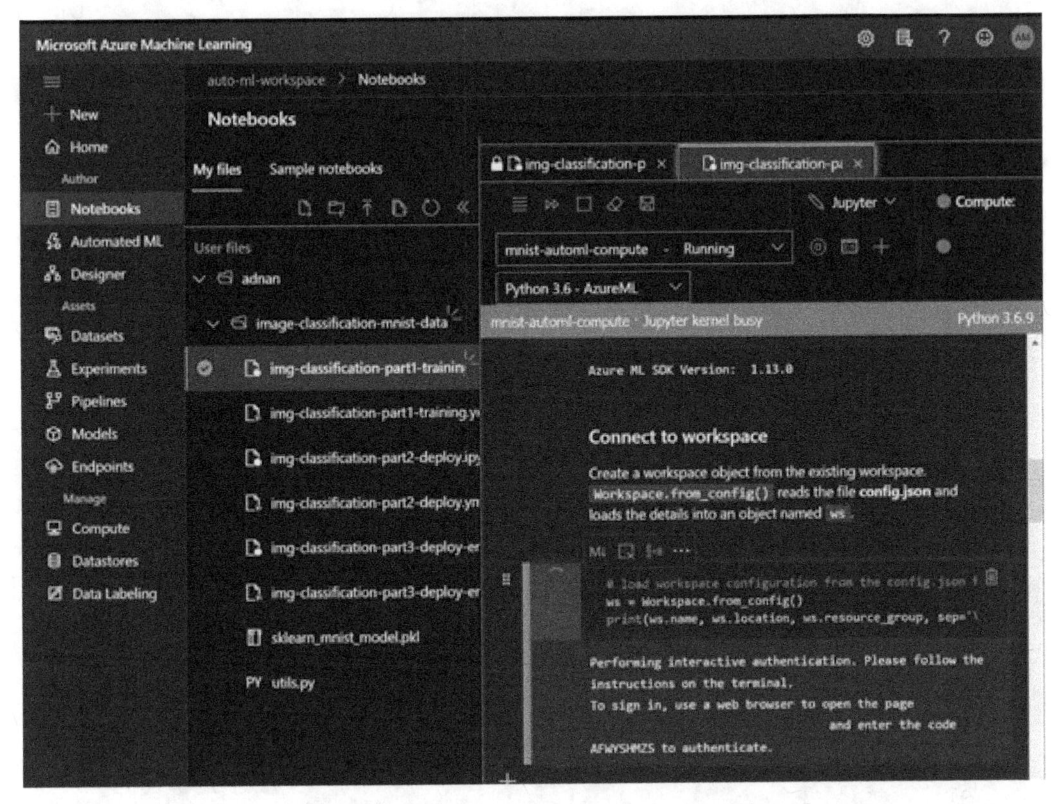

Figure 4.29 – MNIST image classification notebook

10. Now that we are authenticated and all, let's create a compute target. All the `boilerplate` code for configuration is already written for you as part of the Jupyter notebook. By running the cell shown in the following figure, you can now connect to the `ComputeTarget` programmatically and provision the node:

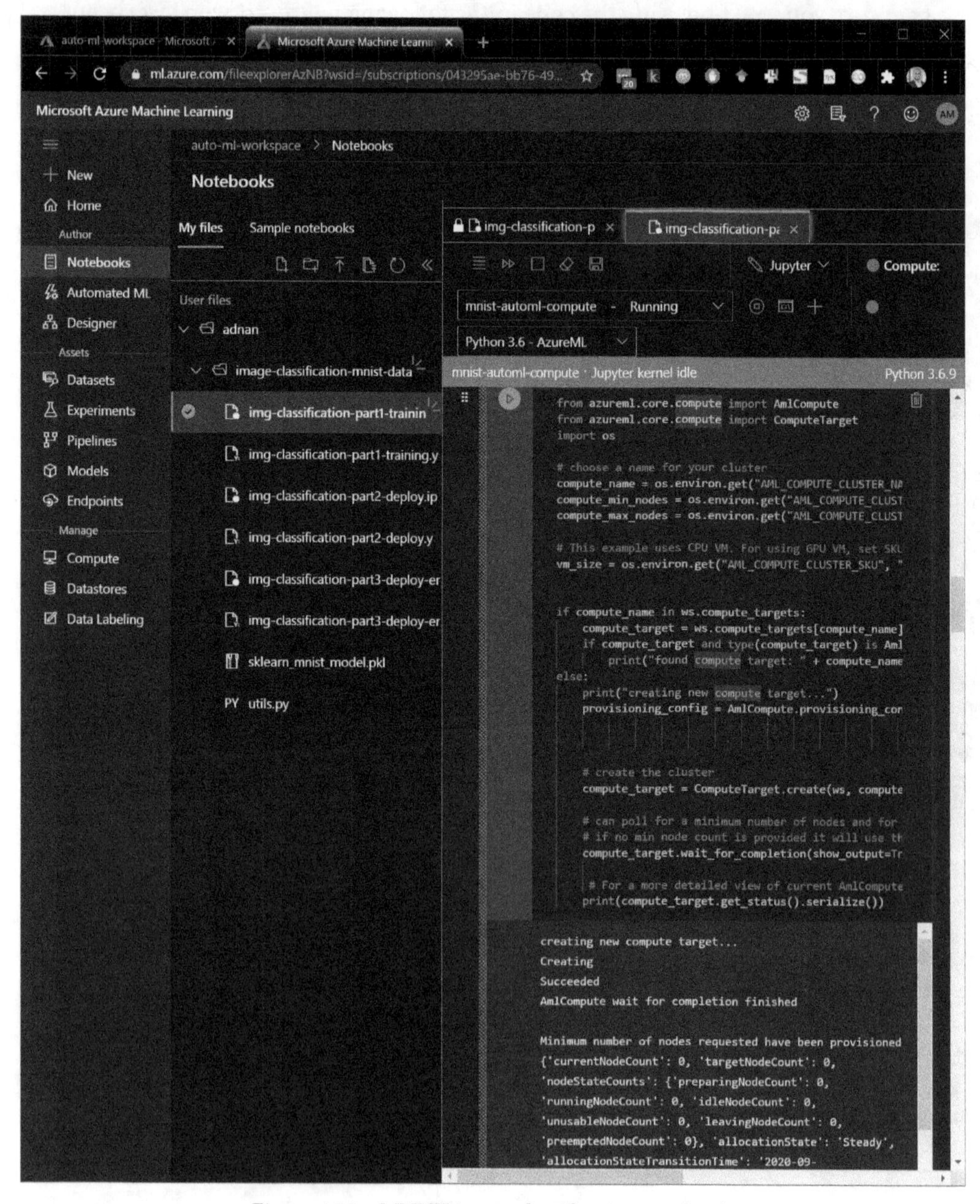

Figure 4.30 – MNIST image classification notebook

11. Now it's time to programmatically download the MNIST dataset for training. Yann LeCun (Courant Institute, NYU) and Corinna Cortes (Google Labs, New York) hold the copyright of the MNIST dataset, which is a derivative work from the original MNIST datasets. The MNIST dataset is made available under the terms of the Creative Commons Attribution-Share Alike 3.0 license. Once downloaded, you can also visualize it as shown in the following notebook screenshot:

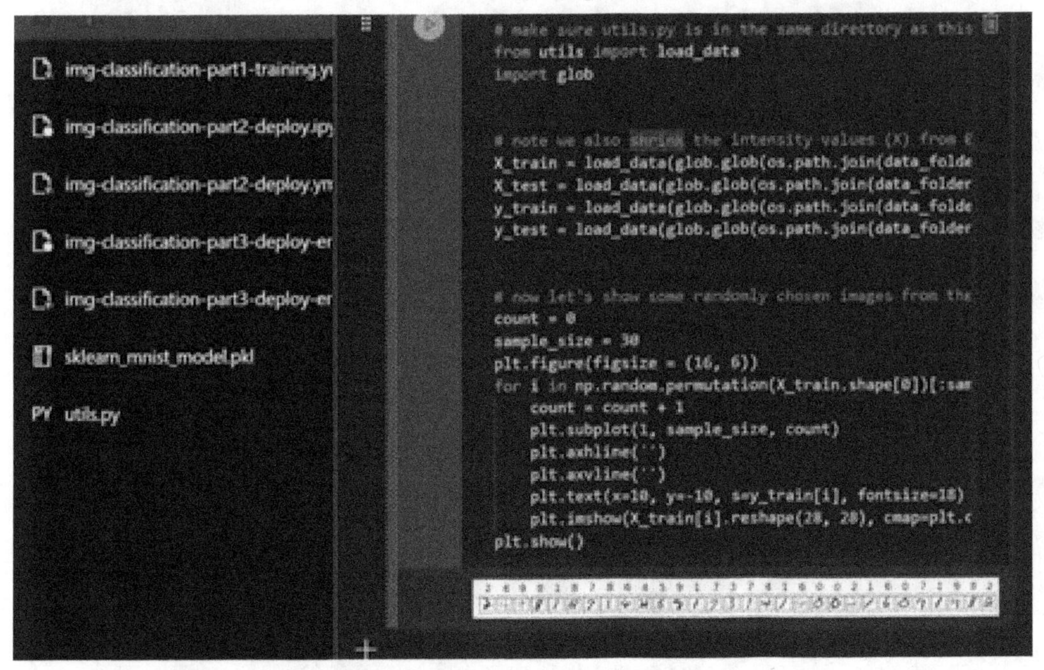

Figure 4.31 – MNIST image classification notebook

12. As you probably have remembered from our earlier adventures with MNIST, we will create the training and testing split for the dataset as shown in the following figure and train a logistic regression model. We'll now also create an estimator, which is used to submit the run to the cluster:

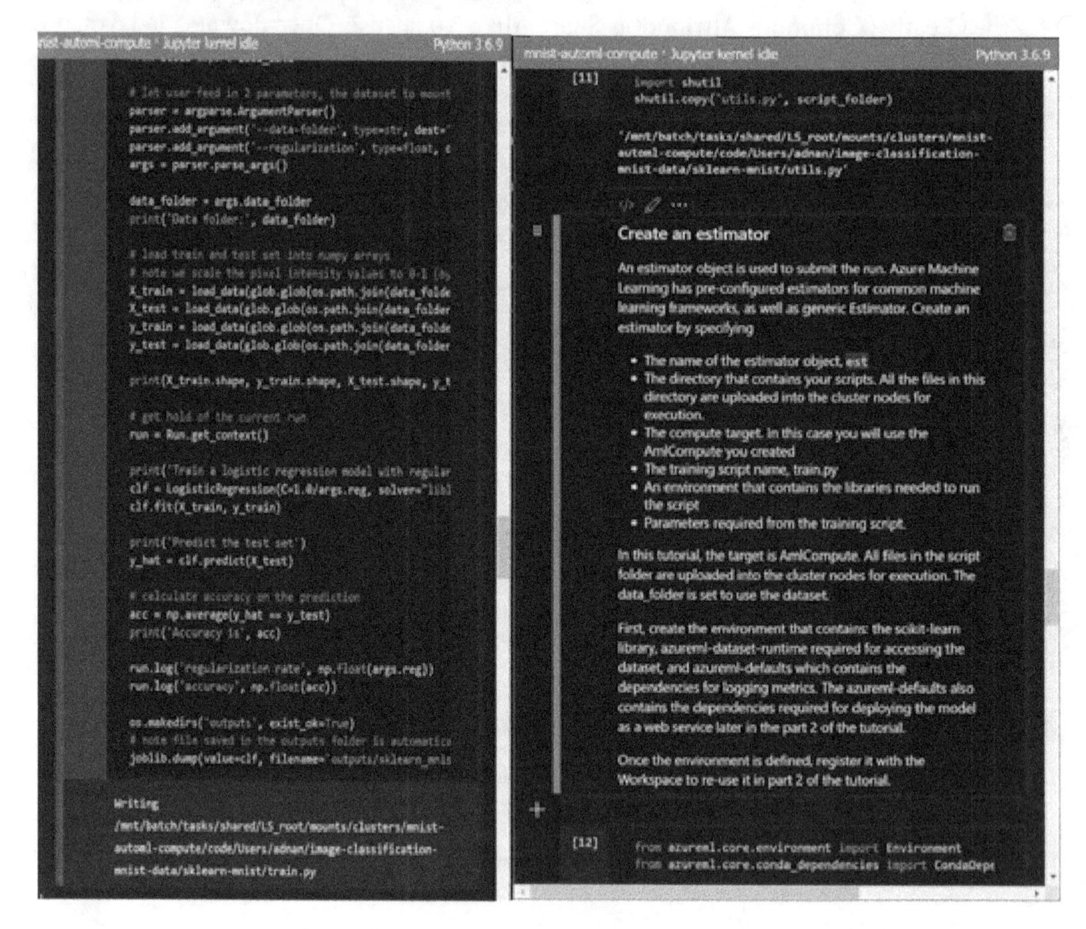

Figure 4.32 – MNIST image classification notebook

When working in Azure Machine Learning, one cardinal rule to remember is that all experiments and associated runs are interlinked to maintain the consistency of the ecosystem. This helps a lot since, regardless of where you run your experiment (in a notebook, JupyterLab, as custom code, and so on), you can pull up the runs and see the details. More on this soon.

In the following figure, you see the code demonstrating the creation of an estimator, and then submitting the job to the cluster by calling the `submit` function of the `experiment` object:

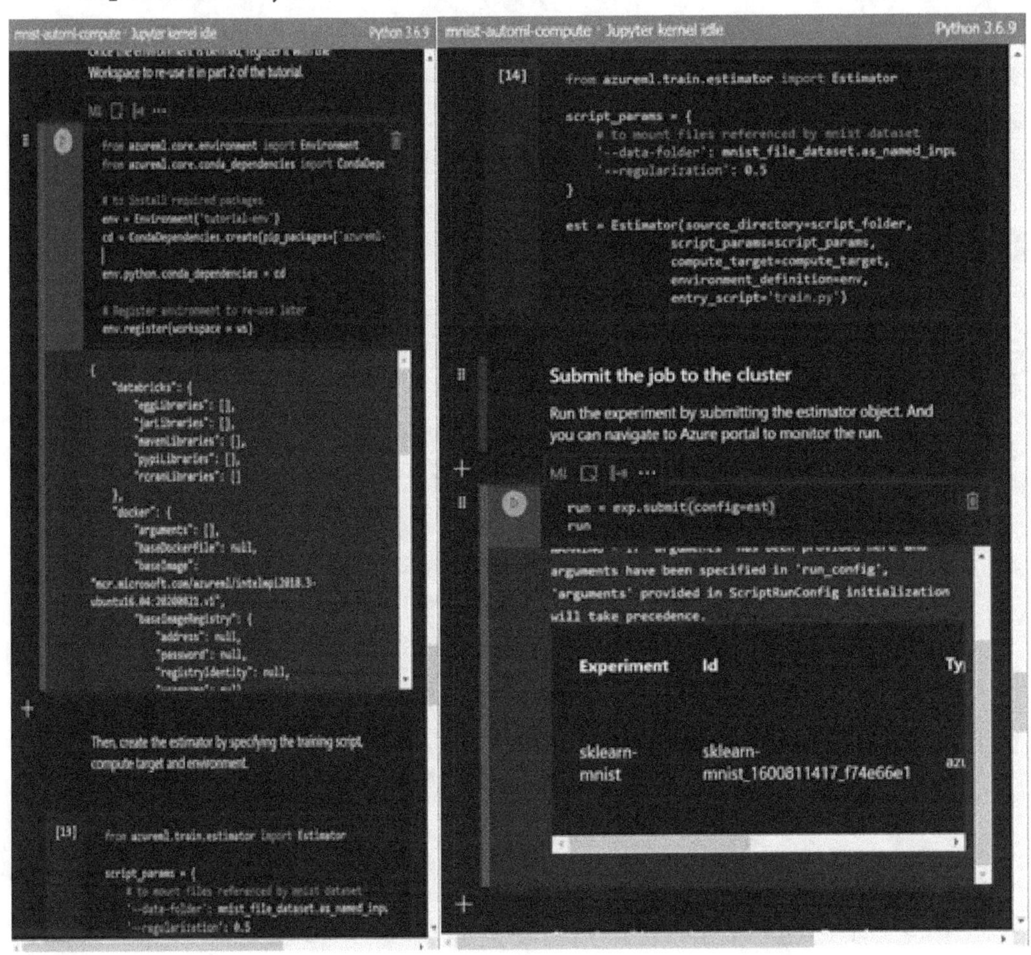

Figure 4.33 – MNIST image classification notebook

13. At this point, running the cells in the following figure demonstrates how you can visualize the details of `experiment` using the Jupyter widget `wait_for_completion` method to see the status of the job. This shows the job running on the remote cluster, and the corresponding build log as part of the widget. You can see the **Run** details as part of the widget as shown in the following figure:

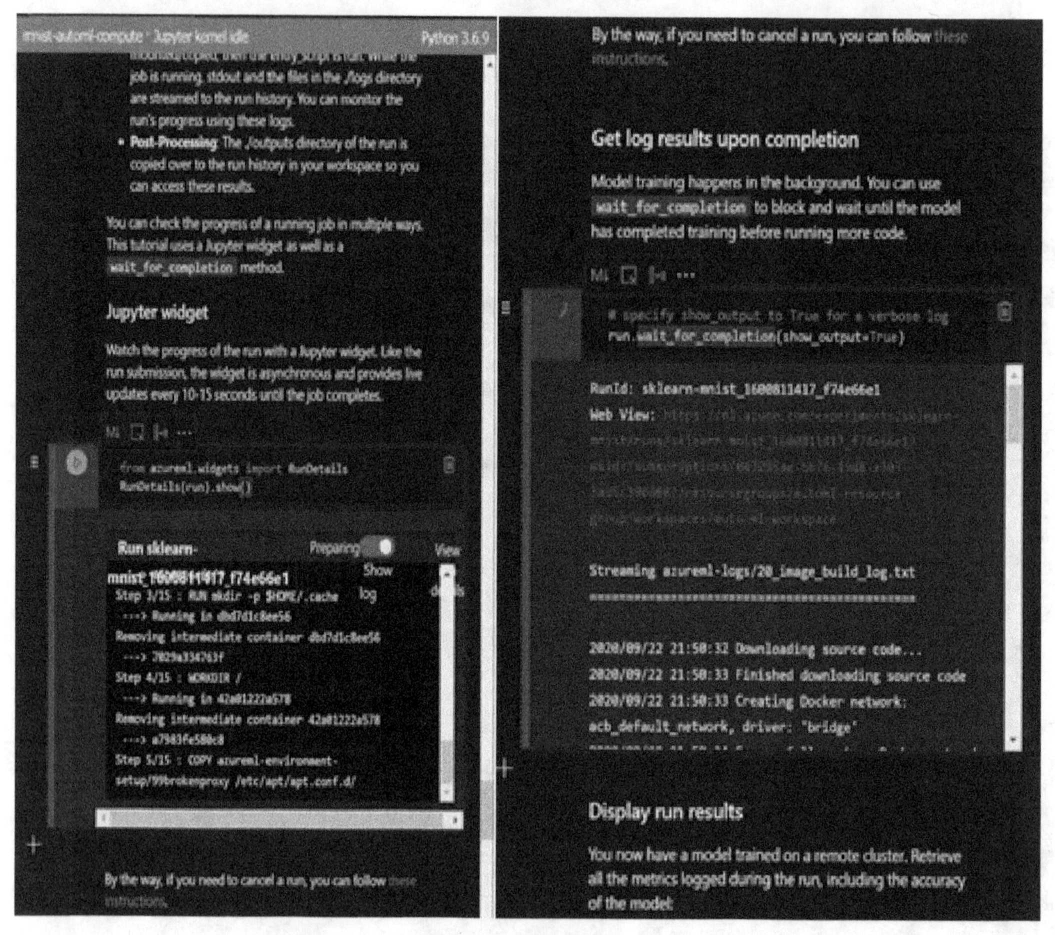

Figure 4.34 – MNIST image classification notebook

As the job runs on the remote cluster, and we see the results streaming in, you can observe the training unfold in front of your eyes with the corresponding percentage indicator:

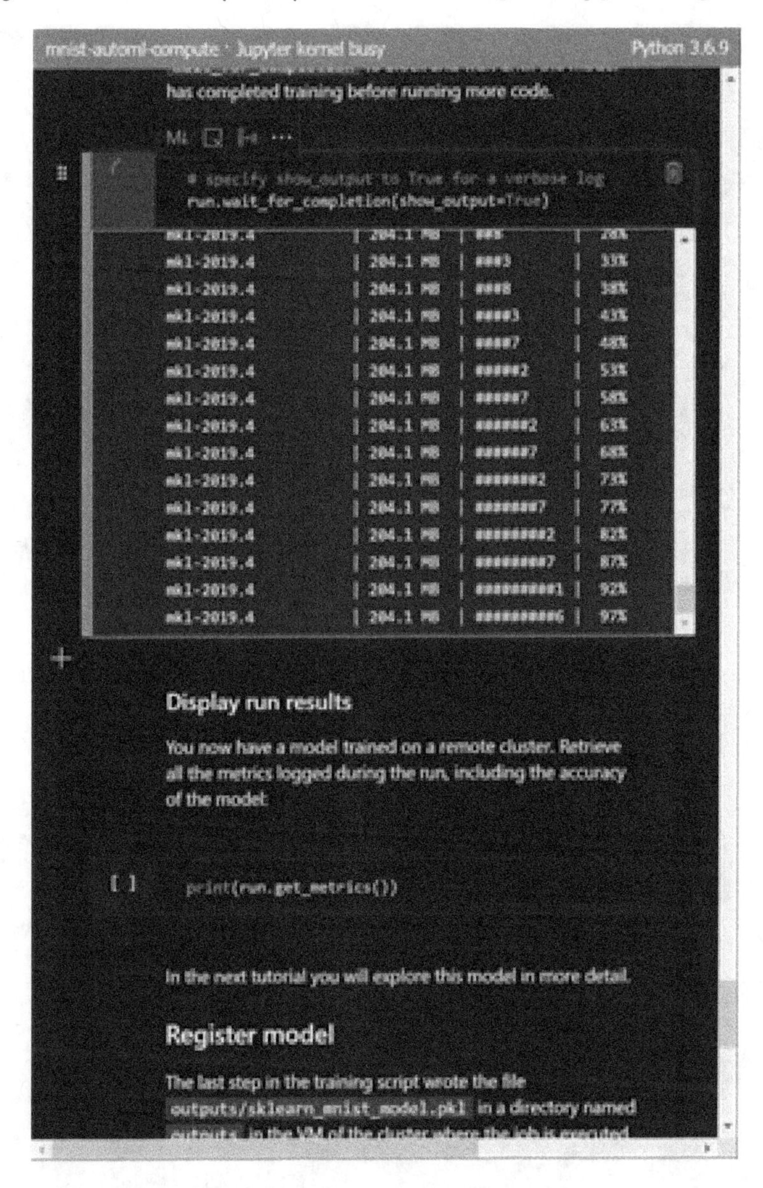

Figure 4.35 – MNIST image classification notebook

As the job completes, you will see the time taken and the run ID displayed in the widget, as shown in the following screenshot:

Get log results upon completion

Model training happens in the background. You can use `wait_for_completion` to block and wait until the model has completed training before running more code.

```
[14]  # specify show_output to True for a verbose log
      run.wait_for_completion(show_output=True)

2020/09/23 02:00:41 The following dependencies were found:
2020/09/23 02:00:41
- image:
    registry: 934b9f82f47a4ac8bede2c0ab17f6a6a.azurecr.io
    repository: azureml/azureml_0a6838d76e7468052f3a857ca80cfaa3
    tag: latest
    digest: sha256:09cdccf921b046044cb49ecf0464d04435ff952cadc1c054a5ae28b913433103
  runtime-dependency:
    registry: mcr.microsoft.com
    repository: azureml/intelmpi2018.3-ubuntu16.04
    tag: 20200821.v1
    digest: sha256:8cee6f674276dddb23068d2710da7f7f95b119412cc482675ac79ba45a4acf99
  git: {}

Run ID: ca1 was successful after 5m2s
```

Figure 4.36 – MNIST image classification notebook

You can also see the details of the experiment in the web UI by clicking on the details page that follows. You can see detailed documentation about how experiments work by looking at the docs page link shown in the following screenshot:

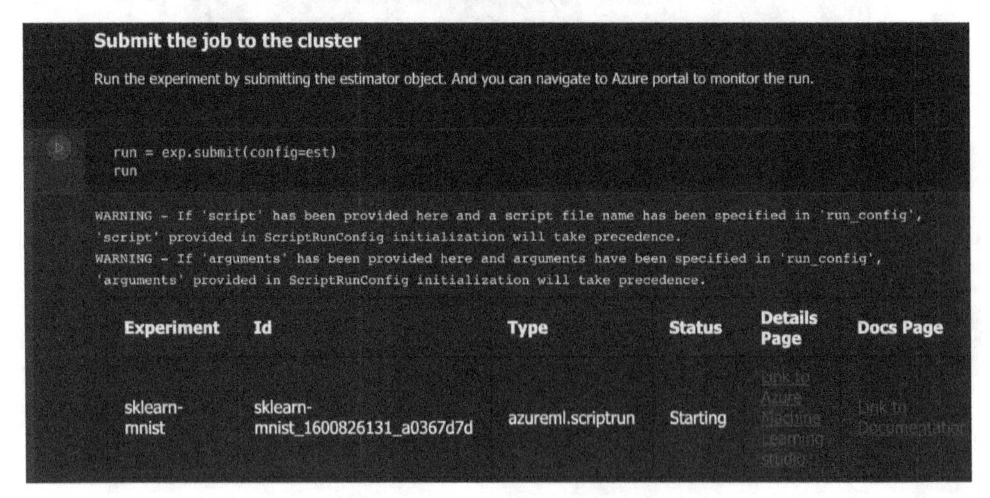

Figure 4.37 – MNIST image classification notebook

Once the training is complete, you will see the results metrics logged during the run, and can also register the model. This means you get the corresponding `.pkl` file for the model:

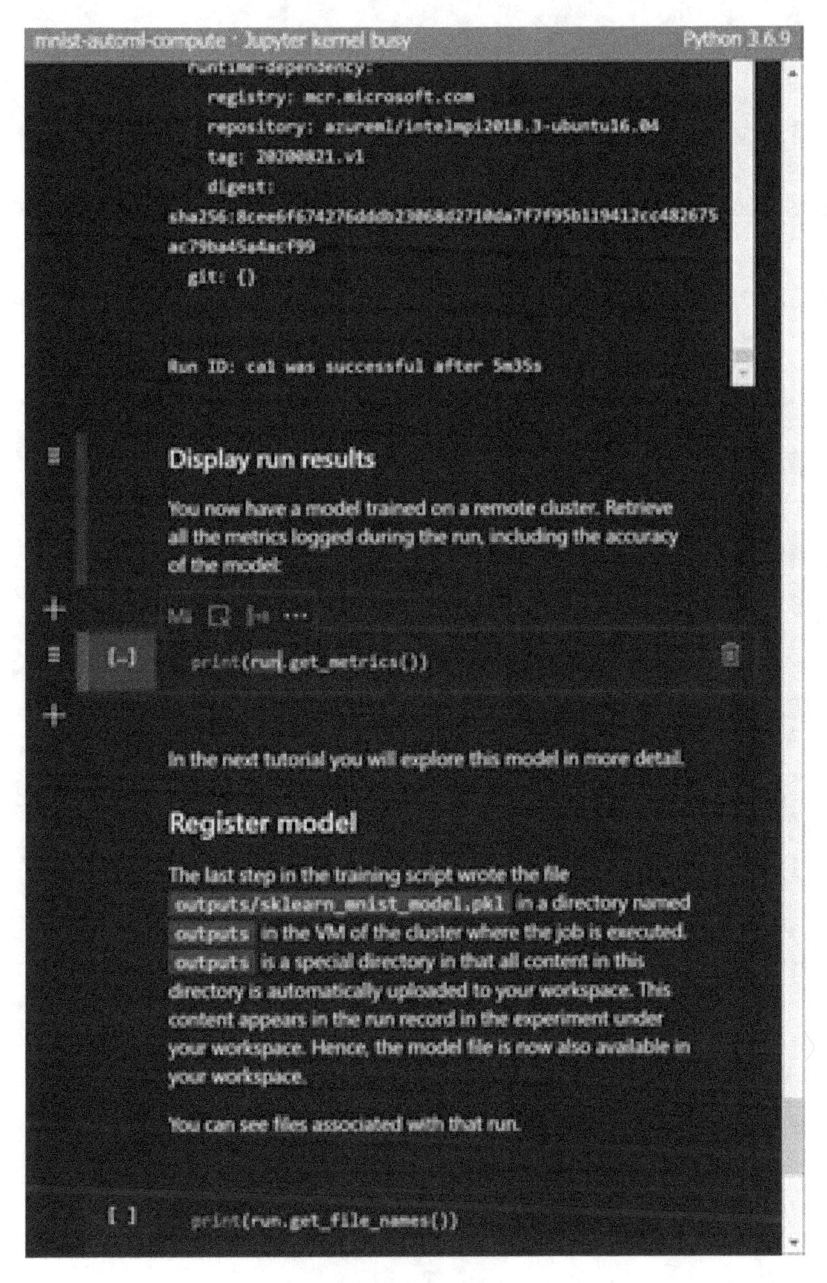

Figure 4.38 – MNIST image classification notebook

Now you can retrieve all the metrics logged during the run by calling `run.get_metrics()`, including the accuracy of the model. In this case, the accuracy is 0.9193:

Display run results

You now have a model trained on a remote cluster. Retrieve all the metrics logged during the run, including the accuracy of the model:

```
[18]  print(run.get_metrics())

{'regularization rate': 0.5, 'accuracy': 0.9193}
```

Figure 4.39 – MNIST image classification notebook

At this point, the notebook automatically saves the model file (`.pkl`) as the output in the output folder. You can see the files by calling the `run.get_file_names()` method. In the next steps, we will use this model file to create a web service and invoke it:

Register model

The last step in the training script wrote the file `outputs/sklearn_mnist_model.pkl` in a directory named `outputs` in the VM of the cluster where the job is executed. `outputs` is a special directory in that all content in this directory is automatically uploaded to your workspace. This content appears in the run record in the experiment under your workspace. Hence, the model file is now also available in your workspace.

You can see files associated with that run.

```
print(run.get_file_names())

['azureml-logs/20_image_build_log.txt', 'azureml-logs/55_azureml-execution-
tvmps_4ac17b36679f8faa19f3b03f634710f765c4d13ba57a6c3e96b075965b4af794_d.txt', 'azureml-logs/65_job_prep-
tvmps_4ac17b36679f8faa19f3b03f634710f765c4d13ba57a6c3e96b075965b4af794_d.txt', 'azureml-
logs/70_driver_log.txt', 'azureml-logs/75_job_post-
tvmps_4ac17b36679f8faa19f3b03f634710f765c4d13ba57a6c3e96b075965b4af794_d.txt', 'azureml-
logs/process_info.json', 'azureml-logs/process_status.json', 'logs/azureml/109_azureml.log',
'logs/azureml/dataprep/backgroundProcess.log', 'logs/azureml/dataprep/backgroundProcess_Telemetry.log',
'logs/azureml/dataprep/engine_spans_l_8e589ded-fe2b-474a-b7b7-9d60265f5567.jsonl',
'logs/azureml/dataprep/engine_spans_l_964ab3d8-9263-416f-b59d-fc49bd448e5d.jsonl',
'logs/azureml/dataprep/python_span_l_8e589ded-fe2b-474a-b7b7-9d60265f5567.jsonl',
'logs/azureml/dataprep/python_span_l_964ab3d8-9263-416f-b59d-fc49bd448e5d.jsonl',
'logs/azureml/job_prep_azureml.log', 'logs/azureml/job_release_azureml.log',
'outputs/sklearn_mnist_model.pkl']
```

Figure 4.40 – MNIST image classification notebook

Deploying and testing models with Azure Machine Learning

The model is now trained, a `.pkl` file has been created, and the model can be deployed for testing. The deployment part is done in the second notebook, `part2-deploy.ipynb`, as seen in the following figure. To deploy the model, we open up the `part 2-deploy.ipynb` notebook by clicking on the notebook in the left pane. We load the `.pkl` file by calling the `joblib.Load` method. You also see the `run` method in the following screenshot, which receives the raw JSON data, invokes the model's `predict` method, and returns the result:

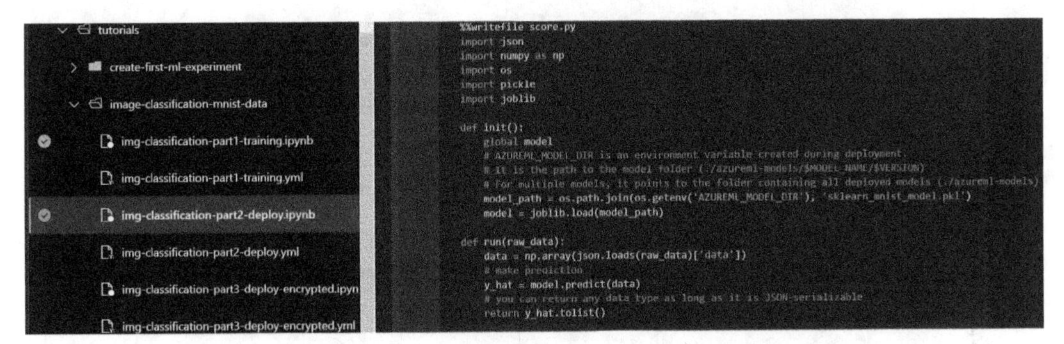

Figure 4.41 – MNIST image classification notebook

In this step, we create a model object by calling the `Model` constructor as shown in the following figure. This model uses the configuration properties from the `Environment` object, and the service name to deploy the endpoint. This endpoint is deployed using **Azure Container Instances** (**ACI**). The endpoint location is available once the deployment is successful:

```
ws = Workspace.from_config()
model = Model(ws, 'sklearn_mnist')


myenv = Environment.get(workspace=ws, name="tutorial-env", version="1")
inference_config = InferenceConfig(entry_script="score.py", environment=myenv)

service_name = 'sklearn-mnist-svc-' + str(uuid.uuid4())[:4]
service = Model.deploy(workspace=ws,
                       name=service_name,
                       models=[model],
                       inference_config=inference_config,
                       deployment_config=aciconfig)

service.wait_for_deployment(show_output=True)
```

```
Running............................
Succeeded
ACI service creation operation finished, operation "Succeeded"
CPU times: user 279 ms, sys: 56.5 ms, total: 336 ms
Wall time: 2min 36s
```

Figure 4.42 – MNIST image classification notebook

1. You can now retrieve the scoring URI, which can be used to invoke the service and get the results from the model:

```
print(service.scoring_uri)
```

Figure 4.43 – MNIST image classification notebook

2. Now you can invoke the web service to get the results:

```
[ ]  import json
     test = json.dumps({"data": X_test.tolist()})
     test = bytes(test, encoding='utf8')
     y_hat = service.run(input_data=test)
```

Figure 4.44 – MNIST image classification notebook

You can also see the corresponding confusion matrix by invoking the `confusion_matrix` method:

```
from sklearn.metrics import confusion_matrix

conf_mx = confusion_matrix(y_test, y_hat)
print(conf_mx)
print('Overall accuracy:', np.average(y_hat == y_test))

[[ 960    0    2    2    1    4    6    3    1    1]
 [   0 1113    3    1    0    1    5    1   11    0]
 [   9    8  919   20    9    5   10   12   37    3]
 [   4    0   17  918    2   24    4   11   21    9]
 [   1    4    4    3  913    0   10    3    5   39]
 [  10    2    0   42   11  768   17    7   28    7]
 [   9    3    7    2    6   20  907    1    3    0]
 [   2    9   22    5    8    1    1  948    5   27]
 [  10   15    5   21   15   26    7   11  852   12]
 [   7    8    2   14   32   13    0   26   12  895]]
Overall accuracy: 0.9193
```

Figure 4.45 – MNIST image classification notebook

This completes the whole cycle of building a model, deploying, and testing it within Azure Machine Learning. We will continue with automated ML in the next chapter.

Summary

In this chapter, you learned how to get started with the Microsoft Azure platform, the ML services ecosystem capabilities, and learned about Microsoft's AI and ML offerings. You were also briefed on different capabilities within the Azure platform, such as collaborative notebooks, drag and drop ML, MLOPS, RStudio integration, reinforcement learning, enterprise-grade security, automated ML, data labeling, autoscaling compute, integration with other Azure services, responsible ML, and cost management. Finally, to test your newly discovered Azure superpowers, you configured, built, deployed, and tested a classification web service using an Azure Machine Learning notebook.

In the next chapter, we will further dive into using the automated ML features of the Azure Machine Learning service.

Further reading

For more information on the following topics, you can visit the given links:

- Python notebooks with ML and deep learning examples with Azure Machine Learning:

  `https://github.com/Azure/MachineLearningNotebooks`

- What are compute targets in Azure Machine Learning?

  `https://docs.microsoft.com/en-us/azure/machine-learning/concept-compute-target`

- Use automated ML in an Azure Machine Learning pipeline in Python:

  `https://docs.microsoft.com/en-us/azure/machine-learning/how-to-use-automlstep-in-pipelines`

- A critical overview of AutoML solutions by Bahador Khaleghi:

  `https://medium.com/analytics-vidhya/a-critical-overview-of-automl-solutions-cb37ab0eb59e`

5

Automated Machine Learning with Microsoft Azure

"By far, the greatest danger of artificial intelligence is that people conclude too early that they understand it."

– Eliezer Yudkowsky

The Microsoft Azure platform and its associated toolset are diverse and part of a larger enterprise ecosystem that is a force to be reckoned with. It enables businesses to focus on what they do best by accelerating growth via improved communication, resource management, and facilitating advance actionable analytics. In the previous chapter, you were introduced to the Azure Machine Learning platform and its services. You learned how to get started with Azure machine learning, and you took a glimpse at the end-to-end machine learning life cycle using the power of the Microsoft Azure platform and its services. That was quite literally (in the non-literal sense of the word) the tip of the iceberg.

In this chapter, we will get started by looking at **Automated Machine Learning** (**AutoML**) in Microsoft Azure. You will build a classification model and perform time series prediction using Azure's AutoML capabilities. This chapter will equip you with the skills you'll need to build and deploy AutoML solutions.

In this chapter, we will cover the following topics:

- AutoML in Microsoft Azure
- Time series prediction using Azure AutoML and JupyterLab

Let's get started!

AutoML in Microsoft Azure

AutoML is treated as a first-class citizen in the Azure platform. The fundamental ideas behind feature engineering, network architecture search, and hyperparameter tuning are the same as what we discussed in *Chapter 2, Automated Machine Learning, Algorithms, and Techniques*, and *Chapter 3, Automated Machine Learning with Open Source Tools and Libraries*. However, the layer of abstraction that's used to democratize these skills makes them much more appealing to non-machine learning experts.

The key principles of AutoML in the Azure platform are shown in the following diagram. User input such as datasets target metrics, and constraints (how long to run the job, what the allocated budget is for compute, and so on) drive the AutoML "engine", which completes iterations to find the best model and rank it according to the score of **Training Success**:

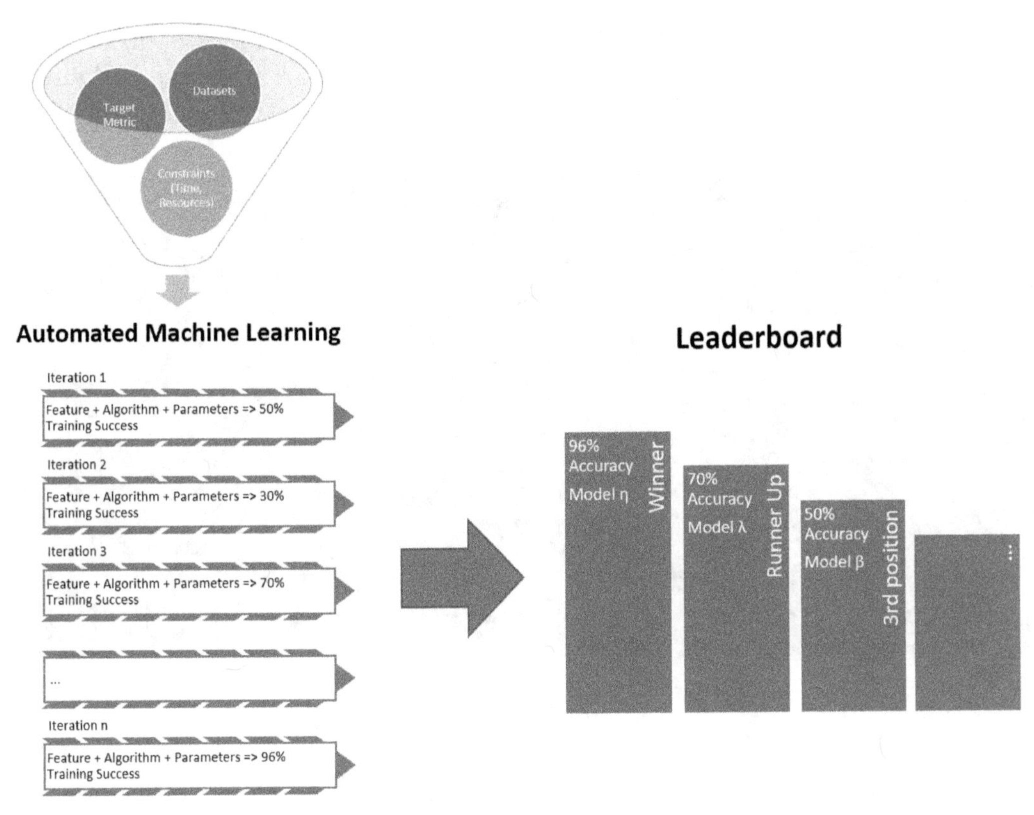

Figure 5.1 – Azure AutoML workflow – how AutoML works

In this section, we'll provide a step-by-step walkthrough of the AutoML approach. In *Chapter 4, Getting Started with Azure Machine Learning*, you saw the main page for Azure machine learning. There, we created a classification model and tested it using a notebook:

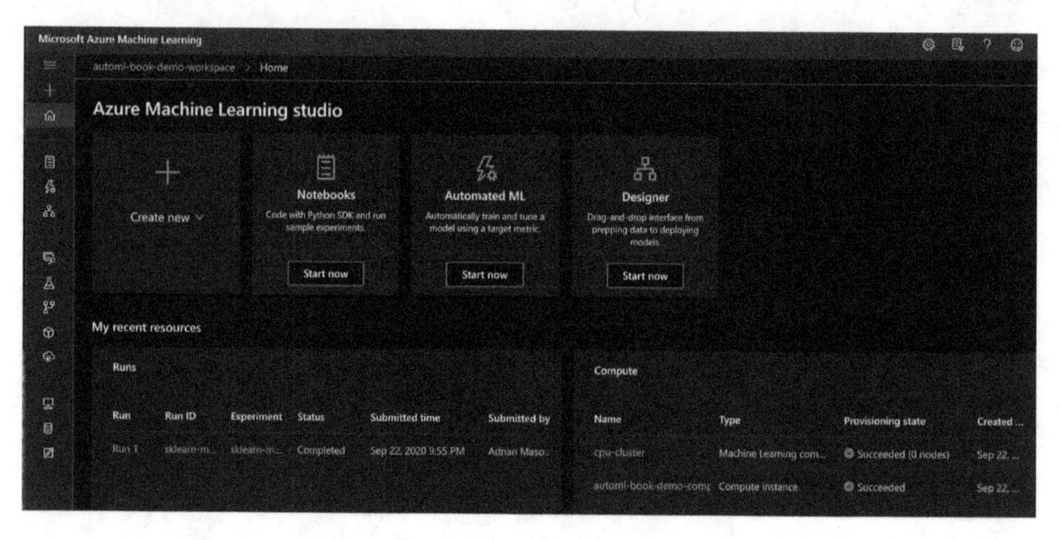

Figure 5.2 – Azure Machine Learning portal

Now, let's explore how AutoML-based model development works when it comes to training and tuning a model:

1. From the Azure portal, click on **Automated ML | Start now**. You will be taken to the following screen, where you can create a new Automated ML run:

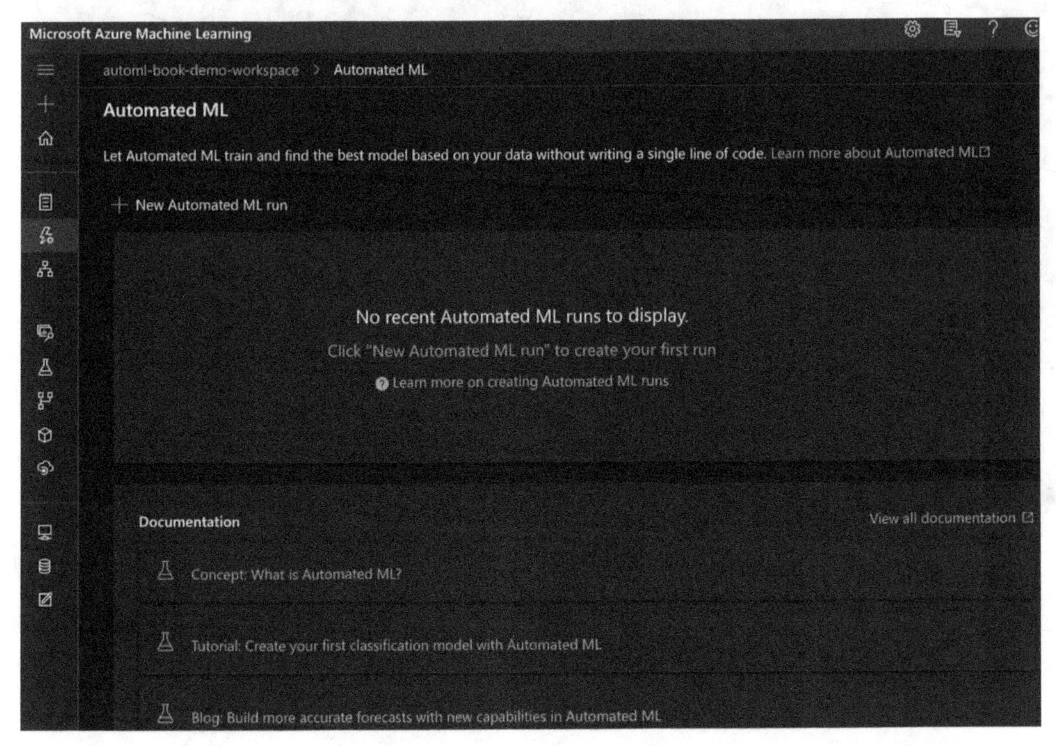

Figure 5.3 – Azure Machine Learning – Creating an Automated ML run

2. The first step of creating an automated ML run is selecting a dataset to work with. Here, you can either create your own dataset or – better yet – select an existing one from the repository of public datasets that Azure provides:

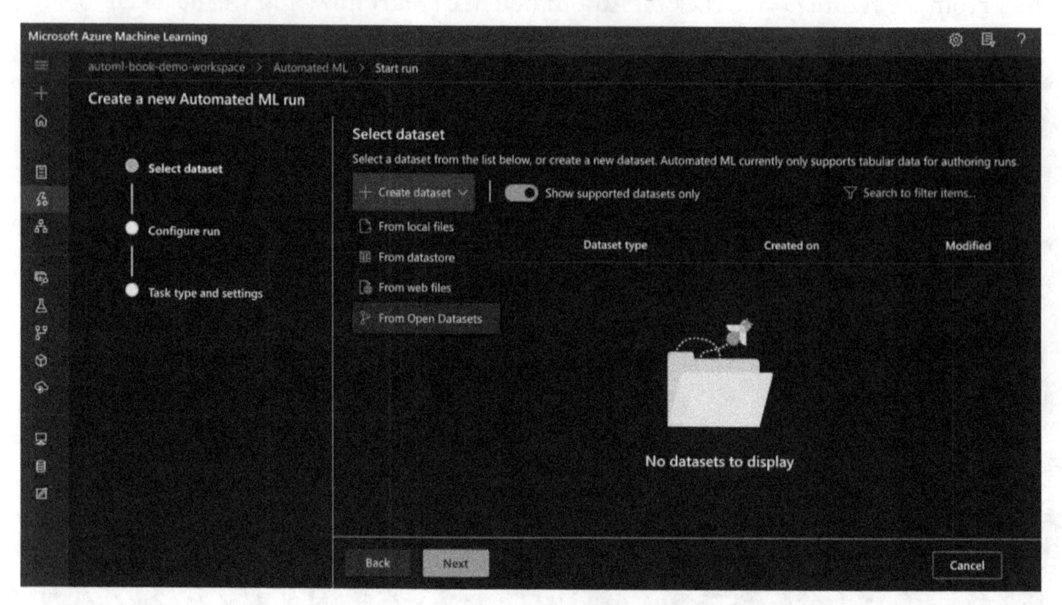

Figure 5.4 – AutoML dataset selection page

3. A dataset can be created from open datasets. In this case, we will use our tried and tested MNIST dataset to create the AutoML run, as shown in the following screenshot:

MNIST dataset

Yann LeCun (Courant Institute, NYU) and Corinna Cortes (Google Labs, New York) hold the copyright for MNIST dataset, which is a derivative work from the original NIST datasets. The MNIST dataset has been made available under the terms of the Creative Commons Attribution-Share Alike 3.0 license.

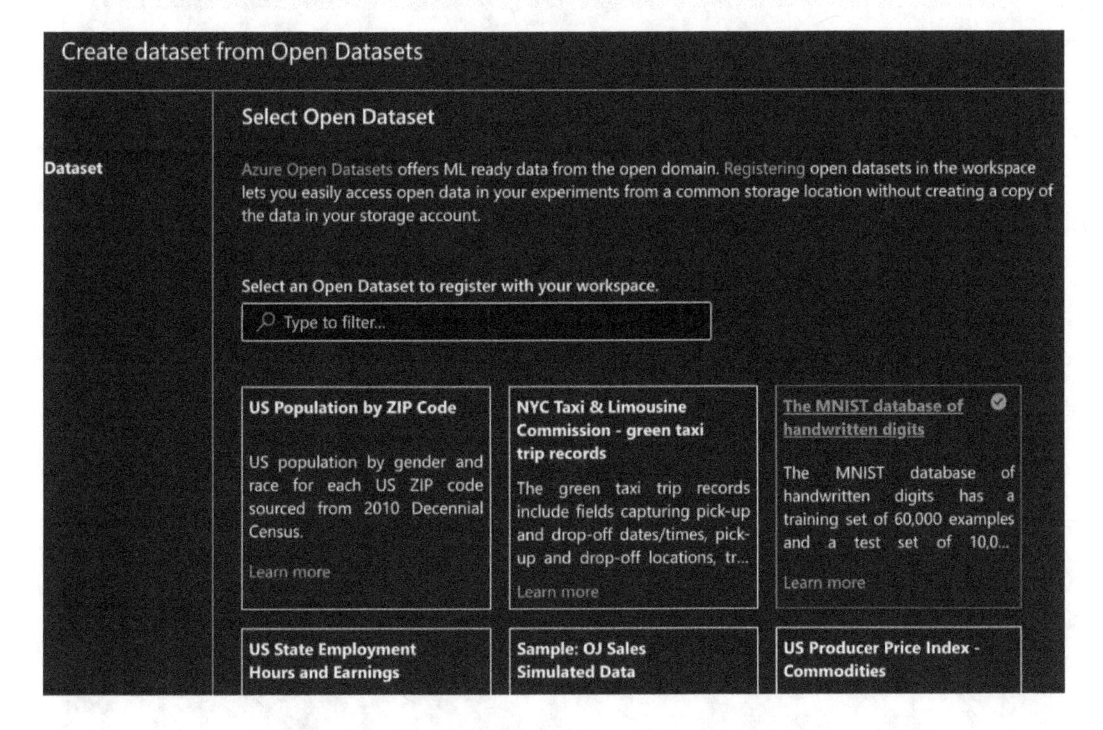

Figure 5.5 – The Create dataset from Open Datasets page

Once you have selected the dataset, it will appear as part of your run, and you can also preview it. Apart from specifying the dataset's version, you can also specify if you would like to use the entire dataset, or whether it should be registered as a tabular data source or a file type data source:

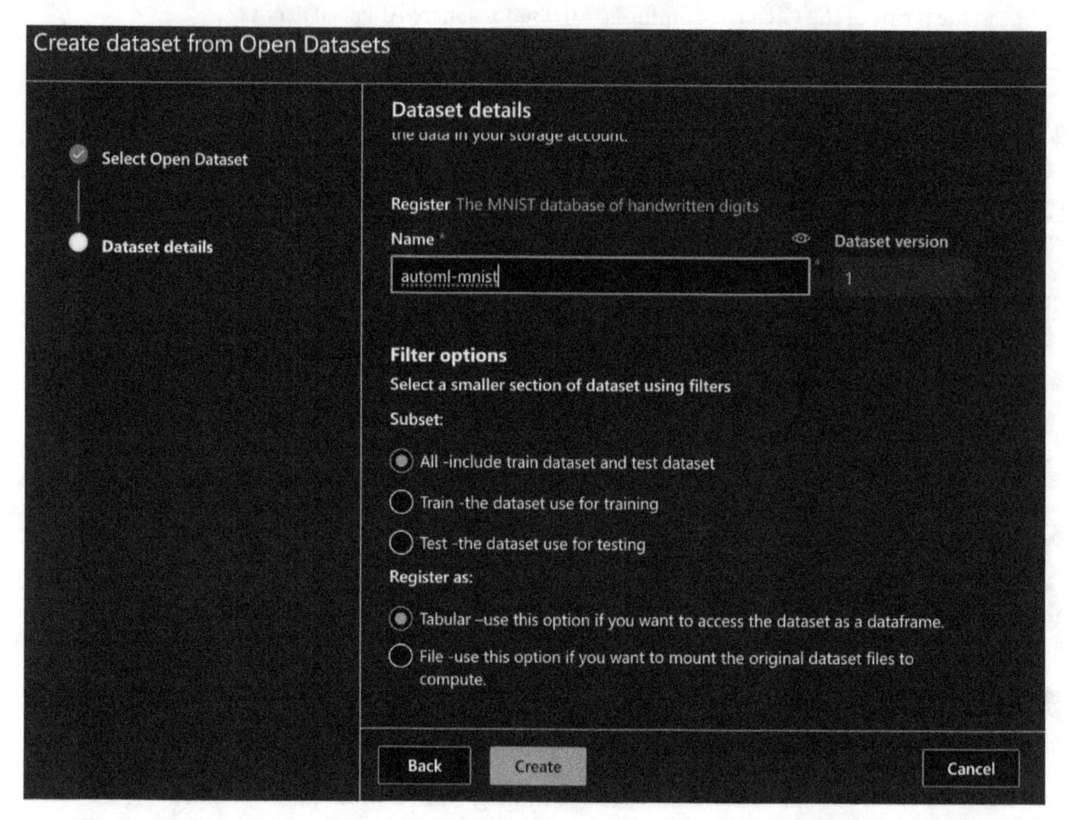

Figure 5.6 – Dataset from the Azure Machine Learning dataset repository of curated datasets

Upon selecting **Create**, you will see the following screen as the dataset becomes part of your run:

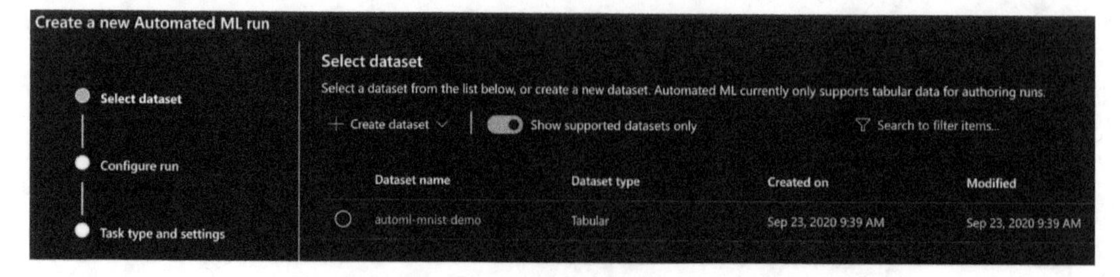

Figure 5.7 – Dataset from the Azure Machine Learning dataset repository of curated datasets

The MNIST dataset can also be seen as part of the data preview if you click on the dataset's name, as shown in the following screenshot:

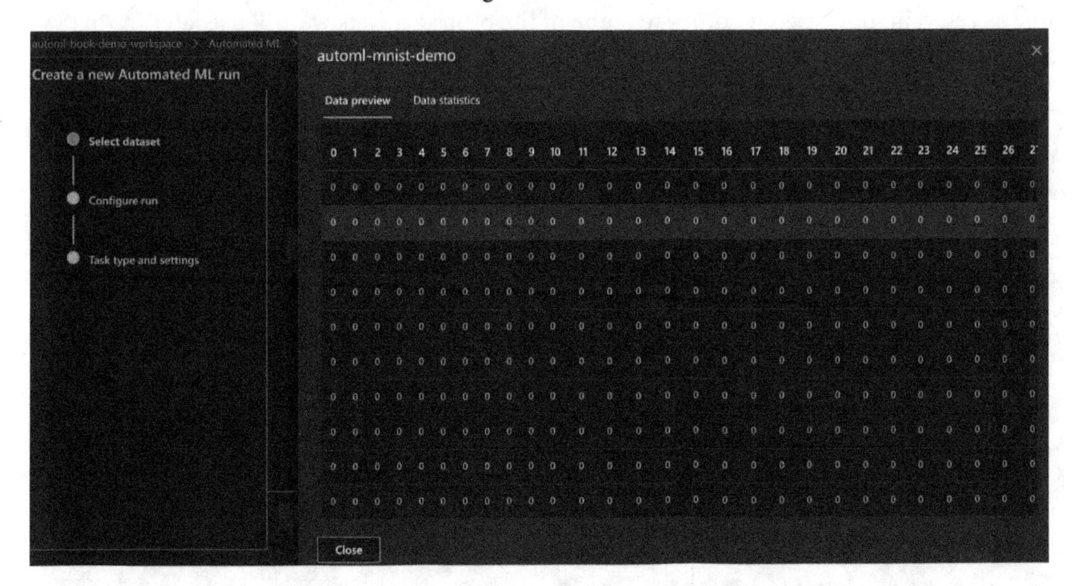

Figure 5.8 – Preview of the dataset from the Azure Machine Learning dataset repository of curated datasets

Let's face it – this preview of the MNIST pixel dataset isn't that exciting, but if you had some more representative data (healthcare, retail, or financial data, and so on), the preview would help us understand that the ingestion process went well and that we are not running the risk of a delimiter fiasco.

Similarly, the data statistics are shown in the following screenshot. If you are pandas-inclined, think of it as the `describe()` feature. Due to its image-based nature, this isn't quite as relevant, but when it comes to some of the other datasets we will use later in this chapter, it comes in quite handy:

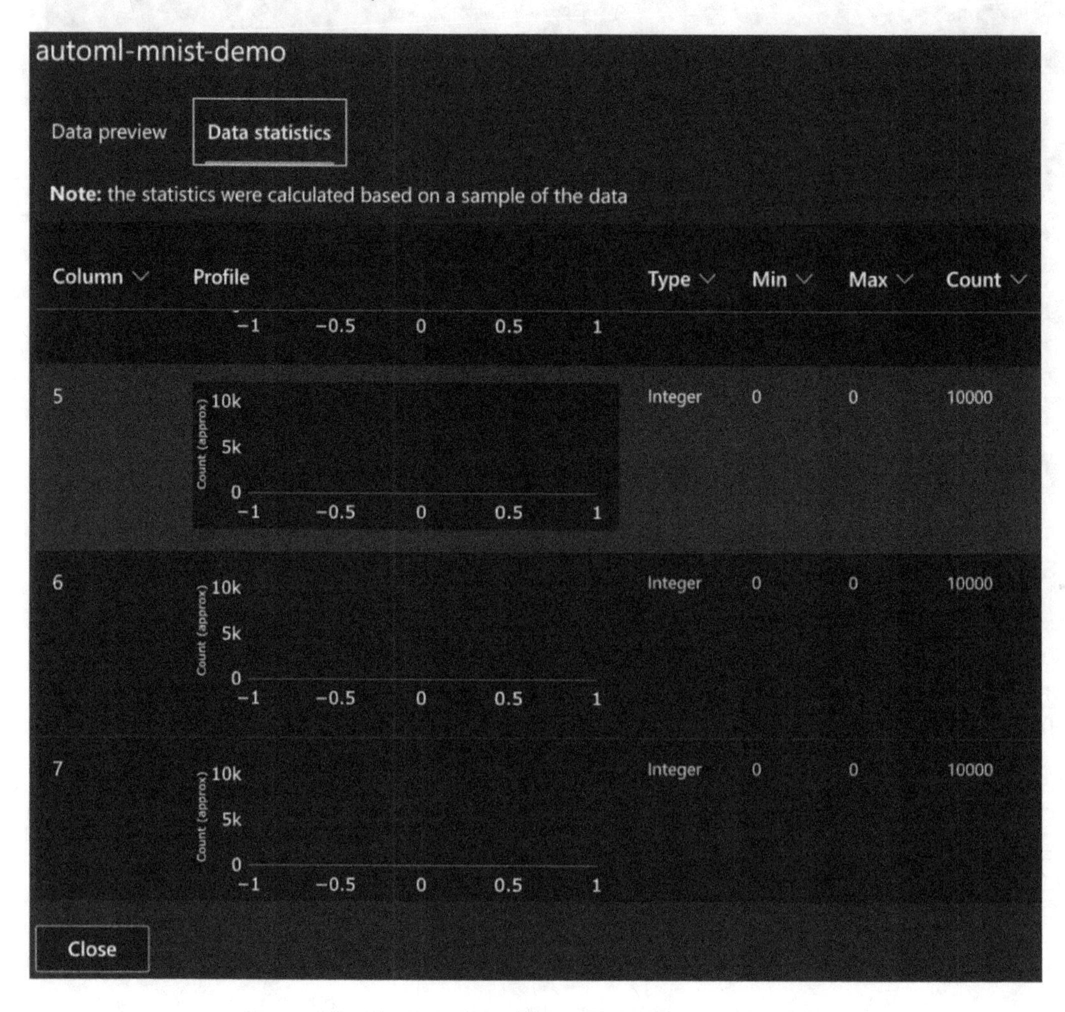

Figure 5.9 – Preview of the data statistics in Azure AutoML

1. Now that we have selected the dataset, we can configure the run by providing the experiment's name, target column (the labeled feature to train on and classify), and the compute cluster, as shown in the following screenshot:

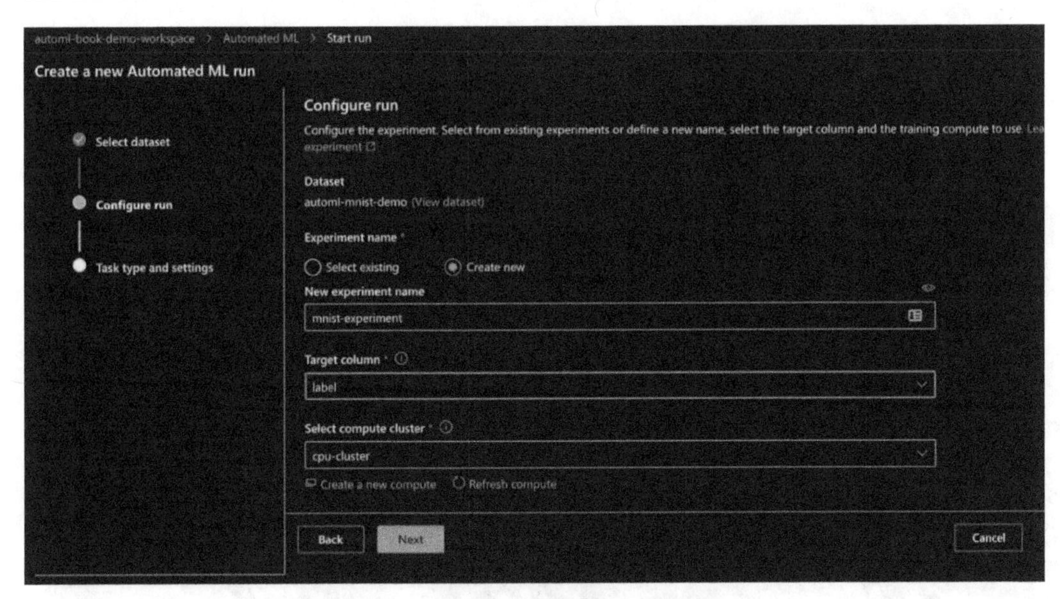

Figure 5.10 – Configuring an AutoML run

2. The third and final step is to select the task type – classification, regression, or time series forecasting. In this case, we are classifying digits based on their associated labels. You will learn how to use the other task types in future examples:

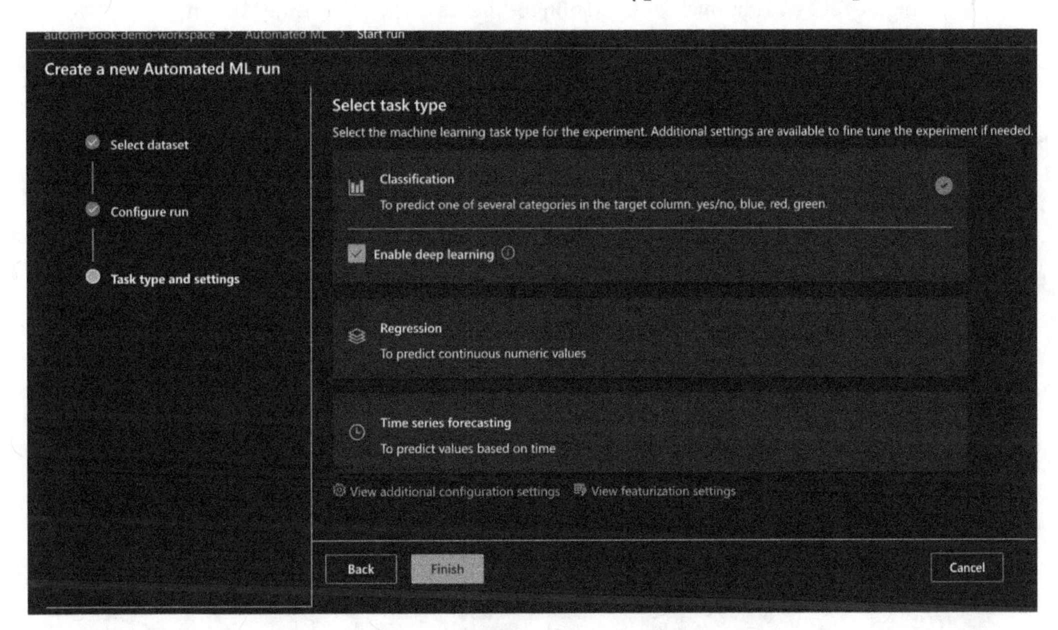

Figure 5.11 – Selecting the task type for the AutoML run

3. It is important to consider the additional configurations. Here, you can select a primary metric, its explainability, any allowed algorithms (by default, all of them are allowed), the exit criteria, and any validation split information, as shown in the following screenshot:

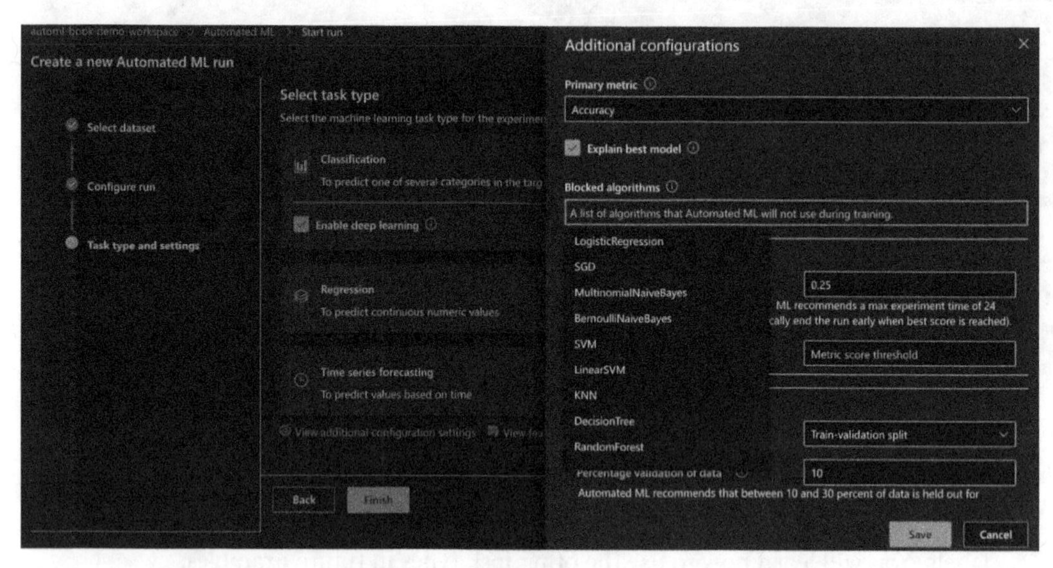

Figure 5.12 – Additional configuration for the task type of the AutoML run

Additional configuration varies based on the task's type. The following screenshot shows the regression configuration elements:

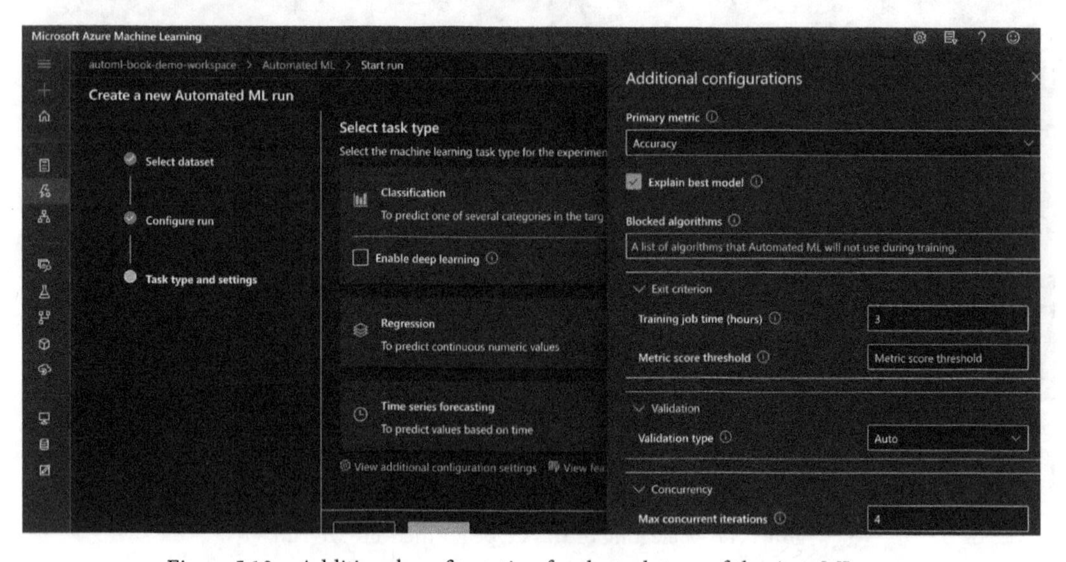

Figure 5.13 – Additional configuration for the task type of the AutoML run

Featurization – that is, selecting and transforming features – is an important factor to keep in mind as you move forward with the dataset. When you click on the **View Featurization Settings** link, Azure machine learning provides you with the following screen. From here, you can select the feature's type, assign a specific data type, and specify what you wish to impute the feature with:

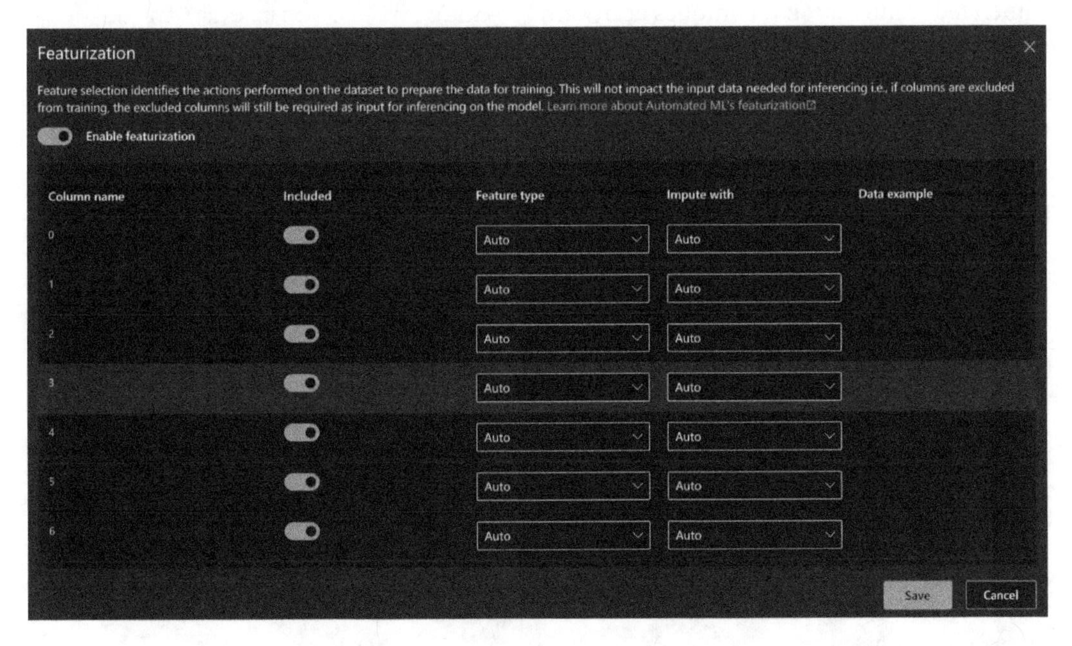

Figure 5.14 – Featurization of the AutoML run

Automatic featurization – that is, turning different data types into numerical vectors – is a typical part of any data science workflow. The following diagram shows the techniques that are applied automatically to the datasets when featurization is turned on (see the blue toggle at the top of the preceding screenshot). The following diagram shows some of the key steps that are taken during auto-featurization. You can find out more about enumerated featurization techniques at `https://docs.microsoft.com/en-us/azure/machine-learning/how-to-configure-auto-features`:

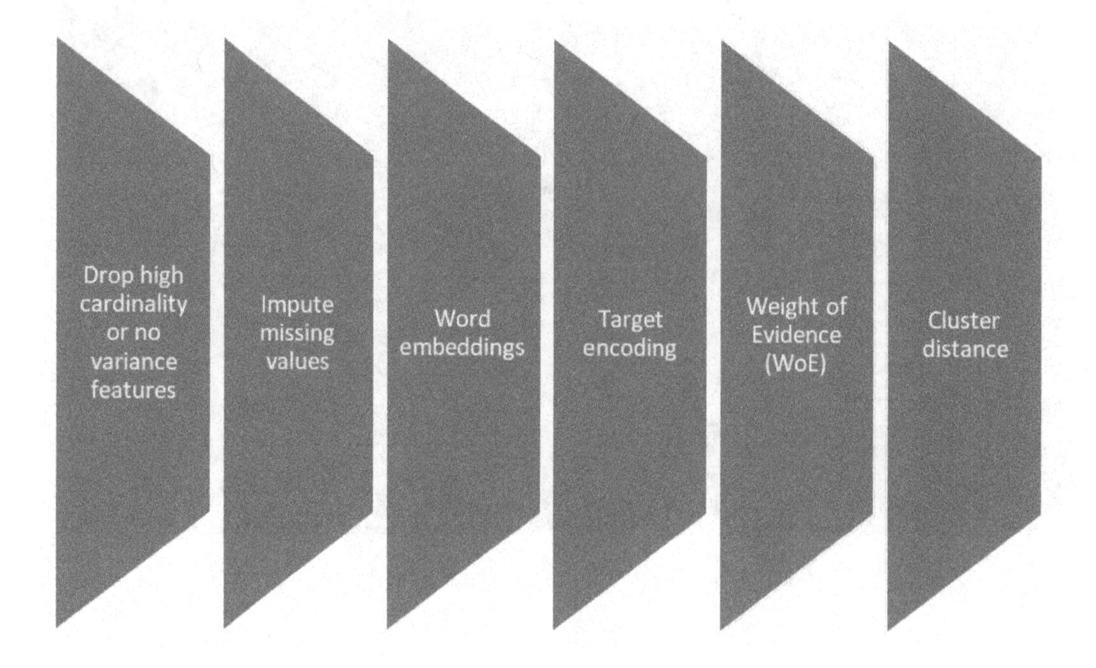

Figure 5.15 – Featurization approaches for the AutoML run

Scaling and **normalization** (also sometimes referred to as **regularization** and **standaridization**) are two important ways of featurization that deal with transforming data into a common range of values. The scaling and normalization techniques that are used in the automatic featurization algorithms can be seen in the following diagram. You can find out more about various enumerated scaling and featurization techniques at `https://docs.microsoft.com/en-us/azure/machine-learning/how-to-configure-auto-features`:

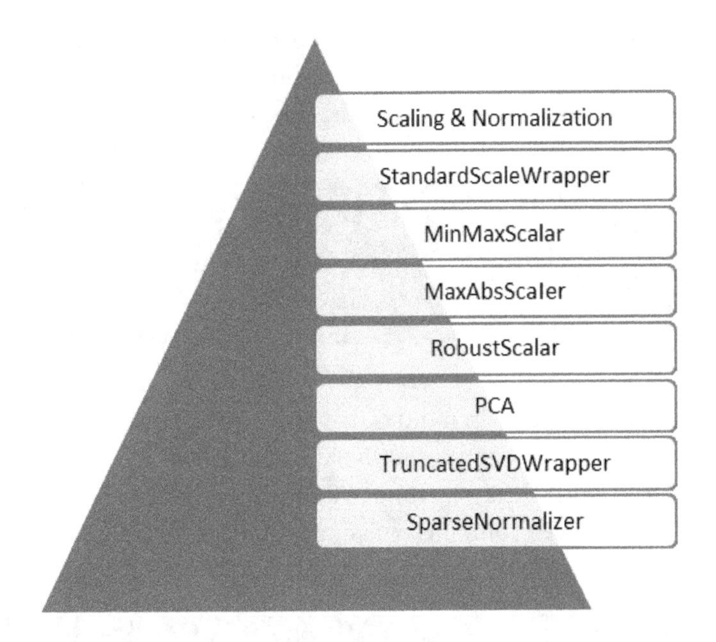

Figure 5.16 – Azure AutoML – scaling and featurization

The topic of featurization is not complete without the mention of guardrails. Data guardrails are part of the AutoML engine which helps identify and address issues with the dataset such as missing feature values, handling high cardinality features (lots of unique values), class imbalance (minority classes and outliers) and so on. The following figure outlines these guardrails you should make yourself familiar with. You can read further details about these guardrails here in the Azure documentation (`https://docs.microsoft.com/en-us/azure/machine-learning/how-to-configure-auto-features`):

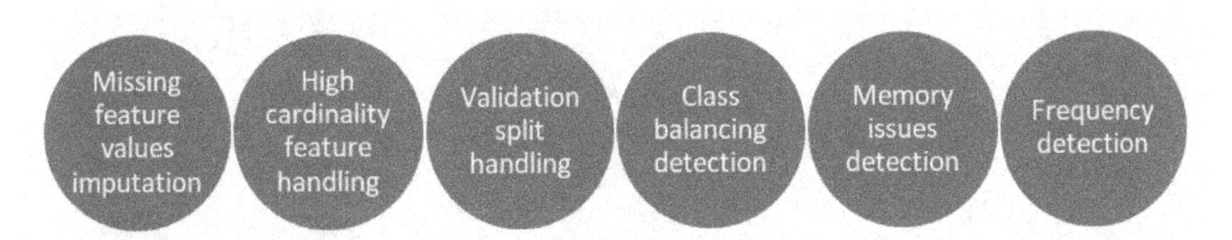

Figure 5.17 – Data guardrails for the AutoML run

4. Now, when you click the **Finish** button, as shown in *Figure 5.10*, after setting up the given parameters for the task type and any additional configuration items, you will see the following screen, which validates the run:

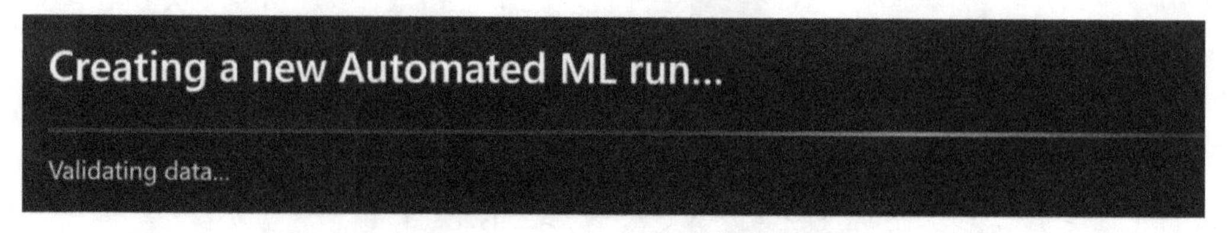

Figure 5.18 – Data guardrails for the AutoML run

One important point to keep in mind is that you need to have good compute resources to run an experiment; otherwise, it will fail. For example, in this experiment, I have set the training time to 0.25 hours; that is, 15 minutes. This is not enough time for the given compute, which means the run is bound to fail, as shown in the following screenshot:

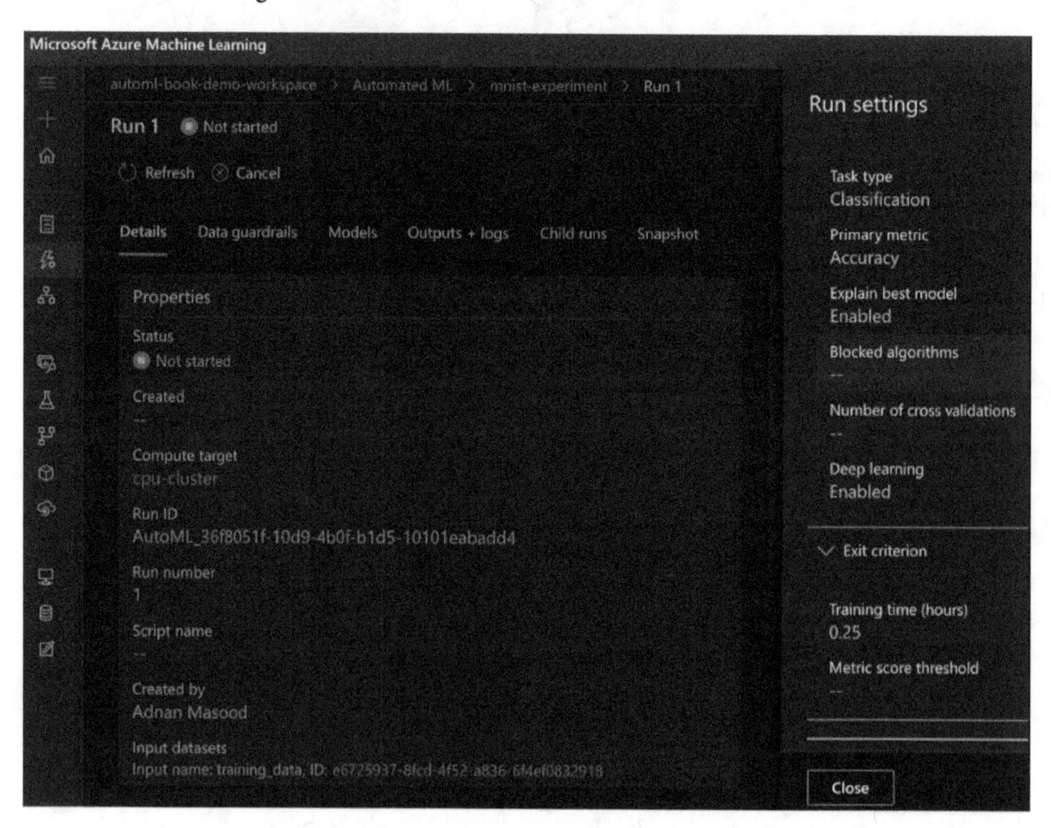

Figure 5.19 – AutoML experiment run settings

The following screenshot shows that since we didn't allocate the right computing resources to run the AutoML experiment, it failed:

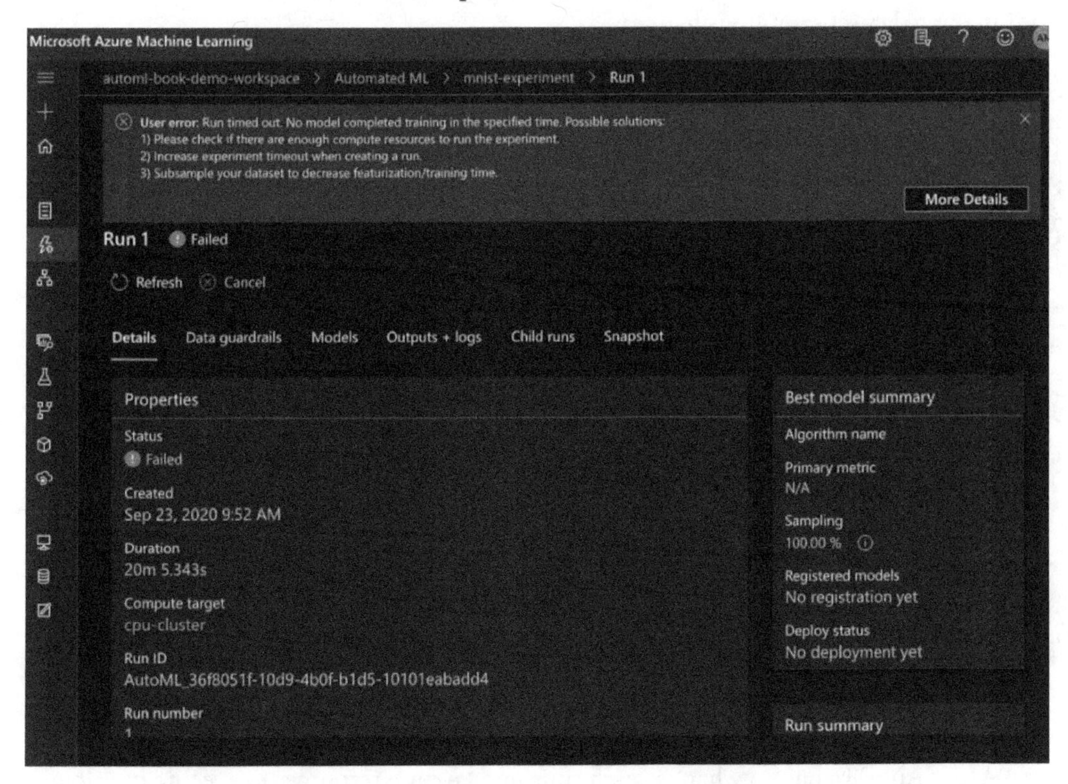

Figure 5.20 – AutoML experiment run failure message

The following error message explains the user error in detail, along with potential solutions, such as adding compute resources, applying an experiment timeout, and making dataset sampling changes:

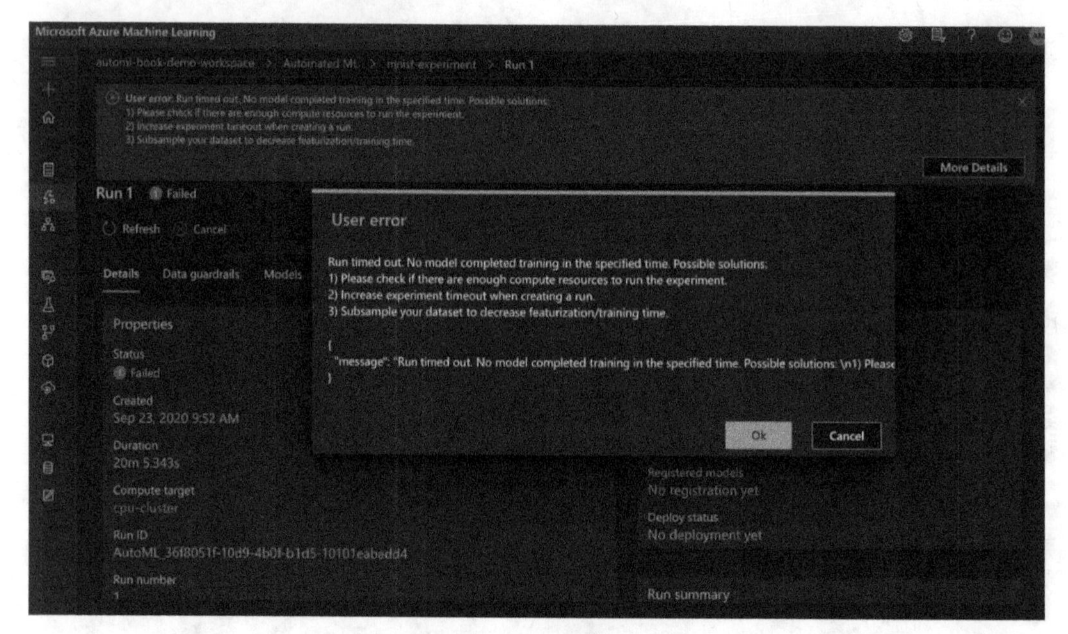

Figure 5.21 – AutoML experiment run failure message

Increasing the time limit to 5 hours will help, as you will see in the following steps. Azure AutoML has now had enough time and enough resources to execute multiple experiments. This has taught you that cheapening out on time and/or resources isn't a good AutoML strategy.

5. The following screen in the AutoML child run shows individual iterations. It clearly demonstrates how different data preprocessing methods, such as `StandardScalerWrapper`, `RobustScaler`, and `MaxAbsScaler/MinMaxScaler`, and forecasting algorithms, such as `RandomForest`, `LightGB`, `ElasticNet`, `DecisionTree`, and `LassoLars`, were used. Runs `54` and `53` in the following screenshot show how ensemble algorithms and their weights can be viewed by clicking on their associated tags:

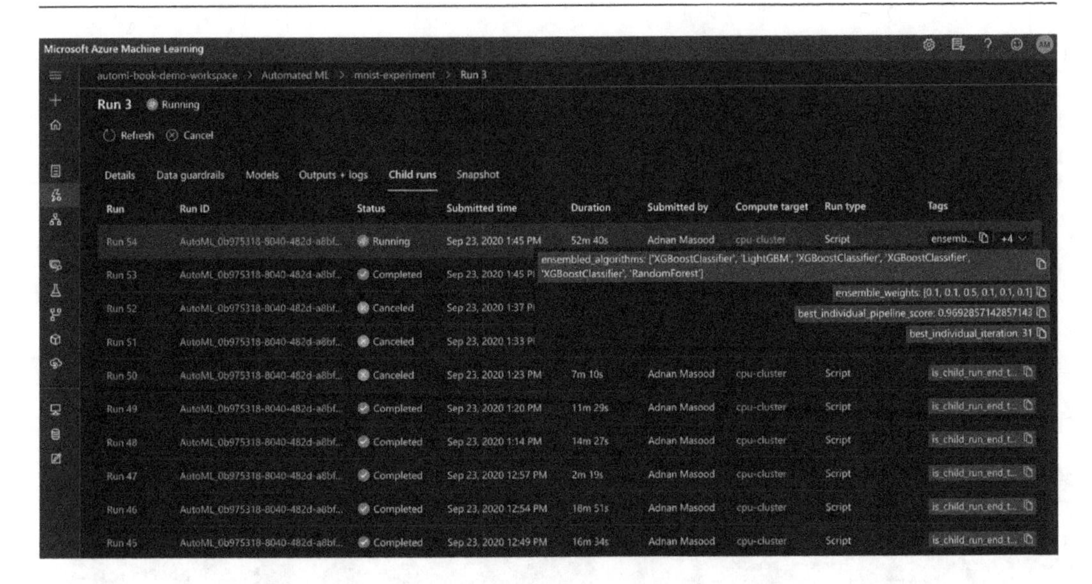

Figure 5.22 – AutoML experiment run details

6. Click on the **Models** tab to see which model provided what degree of accuracy and what run it is associated with, as shown in the following screenshot:

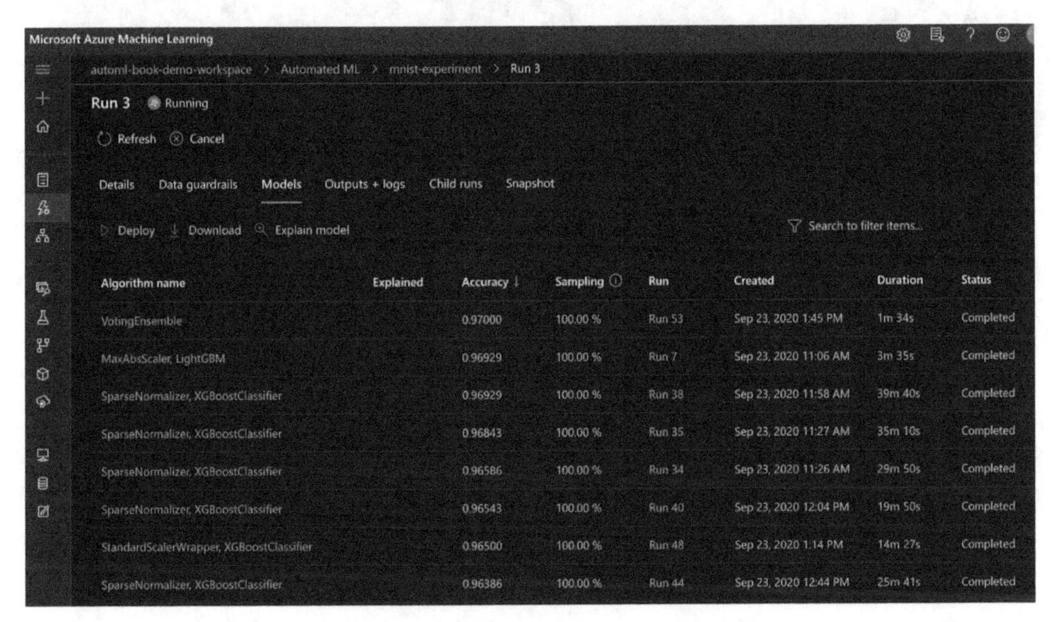

Figure 5.23 – AutoML experiment run details

The run metrics are also a great way to get more detailed information about the associated run. For example, you can see the algorithm's name, the associated accuracy, AUC scores, precision, F1 score, and so on:

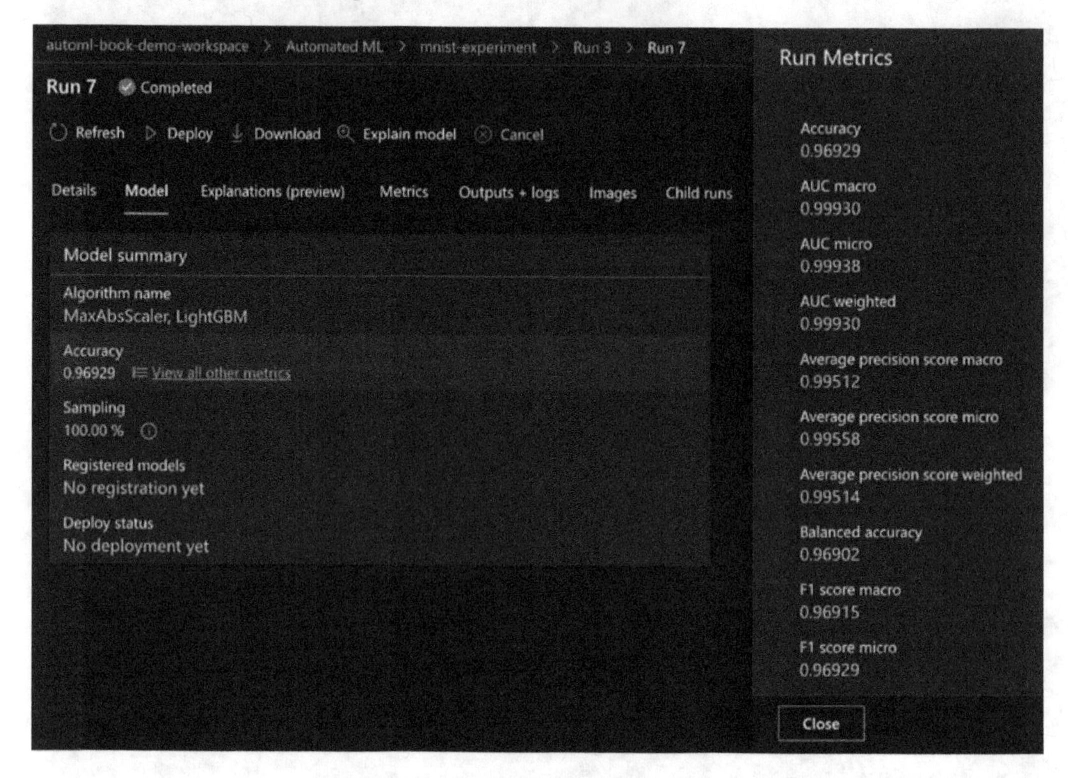

Figure 5.24 – AutoML experiment run details

The data guardrail measures that are taken to protect the quality of the data can be seen by clicking on the corresponding tab, as shown in the following screenshot. This page shows what guardrail techniques have been used to ensure that the input data used to train the model was high quality:

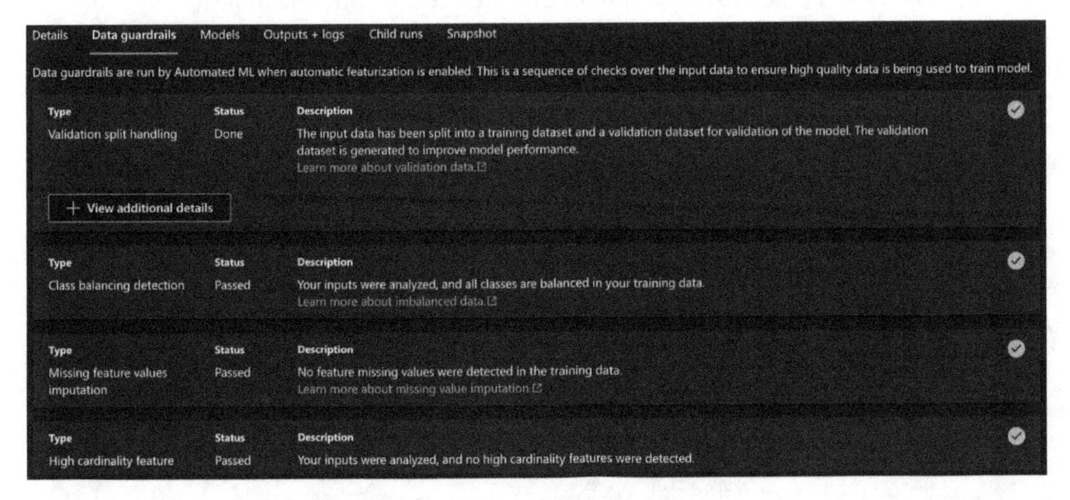

Figure 5.25 – AutoML experiment data guardrails

From the main run summary page, you can view the best model and its summarized outcome. In this case, the soft voting-based `VotingEnsemble()` method was the clear winner. It is among two of the ensemble methods currently supported in Azure AutoML. The other one is `StackEnsemble`, which creates collections from previously run iterations. Ensemble methods are techniques that are used to combine multiple models to get the best results; voting, stacking, bagging, and boosting are some of the categories available for ensemble methods:

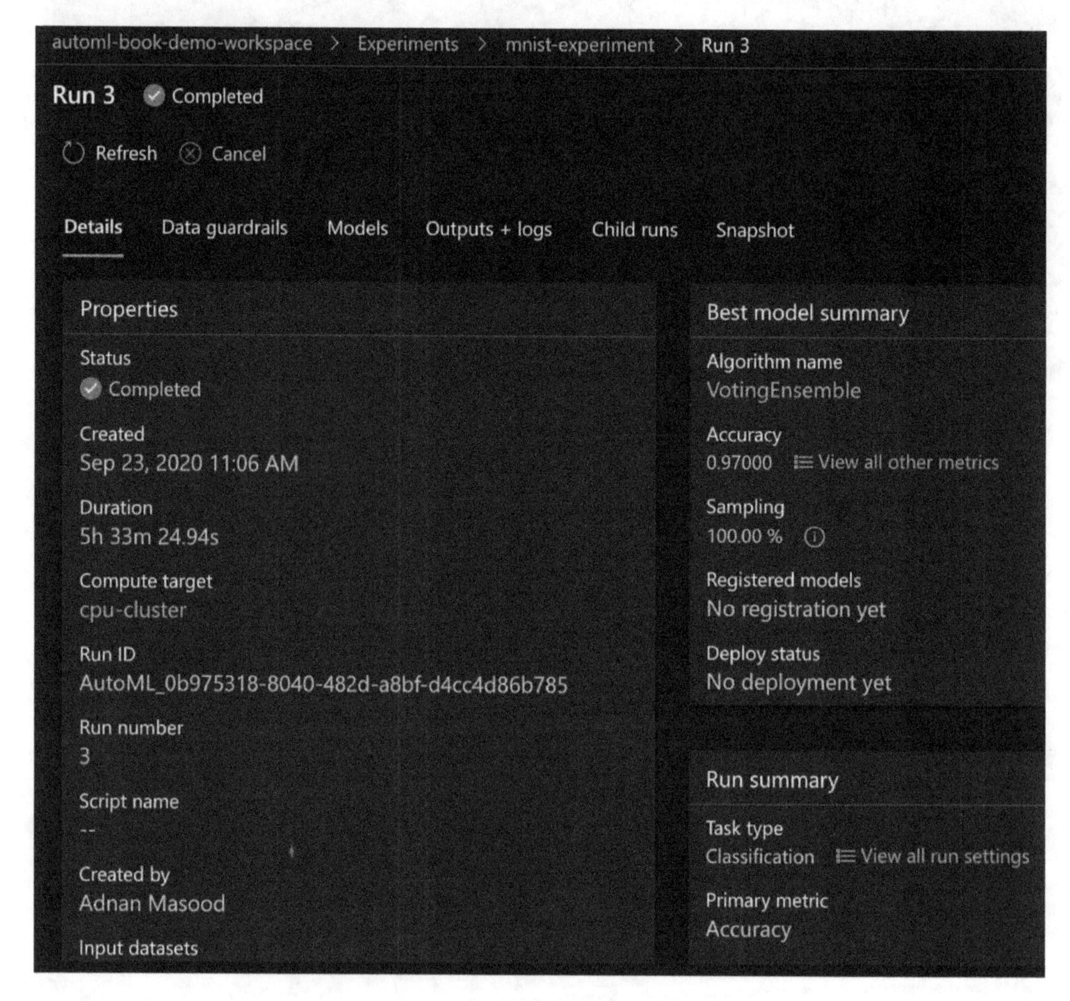

Figure 5.26 – AutoML experiment summary page

Assuming you have followed these experiments so far and tried these steps by yourself, it should be evident that each run has several child runs – that is, individual iterations that each model has. So, when we look at the **Metrics** tab of the **Run** summary page, we not only see the different metrices, but also a precision recall plot:

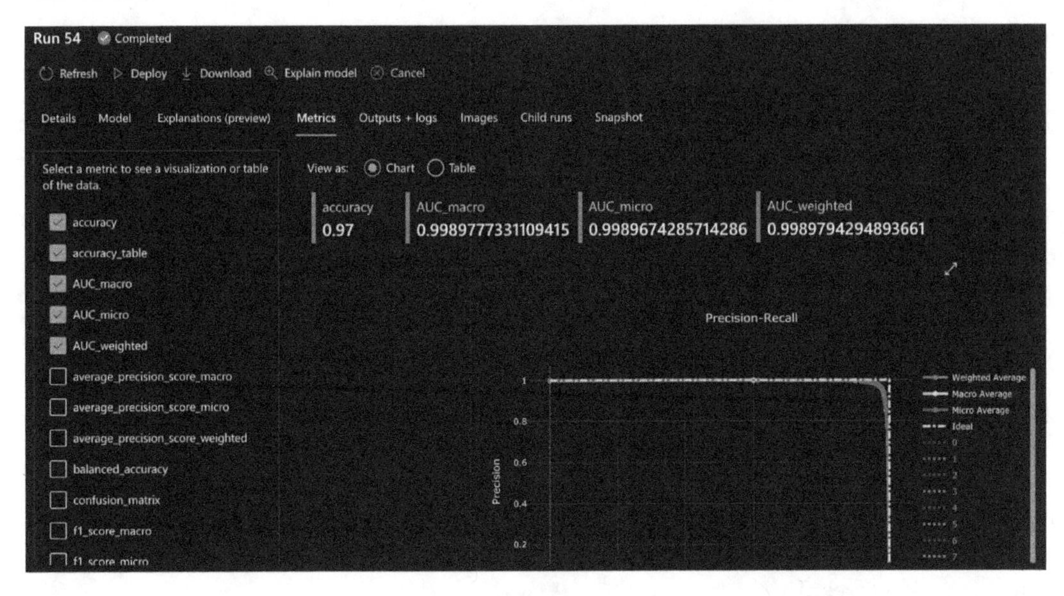

Figure 5.27 – AutoML experiment accuracy metrics and PR curve

Now, let's look at the explanations for the models. The explainability of machine learning models is super important, especially for AutoML. This is because you would like to know, as a subject matter expert, which features played a critical role in the result. In the following screenshot, you can see a tabular explanation of feature importance for the top *k* features, along with a breakdown of how they are used to predict digits from 0-9:

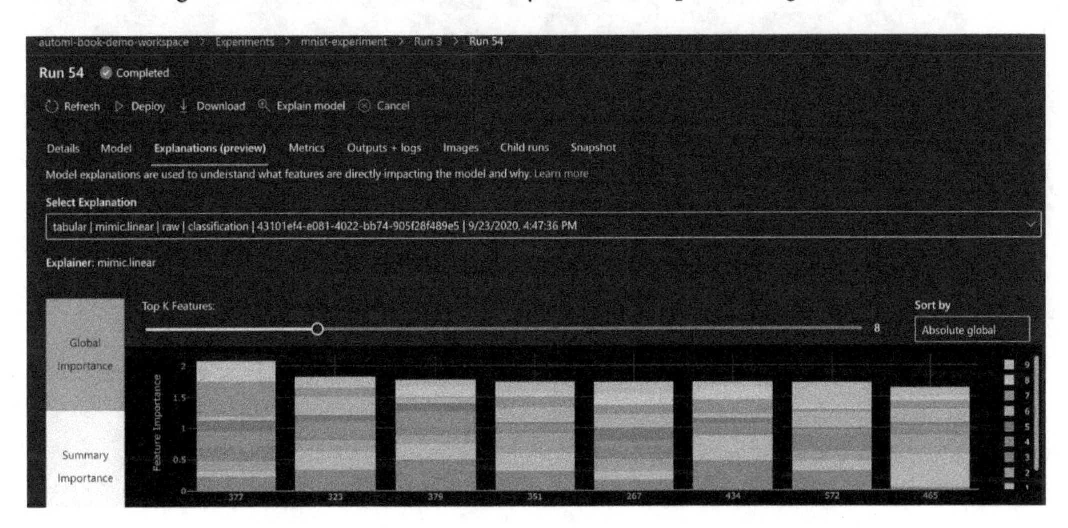

Figure 5.28 – AutoML experiment explanation of features

The preceding screenshot shows which feature played what part in predicting the digits. Feature 377 was significant in predicting digit 7, features 379 and 434 were significant in predicting 9, and so on and so forth. This MNIST dataset might not appear to be relevant to you, but let's imagine you are looking at an HR hiring dataset and gender, race, or age become an important feature. This would raise an alarm since this would go against your corporate policies of sexual bias, racism, or age-related discrimination. It would also likely be against the law and you could get in serious trouble in terms of compliance and reputational damage for having a bigot in the machine. Not to mention, it's unethical (and honestly nonsensical) to discriminate based on attributes that have nothing to do with an employee's capability to get the job done.

This explainability also provides summary importance for features where you can visualize the importance of individual k-features for both global and local features. The swarm chart, shown in the following screenshot, visualizes the same data at a very granular level. It shows a one-to-one mapping between the number of elements in the MNIST dataset and the features they correspond to, similar to what's shown in the tabular representation in the preceding screenshot:

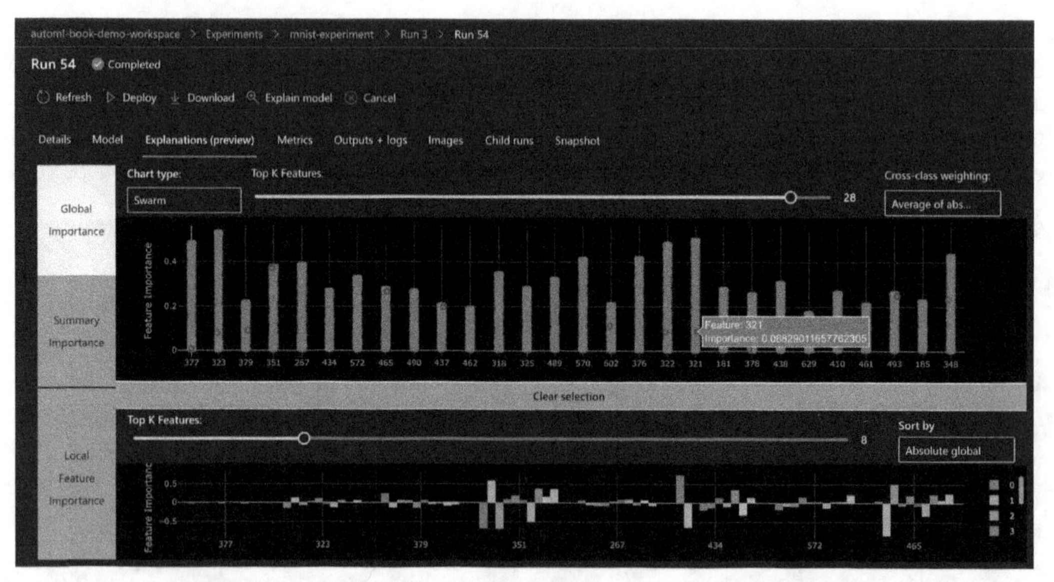

Figure 5.29 – Azure Machine Learning top k features summary importance chart

With this overview of automated ML for classification, let's move on and apply the same techniques to time series forecasting.

Time series prediction using AutoML

Forecasting energy demand is a real problem in the industry where energy providers like to predict the consumer's expected needs in advance. In this example, we will use the New York City energy demand dataset, which is available in the public domain. We will use historic time series data and apply AutoML for forecasting; that is, predicting energy demand for the next 48 hours.

The machine learning notebook is part of the Azure model repository, which can be accessed on GitHub at `https://github.com/Azure/MachineLearningNotebooks/`. Let's get started:

1. Clone the aforementioned GitHub repository on your local disk and navigate to the `forecasting-energy-demand` folder:

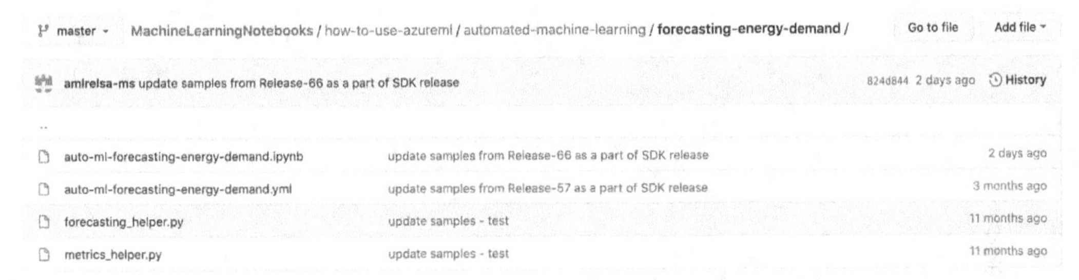

Figure 5.30 – Azure Machine Learning notebooks GitHub repository

2. Click on the **Upload folder** icon and upload the `forecasting-energy-demand` folder to the Azure notebook repository, as shown in the following screenshot:

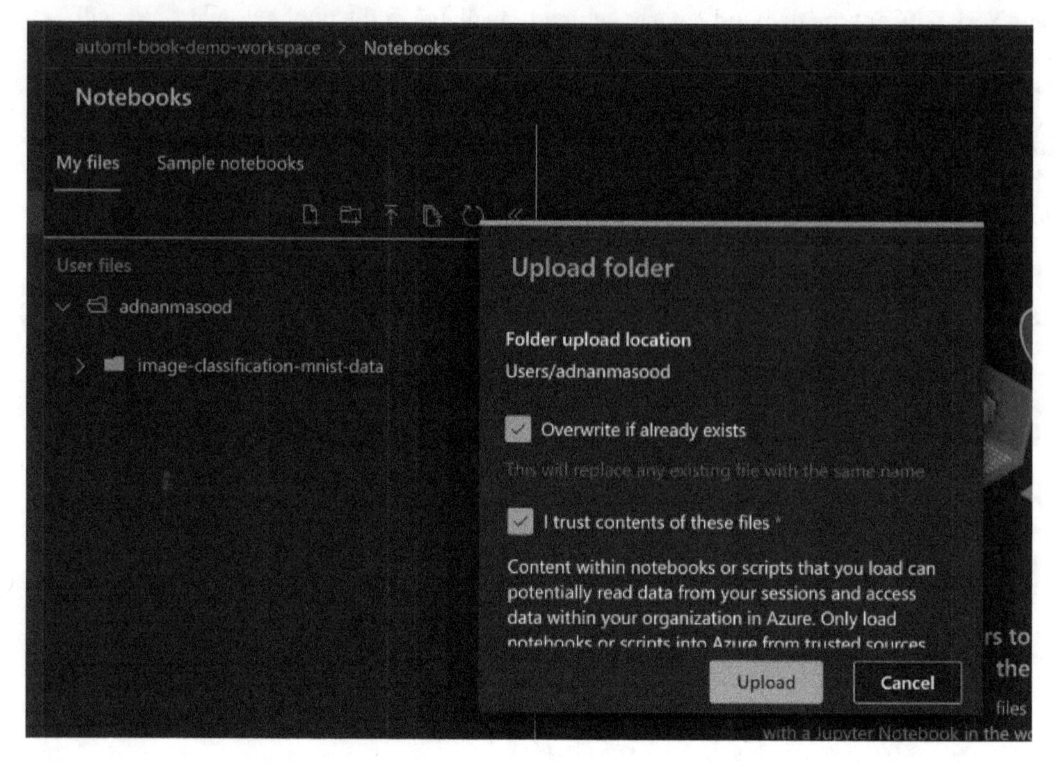

Figure 5.31 – Uploading a folder in the Azure Machine Learning notebook workspace

3. Once the folder has been uploaded (see the files in left-hand pane of the following screenshot), double-click on the, `ipynb` (notebook) file and open it. You will see the following screen:

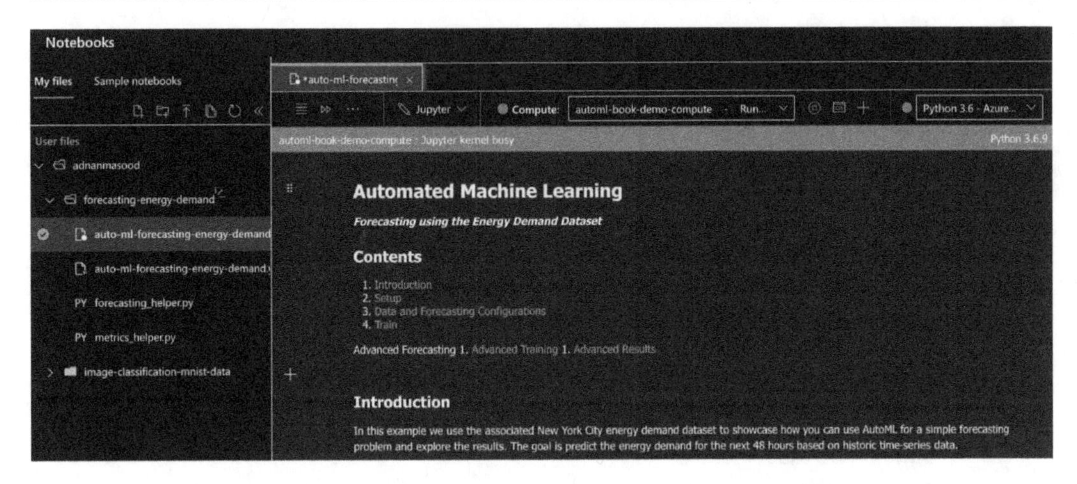

Figure 5.32 – Uploading files in the AutoML notebook workspace

4. Now, open this in JupyterLab by clicking on the respective dropdown, as shown in the following screenshot. It is important to remember that even though you are running the files in JupyterLab, an automated ML experiment is being run in the Azure Machine Learning workspace, and you can always track and view every experiment there. This shows the power of seamless integration with third-party tools:

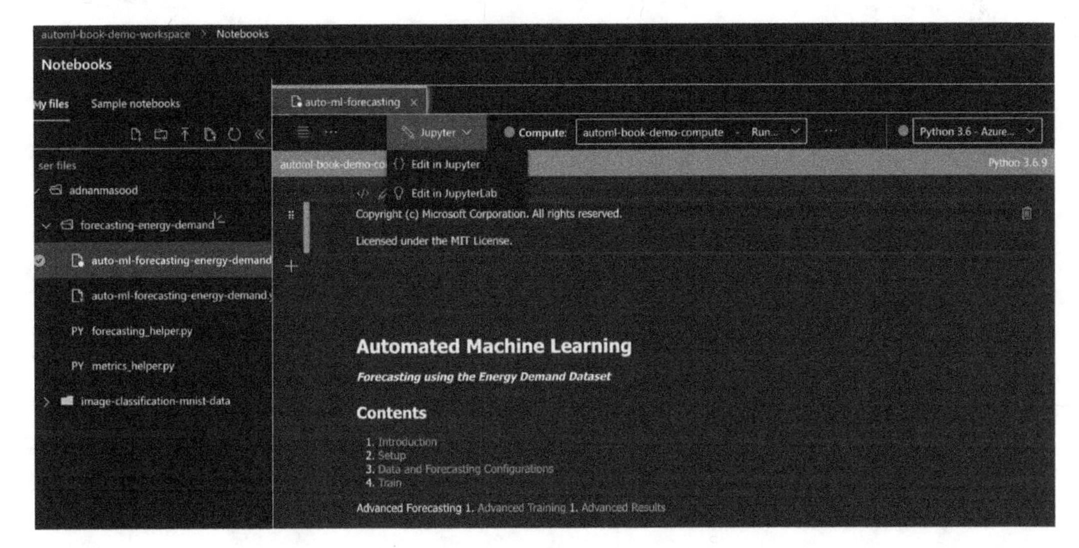

Figure 5.33 – Uploading files in the AutoML notebook workspace and opening them in JupyterLab

Now, the file is running in a very familiar environment, with the kernel being Python 3.6 – the Azure Machine Learning runtime. This seamless integration with notebooks is a powerful feature of Azure Machine Learning:

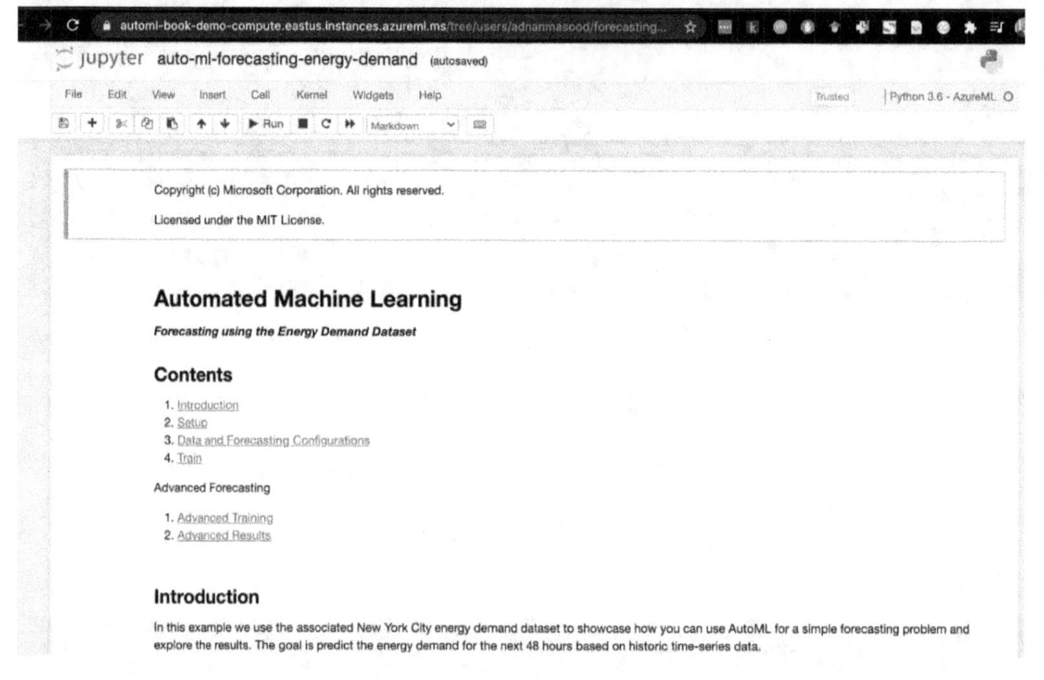

Figure 5.34 – Uploading files in the AutoML notebook workspace and opening them in JupyterLab

Since we are working with time series data, it is useful to note that Azure AutoML offers a variety of native time series as well as deep learning models to support time series-related analytical workloads. The following screenshot shows a list of these algorithms:

Automated ML provides users with both native time-series and deep learning models as part of the recommendation system.

Models	Description	Benefits
Prophet (Preview)	Prophet works best with time series that have strong seasonal effects and several seasons of historical data. To leverage this model, install it locally using `pip install fbprophet`.	Accurate & fast, robust to outliers, missing data, and dramatic changes in your time series.
Auto-ARIMA (Preview)	Auto-Regressive Integrated Moving Average (ARIMA) performs best, when the data is stationary. This means that its statistical properties like the mean and variance are constant over the entire set. For example, if you flip a coin, then the probability of you getting heads is 50%, regardless if you flip today, tomorrow or next year.	Great for univariate series, since the past values are used to predict the future values.
ForecastTCN (Preview)	ForecastTCN is a neural network model designed to tackle the most demanding forecasting tasks, capturing nonlinear local and global trends in your data as well as relationships between time series.	Capable of leveraging complex trends in your data and readily scales to the largest of datasets.

Figure 5.35 – Azure AutoML time series capabilities

Azure automated ML comes with a variety of regression, classification, and time series forecasting algorithms and scoring mechanisms, and you can always add custom metrics. The following screenshot shows a list of Azure AutoML classification, regression, and time series forecasting algorithms and measures:

Classification	Regression	Time Series Forecasting
Logistic Regression*	Elastic Net*	Elastic Net
Light GBM*	Light GBM*	Light GBM
Gradient Boosting*	Gradient Boosting*	Gradient Boosting
Decision Tree*	Decision Tree*	Decision Tree
K Nearest Neighbors*	K Nearest Neighbors*	K Nearest Neighbors
Linear SVC*	LARS Lasso*	LARS Lasso
Support Vector Classification (SVC)*	Stochastic Gradient Descent (SGD)*	Stochastic Gradient Descent (SGD)
Random Forest*	Random Forest*	Random Forest
Extremely Randomized Trees*	Extremely Randomized Trees*	Extremely Randomized Trees
Xgboost*	Xgboost*	Xgboost
Averaged Perceptron Classifier	Online Gradient Descent Regressor	Auto-ARIMA
Naive Bayes*	Fast Linear Regressor	Prophet
Stochastic Gradient Descent (SGD)*		ForecastTCN
Linear SVM Classifier*		

Figure 5.36 – Azure AutoML classification, regression, and time series forecasting algorithms

The following is a list of metrics that are used to measure the accuracy of the aforementioned methods:

Classification	Regression	Time Series Forecasting
accuracy	spearman_correlation	spearman_correlation
AUC_weighted	normalized_root_mean_squared_error	normalized_root_mean_squared_error
average_precision_score_weighted	r2_score	r2_score
norm_macro_recall	normalized_mean_absolute_error	normalized_mean_absolute_error
precision_score_weighted		

Figure 5.37 – Azure AutoML measures for classification, regression, and time series forecasting

5. Skimming over the boilerplate setup code, we can set up the experiment by setting the target column to demand and the time column's name to be the timestamp. Once we've done this, the data is downloaded and made part of the pandas DataFrame, as shown in the following screenshot:

Target column is what we want to forecast.
Time column is the time axis along which to predict.

The other columns, "temp" and "precip", are implicitly designated as features.

```
In [ ]: target_column_name = 'demand'
        time_column_name = 'timeStamp'
```

```
In [ ]: dataset = Dataset.Tabular.from_delimited_files(path = "https://automlsamplenotebookdata.blob.core.windows.net/automl-sample-notebook-dat
        a/nyc_energy.csv").with_timestamp_columns(fine_grain_timestamp=time_column_name)
        dataset.take(5).to_pandas_dataframe().reset_index(drop=True)
```

The NYC Energy dataset is missing energy demand values for all datetimes later than August 10th, 2017 5AM. Below, we trim the rows containing these missing values from the end of the dataset.

```
In [ ]: # Cut off the end of the dataset due to large number of nan values
        dataset = dataset.time_before(datetime(2017, 10, 10, 5))
```

Figure 5.38 – Azure AutoML data loading for the NYC power supply in the notebook

6. Now, let's split up the data into training and testing sets:

Split the data into train and test sets

The first split we make is into train and test sets. Note that we are splitting on time. Data before and including August 8th, 2017 5AM will be used for training, and data after will be used for testing.

```
# split into train based on time
train = dataset.time_before(datetime(2017, 8, 8, 5), include_boundary=True)
train.to_pandas_dataframe().reset_index(drop=True).sort_values(time_column_nam
e).tail(5)
```

```
# split into test based on time
test = dataset.time_between(datetime(2017, 8, 8, 6), datetime(2017, 8, 10, 5))
test.to_pandas_dataframe().reset_index(drop=True).head(5)
```

Figure 5.39 – Data split for the NYC power supply in the notebook

7. One of the key parameters you will have to set as part of this exercise is the forecast horizon; that is, how far in the future you would like to predict for. The automated ML algorithm is smart enough to know what unit to use (hour, days, or months) based on the time series frequency of your dataset. Based on our business problem, we will set the forecast horizon to `48` (hours) and submit the job, as shown in the following screenshot:

```
forecast_horizon = 48
```

```
from azureml.automl.core.forecasting_parameters import ForecastingParameters
forecasting_parameters = ForecastingParameters(
    time_column_name=time_column_name, forecast_horizon=forecast_horizon
)

automl_config = AutoMLConfig(task='forecasting',
                             primary_metric='normalized_root_mean_squared_erro
r',
                             blocked_models = ['ExtremeRandomTrees', 'AutoArim
a', 'Prophet'],
                             experiment_timeout_hours=0.3,
                             training_data=train,
                             label_column_name=target_column_name,
                             compute_target=compute_target,
                             enable_early_stopping=True,
                             n_cross_validations=3,
                             verbosity=logging.INFO,
                             forecasting_parameters=forecasting_parameters)
```

Figure 5.40 – Creating the AutoML configuration for the forecasting job

8. Now that we have created the configuration, let's submit the experiment, as shown in the following screenshot:

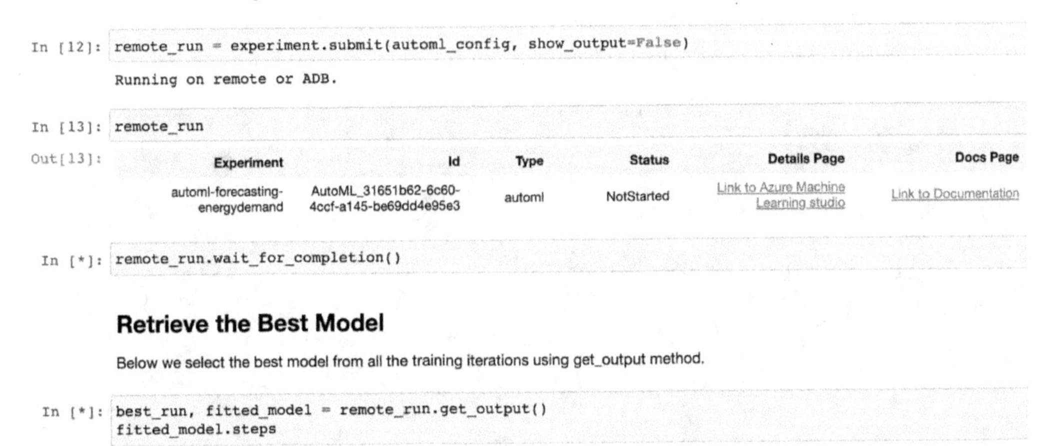

Figure 5.41 – Submitting the AutoML experiment to the remote server for execution

9. To demonstrate the integration of Jupyterlab with the Azure Machine Learning service, click on the **Experiments** tab in the ML service portal, as shown in the following screenshot. Here, you can see that the experiment has been submitted and is now prepared to run with the associated config for the AutoML parameters:

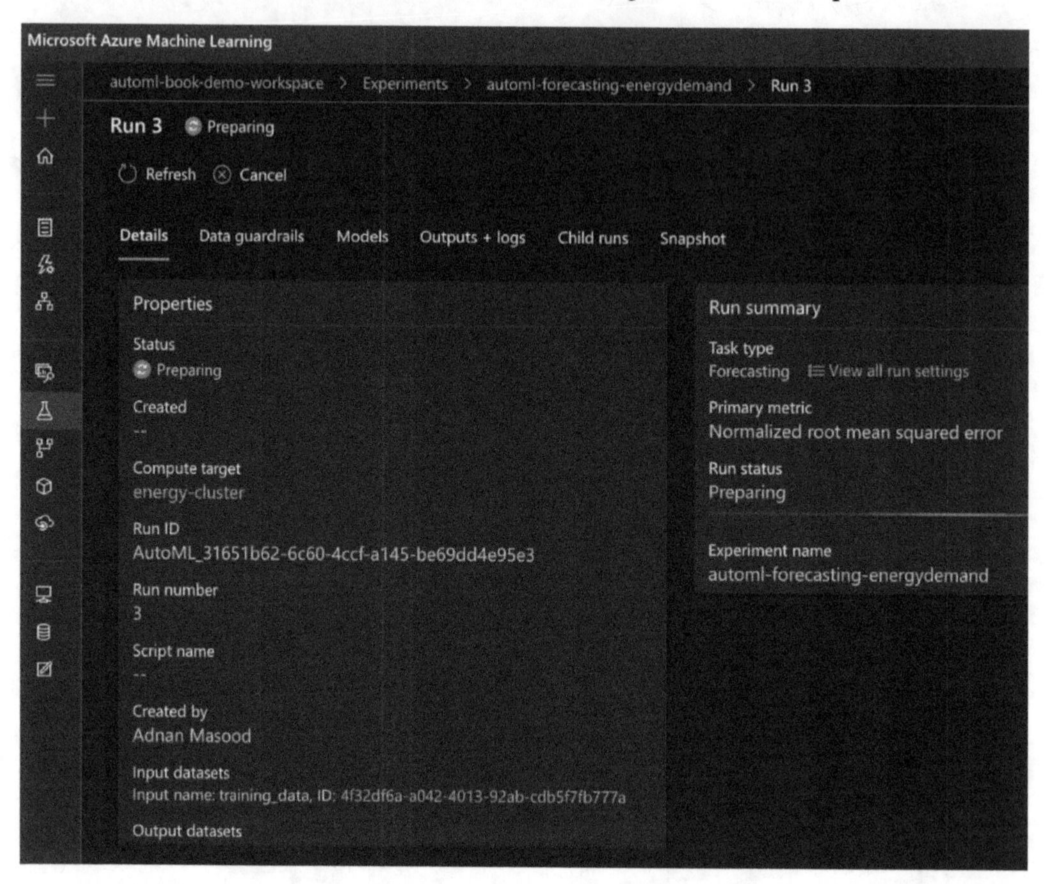

Figure 5.42 – The experiment pane views for the AutoML experiment on the remote server

The AutoML config elements can also be observed as part of the notebook as you wait for the job to complete:

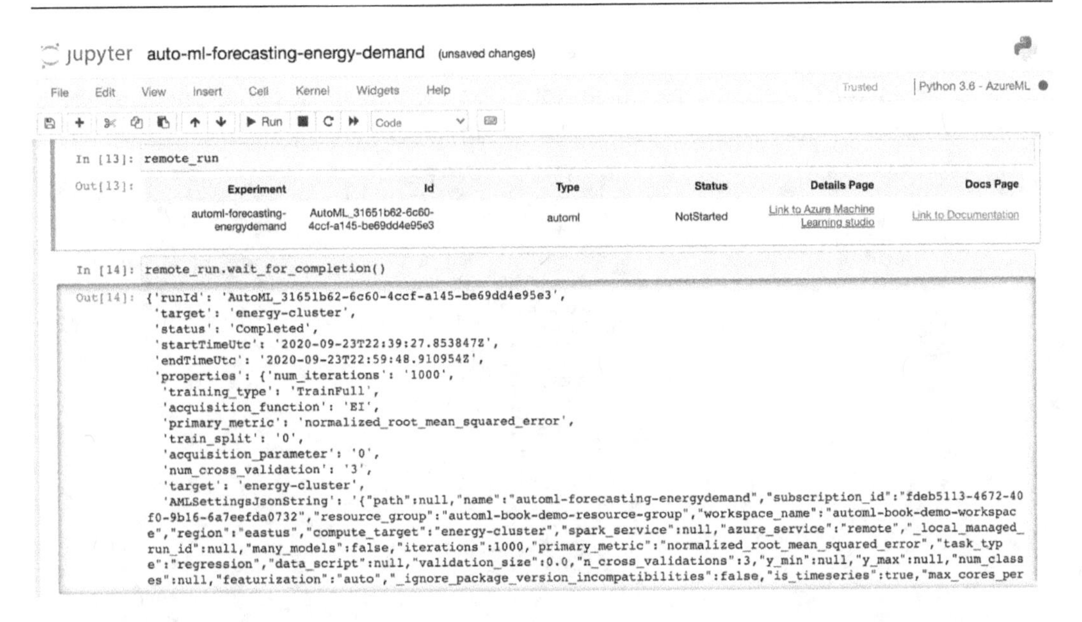

Figure 5.43 – The notebook running the wait_for_completion() method after submitting the job

10. This inherent integration between the notebook and the corresponding experiment can also be seen in the following screenshot. Here, we can see how the **Experiment** notebook is reflected in the experiment console:

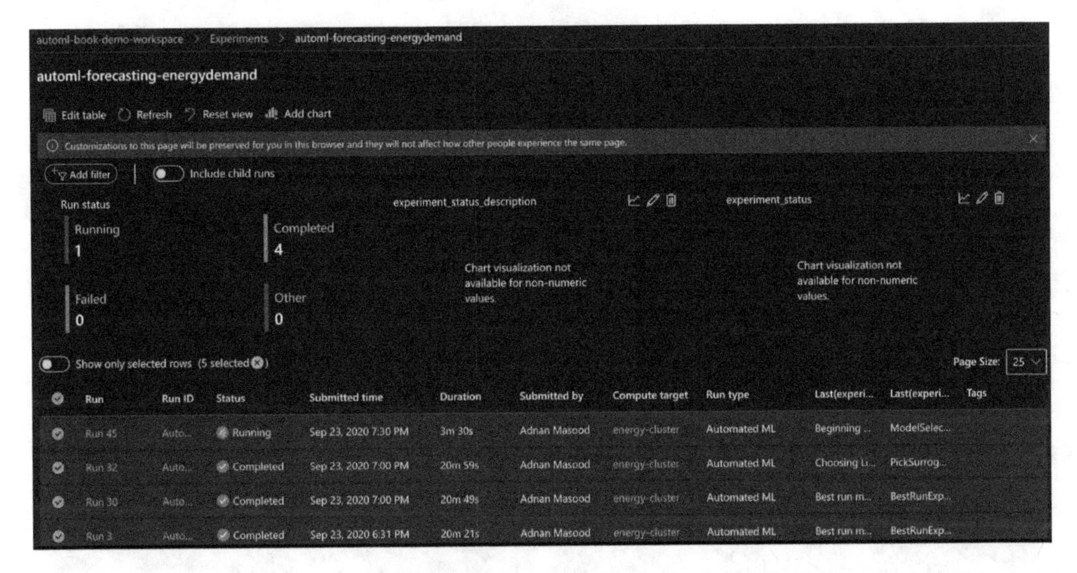

Figure 5.44 – The experiment from the notebook shown in the Experiments pane in Azure Machine Learning

The algorithm's name and error details are outlined for each run and shows a consistent reduction in the error rate. Normalized RMSE and accuracy metrics for MNIST classification are shown int he following screenshot:

Run 3 Completed

Refresh Cancel

Details Data guardrails Models Outputs + logs Child runs Snapshot

Deploy Download Explain model Search to filter items...

Algorithm name	Explained	Normalized root mean s... ↑	Sampling	Run	Created	Duration	Status
VotingEnsemble	View explanation	0.04833	100.00 %	Run 27	Sep 23, 2020 6:57 PM	46s	Completed
MinMaxScaler, DecisionTree		0.05321	100.00 %	Run 20	Sep 23, 2020 6:52 PM	38s	Completed
MinMaxScaler, DecisionTree		0.05447	100.00 %	Run 8	Sep 23, 2020 6:41 PM	35s	Completed
MaxAbsScaler, DecisionTree		0.05640	100.00 %	Run 6	Sep 23, 2020 6:39 PM	33s	Completed
MinMaxScaler, DecisionTree		0.06311	100.00 %	Run 24	Sep 23, 2020 6:56 PM	33s	Completed
RobustScaler, DecisionTree		0.06881	100.00 %	Run 18	Sep 23, 2020 6:50 PM	31s	Completed
RobustScaler, DecisionTree		0.08042	100.00 %	Run 22	Sep 23, 2020 6:54 PM	36s	Completed
RobustScaler, ElasticNet		0.08947	100.00 %	Run 12	Sep 23, 2020 6:45 PM	33s	Completed

Figure 5.45 – The experiment from the notebook shown in the Experiments pane in Azure Machine Learning

The data guardrails types are also notable. In the following screenshot, you can see that they are different from the guardrails we had in the classification exercise. In this case, the data is validated against frequency detection and missing feature value imputation. The AutoML engine is smart enough to learn what types of guardrails need to be applied for different types of experiments and datasets:

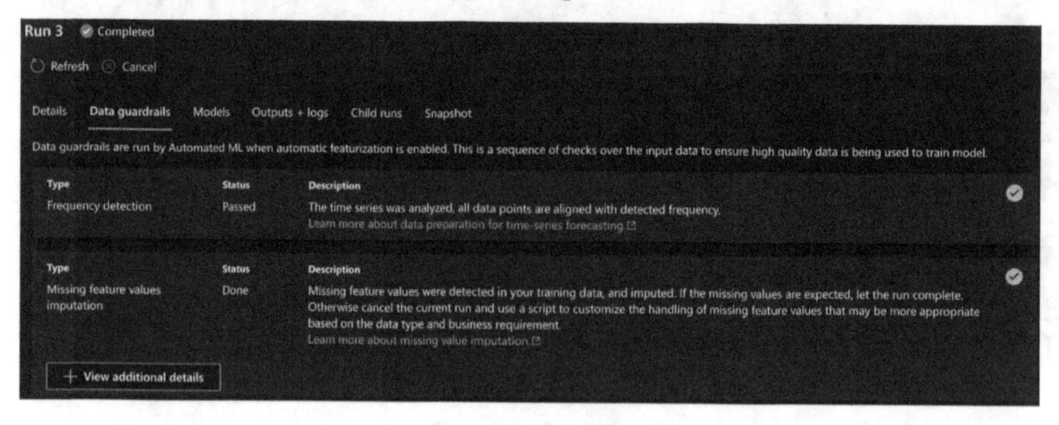

Figure 5.46 – The guardrails for the Experiments pane in Azure Machine Learning

11. Now that the experiment is complete, we can retrieve the best model in the notebook, as shown in the following screenshot (or in the machine learning service console, if you are visually inclined):

Retrieve the Best Model

Below we select the best model from all the training iterations using get_output method.

```
In [15]: best_run, fitted_model = remote_run.get_output()
         fitted_model.steps

Out[15]: [('timeseriestransformer',
           TimeSeriesTransformer(featurization_config=None,
                                 pipeline_type=<TimeSeriesPipelineType.FULL: 1>)),
          ('prefittedsoftvotingregressor',
           PreFittedSoftVotingRegressor(estimators=[('7',
                                                     Pipeline(memory=None,
                                                              steps=[('minmaxscaler',
                                                                      MinMaxScaler(copy=True,
                                                                                   feature_range=(0,
                                                                                                  1))),
                                                                     ('decisiontreeregressor',
                                                                      DecisionTreeRegressor(ccp_alpha=0.0,
                                                                                            criterion='mse',
                                                                                            max_depth=None,
                                                                                            max_features=0.7,
                                                                                            max_leaf_nodes=None,
                                                                                            min_impurity_decrease=0.0,
                                                                                            min_impurity_split=None,
                                                                                            min_samples_leaf=0.001953125,
                                                                                            min_sam...
                                                                                            max_depth=None,
                                                                                            max_features=0.8,
                                                                                            max_leaf_nodes=None,
                                                                                            min_impurity_decrease=0.0,
                                                                                            min_impurity_split=None,
                                                                                            min_samples_leaf=0.018779547644135
         22,
                                                                                            min_samples_split=0.00182615846827
         02607,
                                                                                            min_weight_fraction_leaf=0.0,
                                                                                            presort='deprecated',
                                                                                            random_state=None,
                                                                                            splitter='best'))],
                                                              verbose=False))],
                                        weights=[0.45454545454545453, 0.2727272727272727,
                                                 0.2727272727272727]))]
```

Figure 5.47 – Model retrieval in the notebook

12. You might recall deep feature search or automated feature engineering being introduced in the previous chapters. You can access and retrieve the engineered features from the notebook with the help of the following steps by calling the `get_engineered_feature_names()` method on the model:

Featurization

You can access the engineered feature names generated in time-series featurization.

```
In [16]: fitted_model.named_steps['timeseriestransformer'].get_engineered_feature_names()

Out[16]: ['precip',
          'temp',
          'precip_WASNULL',
          'temp_WASNULL',
          'year',
          'half',
          'quarter',
          'month',
          'day',
          'hour',
          'am_pm',
          'hour12',
          'wday',
          'qday',
          'week']
```

Figure 5.48 – Retrieving engineered features via get_engineered_feature_names

Viewing the featurization summary for these features, both engineered and organic, provides you with the rationale that was used to build these features, as shown in the following screenshot:

View featurization summary

You can also see what featurization steps were performed on different raw features in the user data. For each raw feature in the user data, the following information is displayed:

- Raw feature name
- Number of engineered features formed out of this raw feature
- Type detected
- If feature was dropped
- List of feature transformations for the raw feature

```
In [17]: # Get the featurization summary as a list of JSON
featurization_summary = fitted_model.named_steps['timeseriestransformer'].get_featurization_summary()
# View the featurization summary as a pandas dataframe
pd.DataFrame.from_records(featurization_summary)

Out[17]:
```

	RawFeatureName	TypeDetected	Dropped	EngineeredFeatureCount	Transformations
0	precip	Numeric	No	2	[MedianImputer, ImputationMarker]
1	temp	Numeric	No	2	[MedianImputer, ImputationMarker]
2	timeStamp	DateTime	No	11	[DateTimeTransformer, DateTimeTransformer, DateTimeTransformer, DateTimeTransformer, DateTimeTransformer, DateTimeTransformer, DateTimeTransformer, DateTimeTransformer, DateTimeTransformer, DateTimeTransformer, DateTimeTransformer]

Figure 5.49 – Viewing the engineered features summary via get_featurization_summary()

13. Using the scoring method, we can create the test scores and plot the predicted points on a chart, as shown in the following screenshot:

```
In [21]: from azureml.automl.core.shared import constants
         from azureml.automl.runtime.shared.score import scoring
         from matplotlib import pyplot as plt

         # use automl metrics module
         scores = scoring.score_regression(
             y_test=df_all[target_column_name],
             y_pred=df_all['predicted'],
             metrics=list(constants.Metric.SCALAR_REGRESSION_SET))

         print("[Test data scores]\n")
         for key, value in scores.items():
             print('{}:    {:.3f}'.format(key, value))

         # Plot outputs
         %matplotlib inline
         test_pred = plt.scatter(df_all[target_column_name], df_all['predicted'], color='b')
         test_test = plt.scatter(df_all[target_column_name], df_all[target_column_name], color='g')
         plt.legend((test_pred, test_test), ('prediction', 'truth'), loc='upper left', fontsize=8)
         plt.show()

         [Test data scores]

         normalized_root_mean_squared_error:    0.150
         mean_absolute_percentage_error:    5.491
         normalized_mean_absolute_error:    0.122
         r2_score:    0.743
         normalized_median_absolute_error:    0.097
         root_mean_squared_log_error:    0.064
         normalized_root_mean_squared_log_error:    0.130
         explained_variance:    0.787
         mean_absolute_error:    383.207
         root_mean_squared_error:    473.089
         spearman_correlation:    0.972
         median_absolute_error:    305.623
```

Figure 5.50 – Building a scatter plot for test data scores

The predicted data test scores are in blue, while the actual score is in green:

> **Note**
>
> This image may be black and white to you. You will understand the color reference better when practically working on the example.

```
[Test data scores]

normalized_root_mean_squared_error:   0.150
mean_absolute_percentage_error:   5.491
normalized_mean_absolute_error:   0.122
r2_score:   0.743
normalized_median_absolute_error:   0.097
root_mean_squared_log_error:   0.064
normalized_root_mean_squared_log_error:   0.130
explained_variance:   0.787
mean_absolute_error:   383.207
root_mean_squared_error:   473.089
spearman_correlation:   0.972
median_absolute_error:   305.623
```

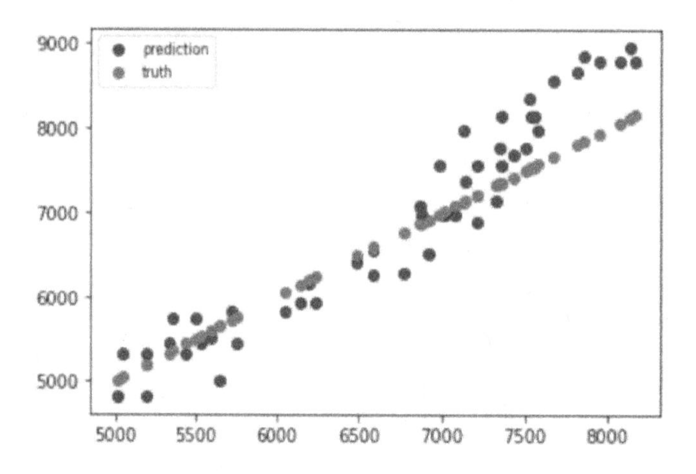

Looking at `X_trans` is also useful to see what featurization happened to the data.

Figure 5.51 – The test data scores and the associated plot

`X_trans` captures the featurization, including the automatic feature engineering changes in the dataset, as shown in the following screenshot:

```
In [22]: X_trans
```

timeStamp	_automl_dummy_grain_col	precip	temp	precip_WASNULL	temp_WASNULL	year	half	quarter	month	day	hour	am_pm	hour12	wday	qday	week	_automl_t
2017-08-08 06:00:00	_automl_dummy_grain_col	0.00	66.17	0	0	2017	2	3	8	8	6	0	6	1	39	32	
2017-08-08 07:00:00	_automl_dummy_grain_col	0.00	66.29	0	0	2017	2	3	8	8	7	0	7	1	39	32	
2017-08-08 08:00:00	_automl_dummy_grain_col	0.00	66.72	0	0	2017	2	3	8	8	8	0	8	1	39	32	
2017-08-08 09:00:00	_automl_dummy_grain_col	0.00	67.37	0	0	2017	2	3	8	8	9	0	9	1	39	32	
2017-08-08 10:00:00	_automl_dummy_grain_col	0.00	68.30	0	0	2017	2	3	8	8	10	0	10	1	39	32	

Figure 5.52 – X_trans showing the time series features for energy forecasting

Even though the explainability of the MNIST dataset wasn't quite as intuitive, in terms of exploring the energy demand dataset, you can visualize different models and see which features have the most impact on the predicted usage. It is quite intuitive that temperature would have a positive correlation with the global importance of power usage. A higher temperature leads to a heavier usage of air conditioning, and hence higher power usage. The time of day and day of the week are also deemed important by the model, as shown in the following chart:

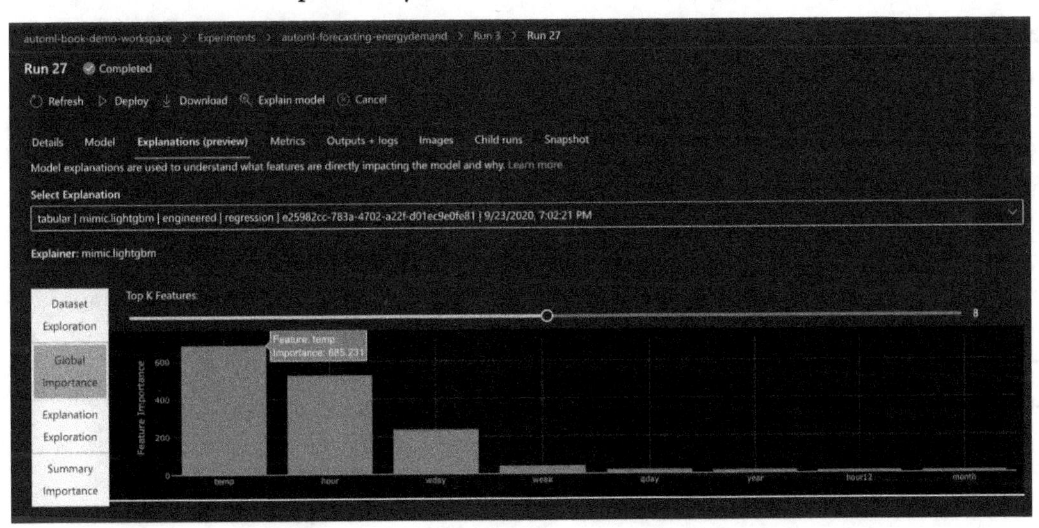

Figure 5.53 – Global importance explainability graph

In the following screenshot, different explanation models (engineered features versus raw) map the result against the different predicted values of Y. Model explanation views help us understand which features directly impact the model:

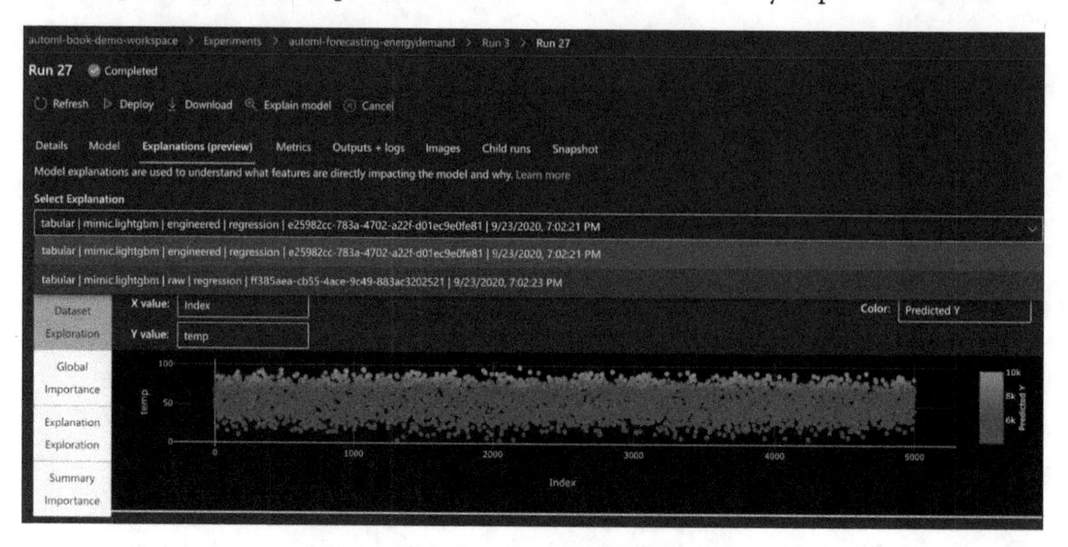

Figure 5.54 – Global importance explainability graph

This concludes our demonstration of time series prediction using AutoML in Azure.

Summary

In this chapter, you learned how to apply AutoML in Azure to a classification problem and a time series prediction problem. You were able to build a model within the Azure Machine Learning environment with an Azure notebook and via JupyterLab. You then understood how the entire workspace relates to the experiments and runs. You also see the visualization during these automated runs; this is where feature importance, the global and local impact of features, and explanations based on raw and engineered features provide an intuitive understanding. Besides your affinity with a tool, it is also important that the platform aligns with your enterprise roadmap. Azure is an overall great platform with a comprehensive set of tools, and we hope you enjoyed exploring its automated ML capabilities.

Further reading

For more information on the topics that were covered in this chapter, please take a look at the following links:

- Azure AutoML:

  `https://docs.microsoft.com/en-us/azure/machine-learning/concept-automated-ml`

- Practical AutoML on Azure:

  `https://github.com/PracticalAutomatedMachineLearning/Azure`

6
Machine Learning with AWS

"Whatever you are studying right now, if you are not getting up to speed on deep learning, neural networks, etc., you lose. We are going through the process where software will automate software, automation will automate automation."

– Mark Cuban

In the previous chapter, you were introduced to the Azure **Machine Learning** (**ML**) landscape and how to do automated ML in the Azure platform. In this chapter, you will learn how to get started with ML using **Amazon Web Services** (**AWS**), along with different offerings and a detailed understanding of the ginormous AWS cloud stack.

The theme of this chapter is to get started with an introduction to AWS ML capabilities to give a wider perspective of this large ecosystem; not only AWS as a hyperscaler but also the breadth of the field itself. Many use cases and permutations require specialized solutions and there is no one-size-fits-all solution for an enterprise's AI and ML needs. That is why this breadth of knowledge about every cloud offering is important as you embark on your automated ML journey.

In this chapter, we will cover the following topics:

- Overview of ML in the AWS landscape
- Getting started with AWS ML
- Working with AWS SageMaker Autopilot

ML in the AWS landscape

Gartner is among a few major advisory companies that regularly review the landscape of technology and provide a comprehensive overview of their findings in their Magic Quadrant reports. In its latest release, the Magic Quadrant contains Anaconda and Altair as niche players, Microsoft, DataRobot, KNIME, Google, H2O.ai, RapidMiner, and Domino as visionaries, IBM as a challenger, and Alteryx, SAS, Databricks, MathWorks, TIBCO, and Dataiku as leaders in the data science and ML space.

It is surprising for us to not see AWS mentioned here. There are six companies in the leadership quadrant due to their consistent record of data science and AI solution deliveries, and seven are classified as visionaries. However, AWS not making it to the visionaries and/or the leaders quadrant is attributed to the announcement delay. The AWS flagship AI products SageMaker Studio and SageMaker Autopilot were announced after the deadline for Gartner submission; hence they didn't make the cut. It is surprising not to see AWS on the list because of the breadth of the AWS solution landscape. As a cloud services provider with a headstart, AWS dominates the market share greater than the sum of its closest three competitors.

AWS provides a comprehensive set of tools to developers, data scientists, ML engineers, and enthusiasts to work with AI and ML. These tools range from frameworks and infrastructure components, ML services, AI services, **Integrated Development Environments (IDEs)**, and APIs to training and tutorials to get you started in the ever-growing world of AWS offerings. The following is a bird's-eye view of the AWS ML stack:

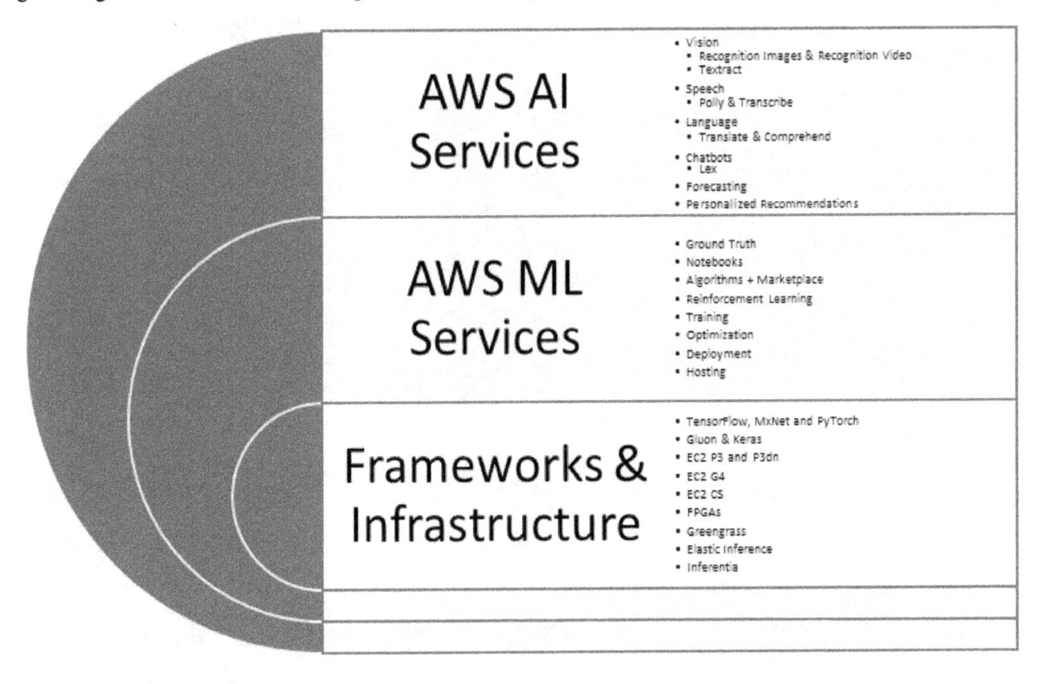

Figure 6.1 – Amazon ML stack – set of AI and ML services and capabilities – courtesy of Amazon re:Invent

Due to both the depth and breadth of these myriad offerings, each of these services deserves at least a chapter of its own, but then we would not be able to meet those slippery deadlines. Therefore, in the interest of time and brevity, we will only focus on the automated ML part, that is, Amazon SageMaker and its Autopilot offering. Please see the *Further reading* section for links to read more about Amazon's ML stack.

SageMaker is a fully managed, cloud-based ML platform that provides AI operationalization capabilities by enabling an end-to-end ML workflow, without the metaphorical tears. In the following figure, you will see the components for end-to-end ML with AWS that are used to prepare, build, train, tune, deploy, and manage models. You will also notice how SageMaker Autopilot offers capabilities across the spectrum to automatically build and train models, as well as to prepare, build, and maintain these models throughout the life cycle:

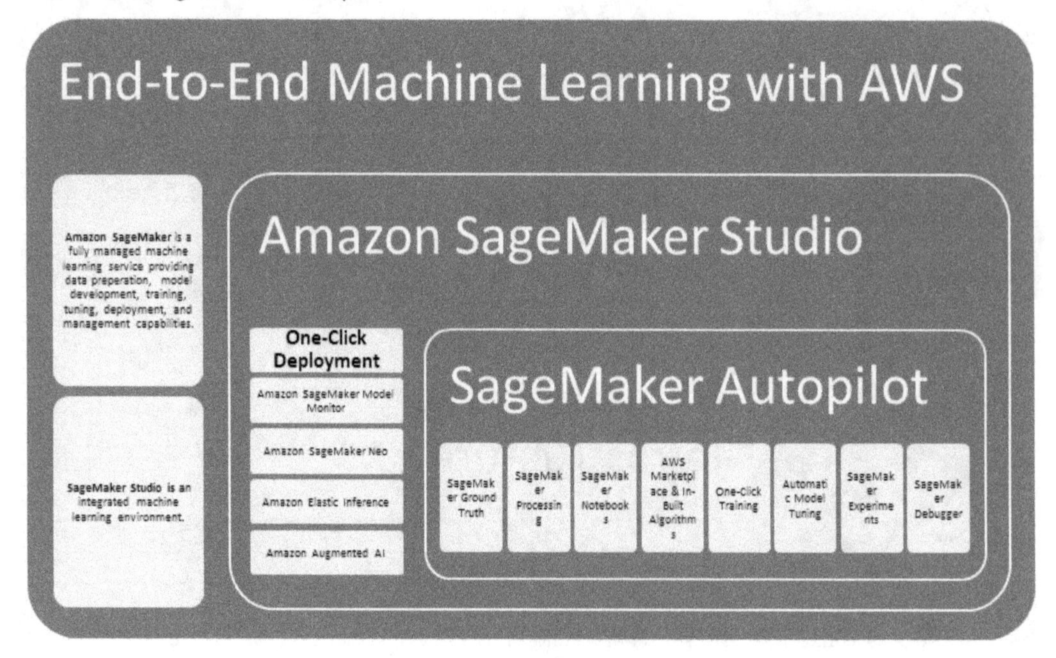

Figure 6.2 – End-to-end ML with AWS

Even though our focus is on automated ML capabilities, it is worthwhile for you to explore the Amazon SageMaker landscape. One of the key offerings is Amazon SageMaker Studio, a web-based ML IDE to prepare, build, deploy, and operationalize models. When SageMaker is mentioned, this IDE is what most people think of, but you will see that it is part of a larger ecosystem.

Notebooks are the Swiss Army knives of data scientists. Amazon SageMaker Studio notebooks provide the comforting environment that data scientists have come to know and love, for the most part. Amazon SageMaker Ground Truth provides training datasets, while Amazon **Augmented AI** (**A2I**) enables **Human-in-the-Loop** (**HITL**), where a carbon-based lifeform is required to review the ML predictions, especially ones with low confidence scores. Amazon SageMaker Experiments is similar to what you have explored earlier, in the other hyperscaler offerings. It helps to track data, enables reconstruction and sharing of an experiment, and provides tracing information for auditing purposes. Amazon SageMaker comes with a wide variety of built-in algorithms for classification, regression, text analysis, topic modeling, forecasting, clustering, and many more use cases, as seen in the following figure:

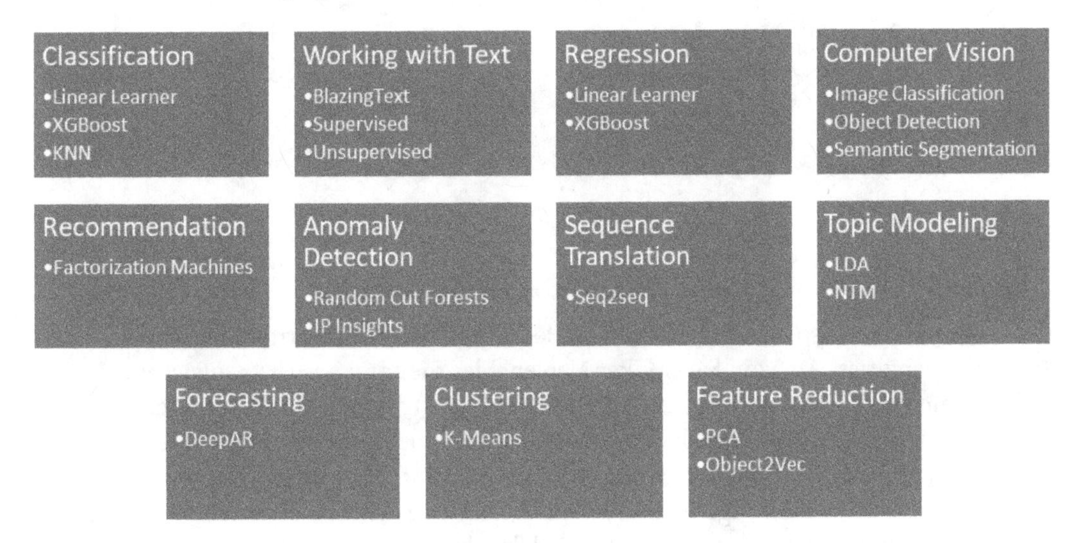

Figure 6.3 – Amazon SageMaker built-in algorithms

Amazon SageMaker Debugger helps inspect parameters and data, while Amazon SageMaker Model Monitor keeps an eye on model behavior in production. Model monitoring has gained loads of attention recently since data drift can significantly impact model quality, and therefore predictions. Online learning can be perilous; let Tay serve as a lesson learned. Some of the features of Amazon SageMaker for the democratization of AI can be seen in the following figure:

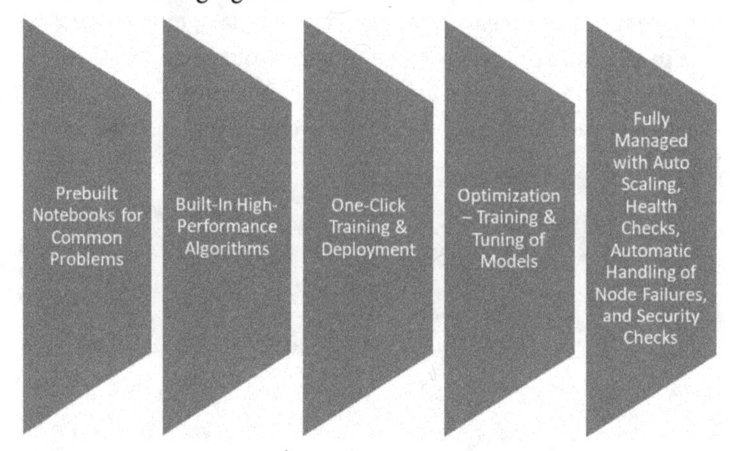

Figure 6.4 – Different types of Amazon SageMaker capabilities – courtesy of Amazon re:Invent

Amazon SageMaker also offers reinforcement learning, batch transformation, and elastic inference capabilities. Amazon SageMaker Neo enables the *train once, run anywhere* capability and helps separate training and inference infrastructure. Neo is powered by the Apache-licensed Neo-AI-DLR common runtime, supports the usual suspects (TensorFlow, MXNet, PyTorch, ONNX, and XGBoost), and even claims to speed them up. Lastly, we have Amazon SageMaker Autopilot, our focus for this book, where citizen data scientists can build, train, and test ML models – bringing us one step closer to the democratization of AI.

We will discuss more on SageMaker Autopilot in the second half of this chapter. First, let's explore writing some code in AWS SageMaker.

Getting started with AWS ML

In this section, we will do a walkthrough of the AWS Management Console and show you how to use AWS SageMaker with step-by-step instructions. Let's dive in. The AWS ML environment is fairly intuitive and easy to work with:

1. To start, first open up the AWS Management Console by visiting `aws.amazon.com` in your browser. Now, click on **Sign in to the Console**, or log back in (if you are a returning user):

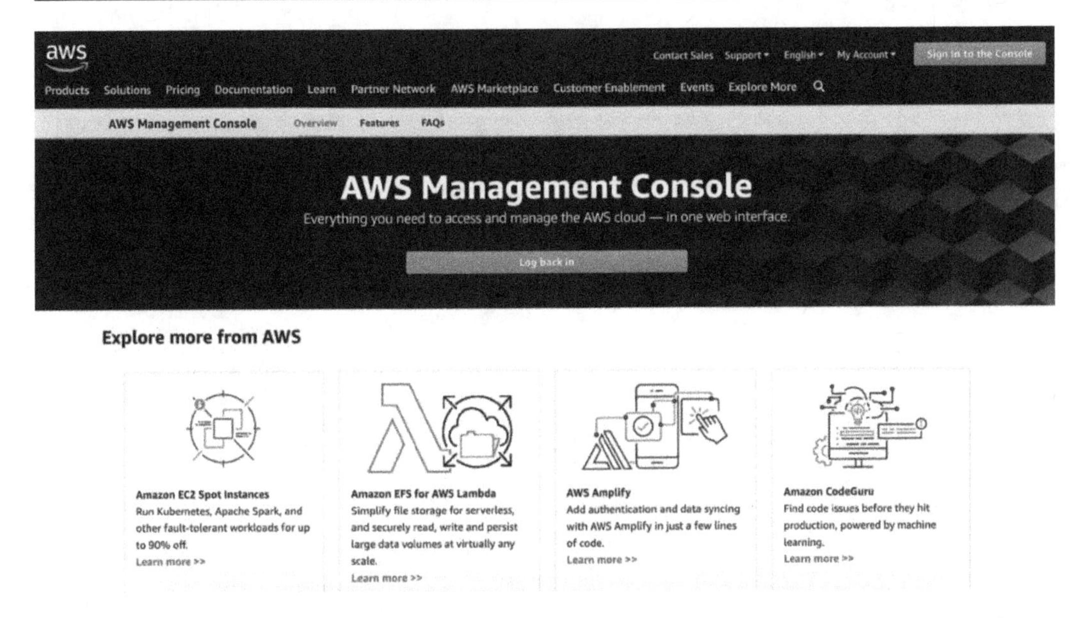

Figure 6.5 – AWS Management Console

2. Enter your root (account) user's email address in the **Root user email address** field to proceed:

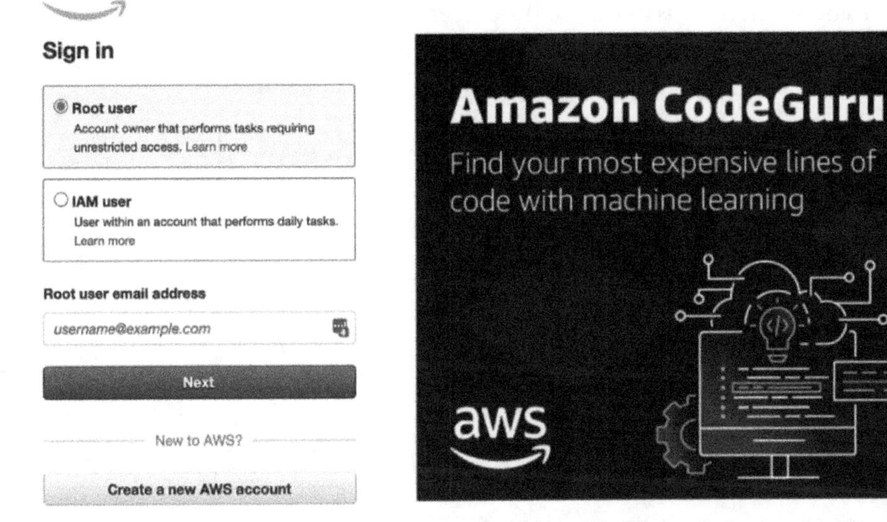

Figure 6.6 – AWS Management Console login

3. Upon successful login, you will be taken to the following screen, the AWS Management Console:

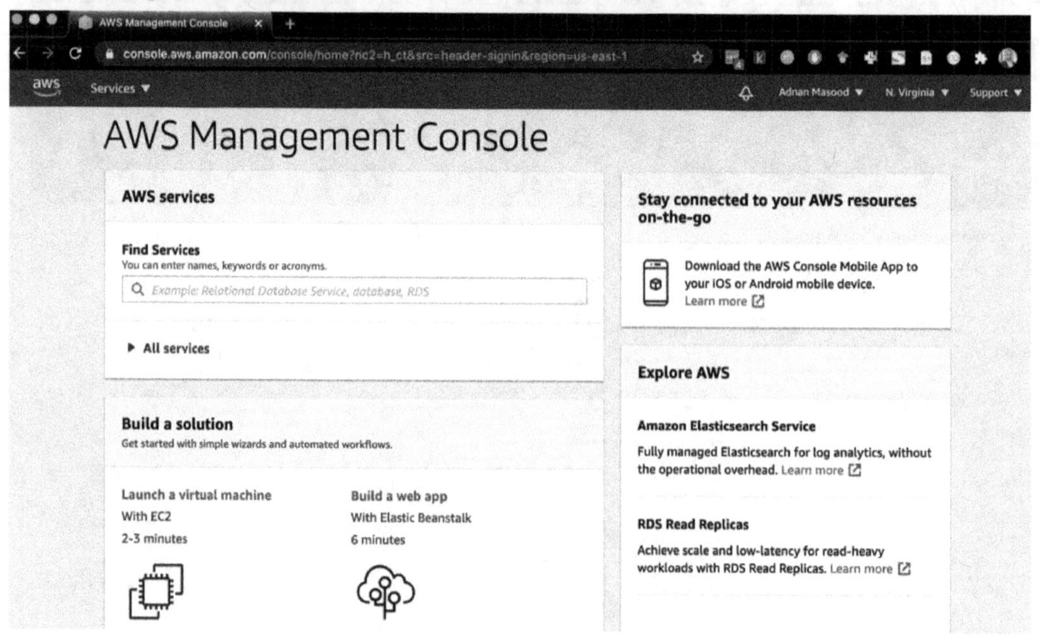

Figure 6.7 – AWS Management Console

4. AWS has a collection of tons of different services. In the AWS Management Console, find the services search box, then type `sagemaker` to find the **Amazon SageMaker** service, as shown in the following screenshot, and click on it:

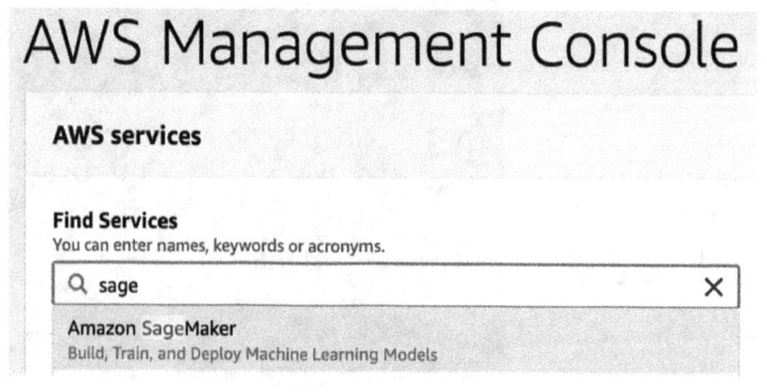

Figure 6.8 – SageMaker search in the AWS Management Console

5. This will take you to the SageMaker home page shown in the following screenshot. Here, you can read up on the different offerings, such as Ground Truth, notebooks, jobs processing, and so on:

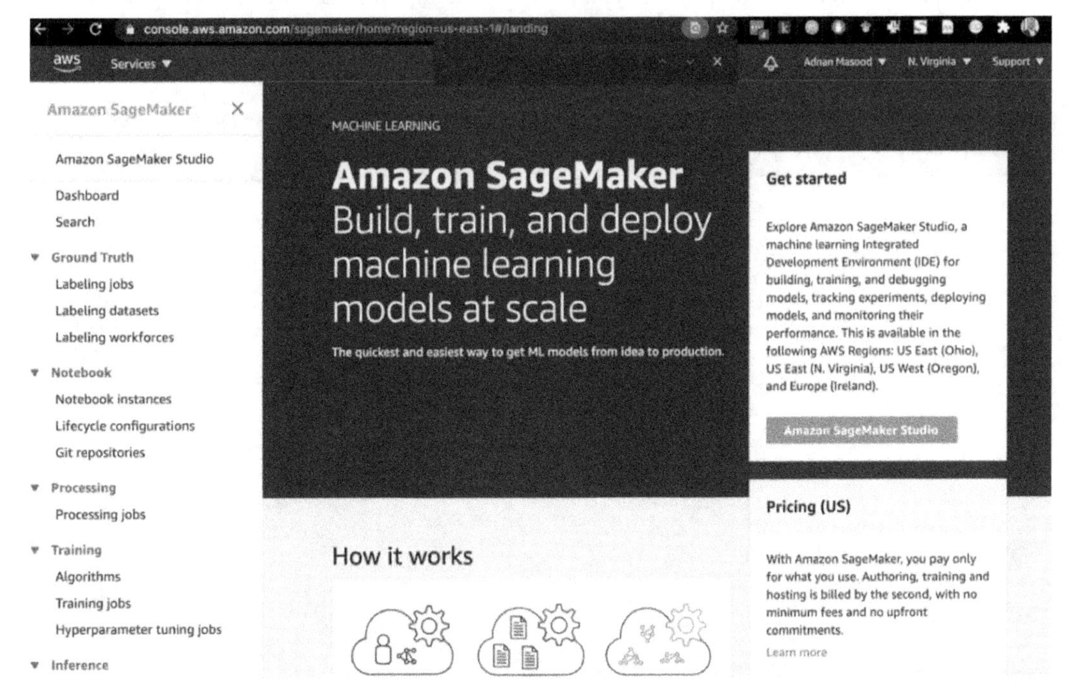

Figure 6.9 – Amazon SageMaker home page

The AWS team has put a copious amount of work into building the documentation, training videos, and partner training programs to get developers up to speed. You can see some of these courses in the *Further reading* section at the end of the chapter. For example, here, when you click on the top link in the left-hand pane, it will show you information regarding how you can build, train, and deploy models using Amazon SageMaker Studio. Contextual documentation, pretty neat eh!

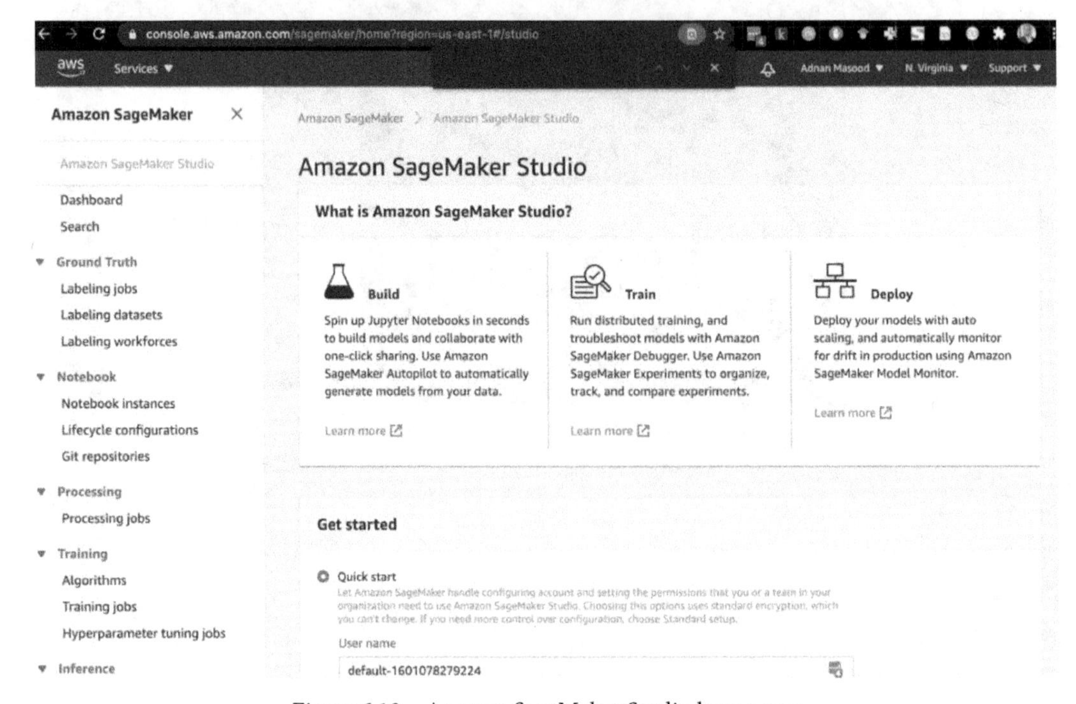

Figure 6.10 – Amazon SageMaker Studio home page

6. Now, let's explore SageMaker Studio, where we will be developing a classification model for our favorite dataset. In the **Get started** section, you will create a username and define an execution **Identity and Access Management** (**IAM**) role. An IAM role gives you granular permissions as to what you can and cannot do on the AWS platform. Click on the **Execution role** dropdown to select the role you intend to use (or create a new role):

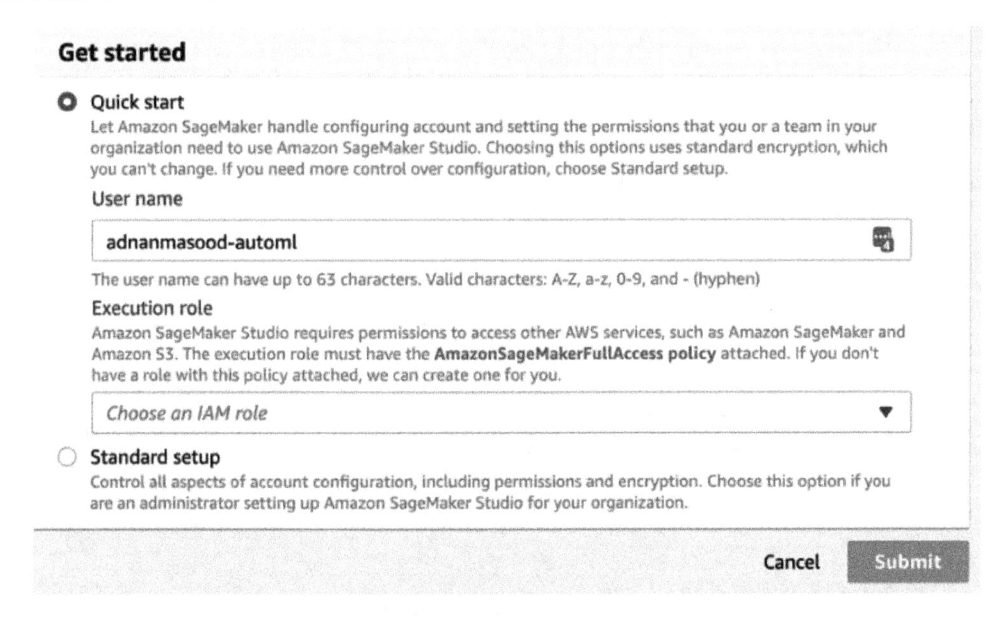
Get started
Quick start
Let Amazon SageMaker handle configuring account and setting the permissions that you or a team in your organization need to use Amazon SageMaker Studio. Choosing this options uses standard encryption, which you can't change. If you need more control over configuration, choose Standard setup.
User name
adnanmasood-automl
The user name can have up to 63 characters. Valid characters: A-Z, a-z, 0-9, and - (hyphen)
Execution role
Amazon SageMaker Studio requires permissions to access other AWS services, such as Amazon SageMaker and Amazon S3. The execution role must have the AmazonSageMakerFullAccess policy attached. If you don't have a role with this policy attached, we can create one for you.
Choose an IAM role
Standard setup
Control all aspects of account configuration, including permissions and encryption. Choose this option if you are an administrator setting up Amazon SageMaker Studio for your organization.
Cancel
Submit

Figure 6.11 – Amazon SageMaker Studio Get started screen

7. You can create an IAM role if you don't have one already and grant the appropriate permissions, as shown in the following screenshot. S3 is one of the AWS storage mechanisms and the following screen allows you to create an IAM role to access S3 buckets. This is a one-time setup process, unless you plan to make changes:

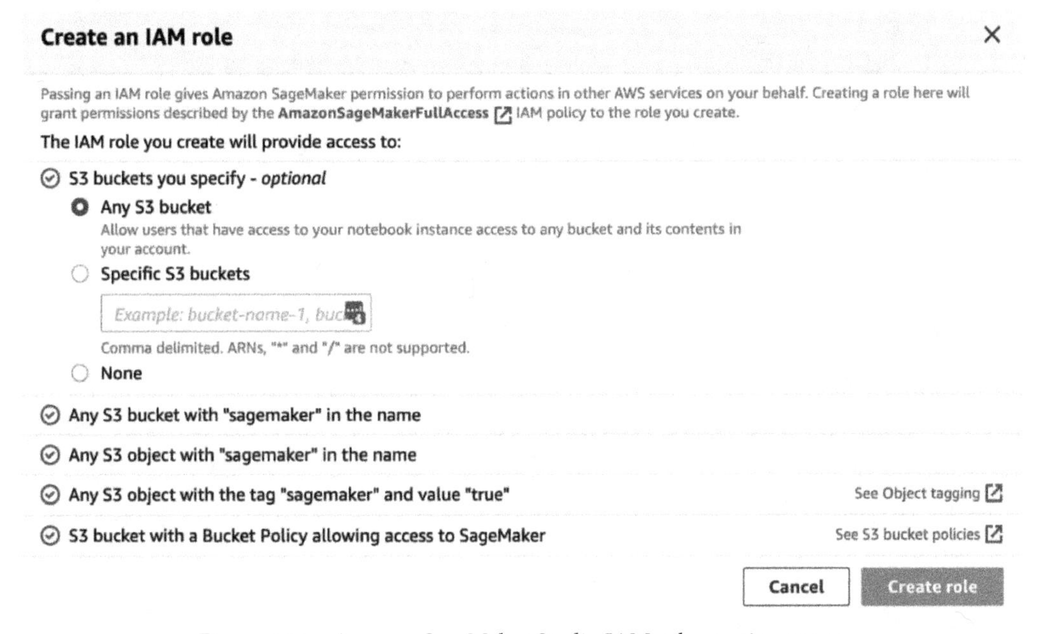
Create an IAM role
Passing an IAM role gives Amazon SageMaker permission to perform actions in other AWS services on your behalf. Creating a role here will grant permissions described by the AmazonSageMakerFullAccess IAM policy to the role you create.
The IAM role you create will provide access to:
S3 buckets you specify - optional
Any S3 bucket
Allow users that have access to your notebook instance access to any bucket and its contents in your account.
Specific S3 buckets
Example: bucket-name-1, buc
Comma delimited. ARNs, "*" and "/" are not supported.
None
Any S3 bucket with "sagemaker" in the name
Any S3 object with "sagemaker" in the name
Any S3 object with the tag "sagemaker" and value "true"
See Object tagging
S3 bucket with a Bucket Policy allowing access to SageMaker
See S3 bucket policies
Cancel
Create role

Figure 6.12 – Amazon SageMaker Studio IAM role creation screen

8. Once you are done creating the IAM role, you will see the following success message. Press **Submit** to navigate to the next screen:

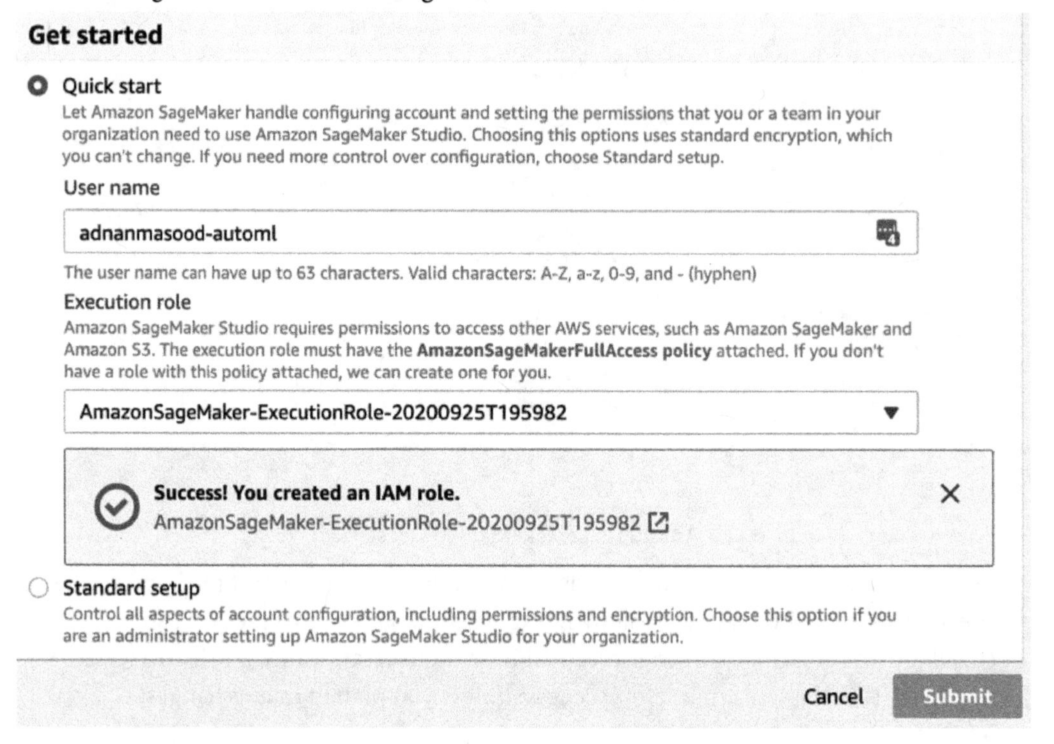

Get started
Quick start
Let Amazon SageMaker handle configuring account and setting the permissions that you or a team in your organization need to use Amazon SageMaker Studio. Choosing this options uses standard encryption, which you can't change. If you need more control over configuration, choose Standard setup.
User name
adnanmasood-automl
The user name can have up to 63 characters. Valid characters: A-Z, a-z, 0-9, and - (hyphen)
Execution role
Amazon SageMaker Studio requires permissions to access other AWS services, such as Amazon SageMaker and Amazon S3. The execution role must have the AmazonSageMakerFullAccess policy attached. If you don't have a role with this policy attached, we can create one for you.
AmazonSageMaker-ExecutionRole-20200925T195982
Success! You created an IAM role.
AmazonSageMaker-ExecutionRole-20200925T195982
Standard setup
Control all aspects of account configuration, including permissions and encryption. Choose this option if you are an administrator setting up Amazon SageMaker Studio for your organization.
Cancel
Submit

Figure 6.13 – Amazon SageMaker Studio IAM role creation success

9. Once the role is created, you will be taken to the SageMaker dashboard, where you will see the available offerings, as shown in the following screenshot:

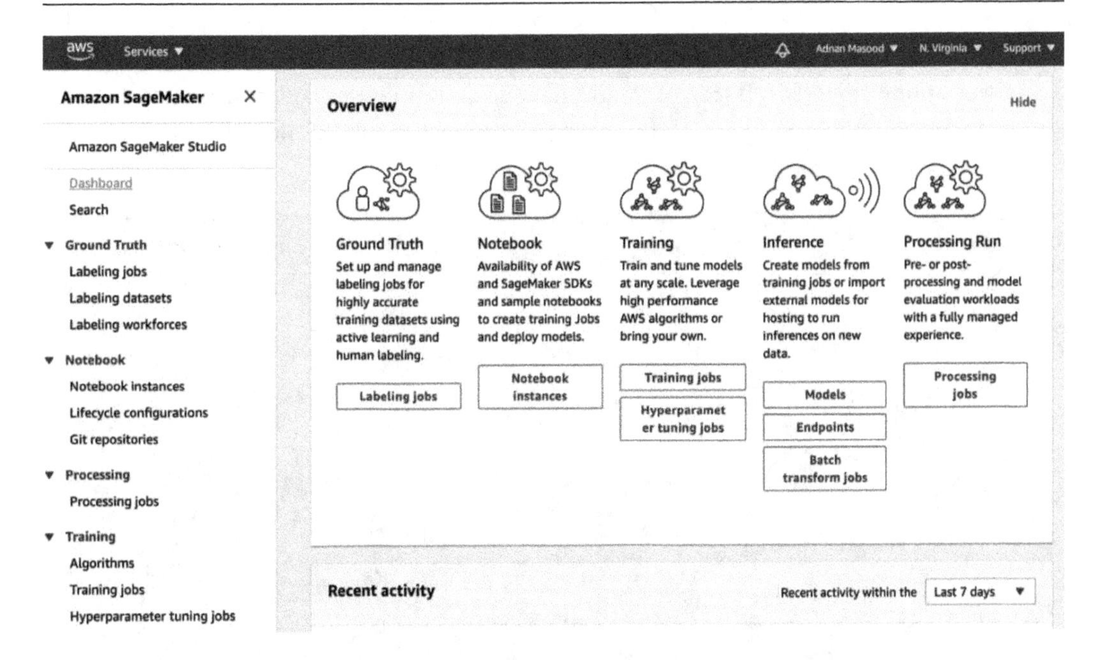

Figure 6.14 – Amazon SageMaker Studio dashboard

10. From this earlier screen, you can now navigate to the **Control Panel** to see your associated user and click on **Open Studio**, which will take you (finally) to SageMaker Studio:

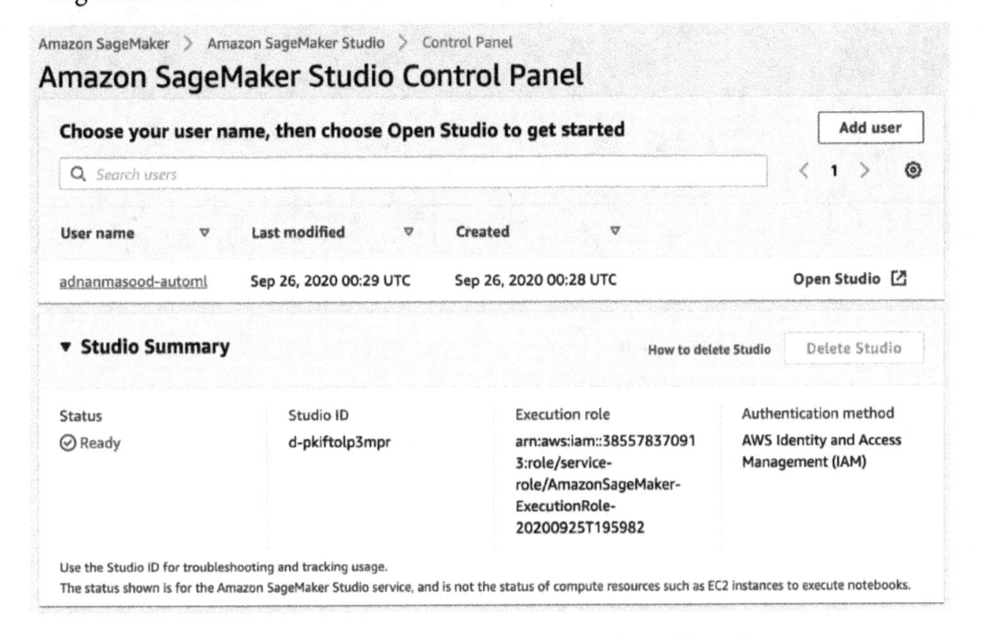

Figure 6.15 – Amazon SageMaker Studio dashboard

The following screenshot shows what SageMaker Studio looks like. It is similar to other online cloud ML IDEs you might have seen in previous, and also in future, chapters about hyperscalers. Here, you can create notebooks, build experiments, and deploy and monitor your ML services:

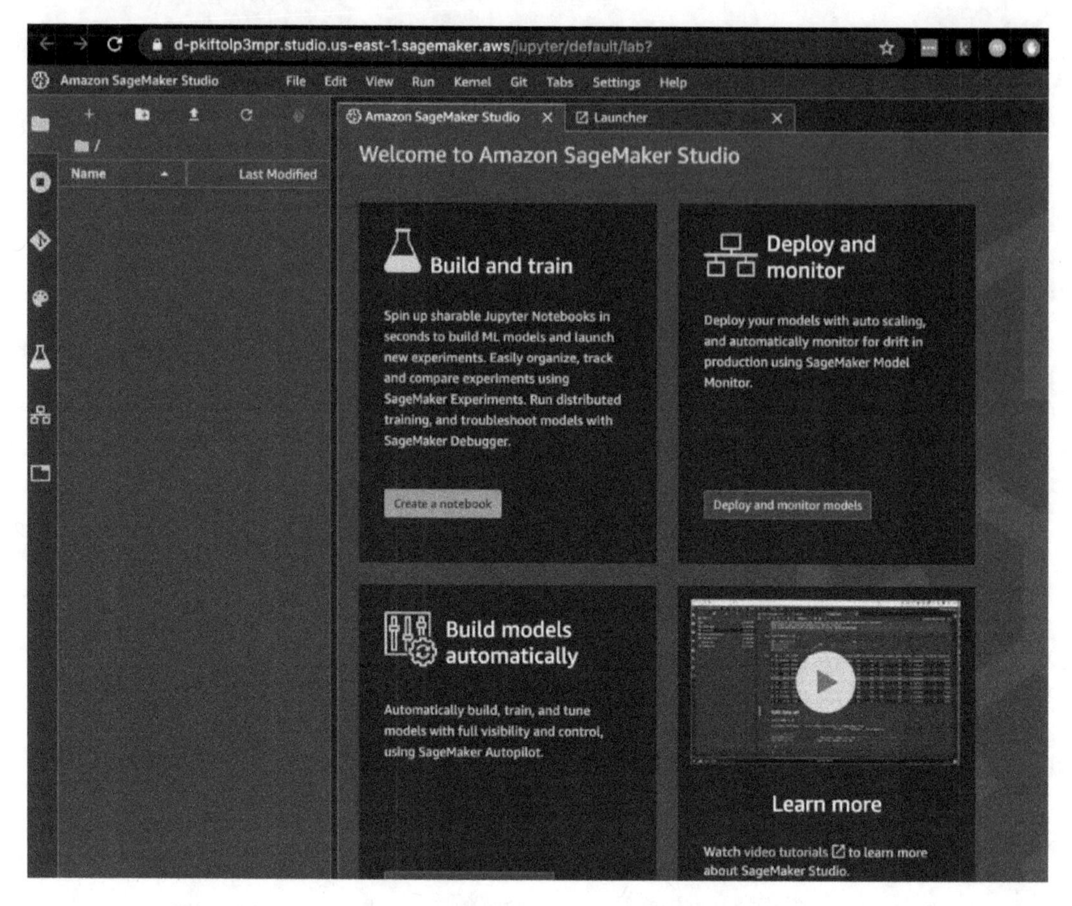

Figure 6.16 – Amazon SageMaker Studio dashboard

11. Click on the **Create a notebook** button and you will see the following screen, which will open up the Jupyter Notebook:

Figure 6.17 – Amazon SageMaker Studio loading

12. Once you see the following launcher screen, you will be able to create notebooks. In our case, we will clone the AWS SageMaker `examples` repository on GitHub (`https://GitHub.com/awslabs/amazon-sagemaker-examples`):

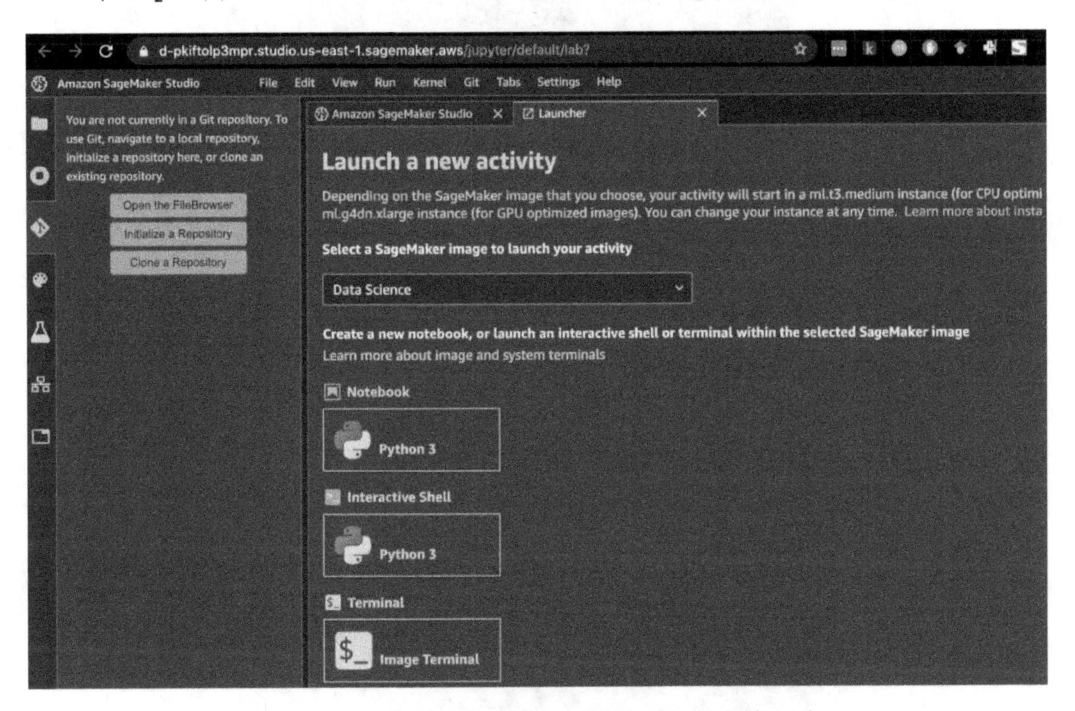

Figure 6.18 – Amazon SageMaker notebook activity launcher

13. Click on **Clone a Repository** and provide the GitHub repository to download it locally, and you will see the following window pop up. Click on **CLONE** to clone the repo:

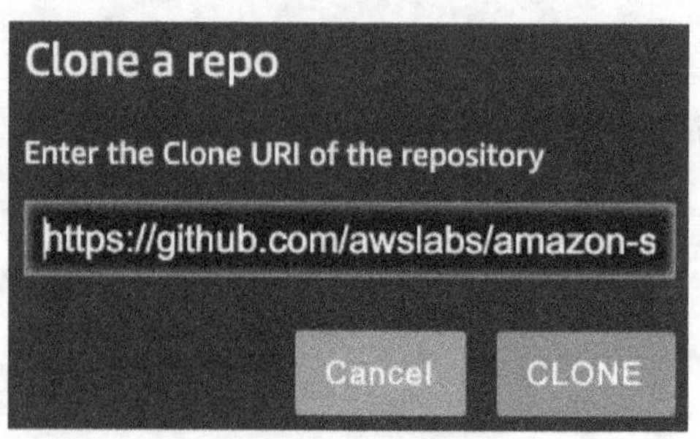

Figure 6.19 – Amazon SageMaker Clone a repo dialog

14. Once the repository is cloned, you will be able to see the following tree in AWS SageMaker:

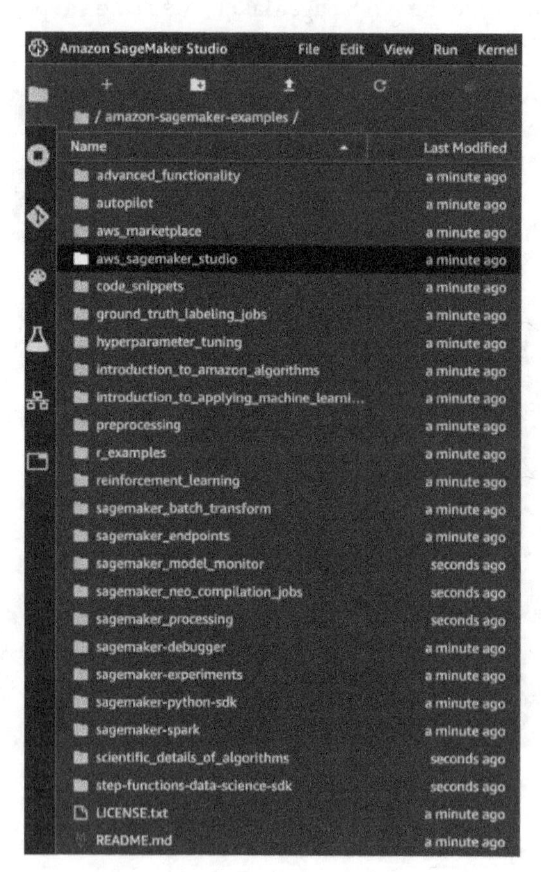

Figure. 6.20 – Amazon SageMaker cloned repository directory view

15. Navigate to the `/aws_sagemaker_studio/getting_started/` folder and open the `xgboost_customer_churn_studio.ipynb` notebook. Upon opening this notebook, you will need to choose a preferred Python kernel to run it. Select the **Python 3 (Data Science)** kernel:

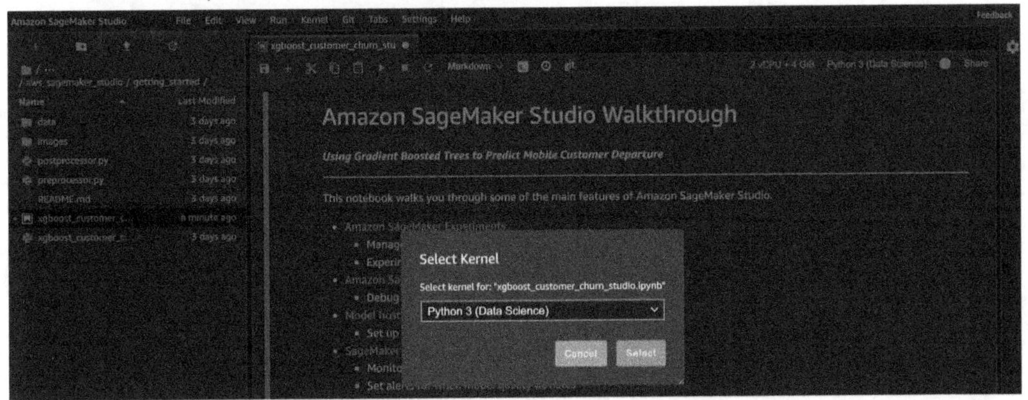

Figure. 6.21 – Amazon SageMaker repository – selecting the preferred kernel

16. Now that you have selected the kernel, you will notice that you still cannot run the notebook. Why? Well, you need a compute resource! Now it's time to select the compute instance you would like to use. In our case, we went with the `ml.t3.medium` general-purpose instance (which costs $0.05 per hour at the time of writing) but you can choose bigger and better machines to make your experiments run faster. The pricing details are found here: `https://aws.amazon.com/sagemaker/pricing/`. Click on **Save and continue** to proceed:

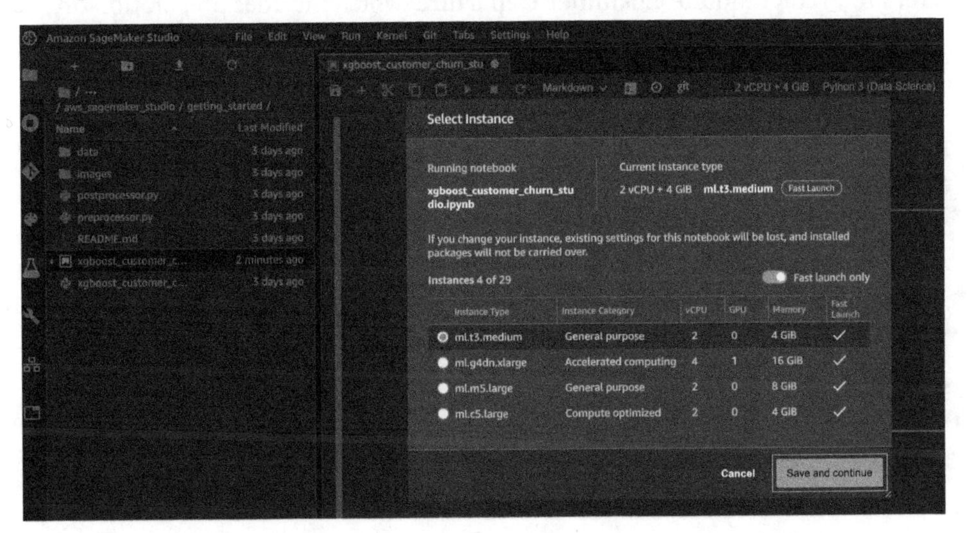

Figure. 6.22 – Amazon SageMaker repository – select the preferred compute instance

17. There might be cases where you need to change the type of compute instance. In that case, you would need to delete the previous instances since your account may not allow multiple instances to run simultaneously. You can see the link for deleting an app in the following screenshot:

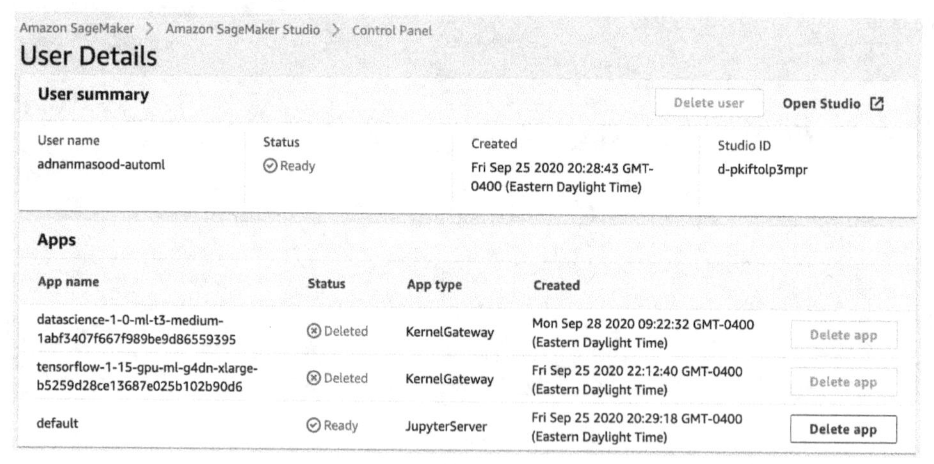

Figure. 6.23 – Amazon SageMaker repository – select the preferred compute instance

18. The compute resource and kernel are identified; now you are all set and ready to run this notebook using the control toolbar shown in the following screenshot and install Python, and other related SDK. This introductory Amazon SageMaker Studio walkthrough is an excellent place to start since it walks you through some of the key features of SageMaker Studio. The key use case here is **Using Gradient Boosted Trees to Predict Mobile Customer Departure**, which includes preparation of the dataset and uploading it to Amazon S3, training with the Amazon SageMaker XGBoost algorithm, building an S3 experiment, debugging, hosting, and monitoring. We will leave the walkthrough as homework for you:

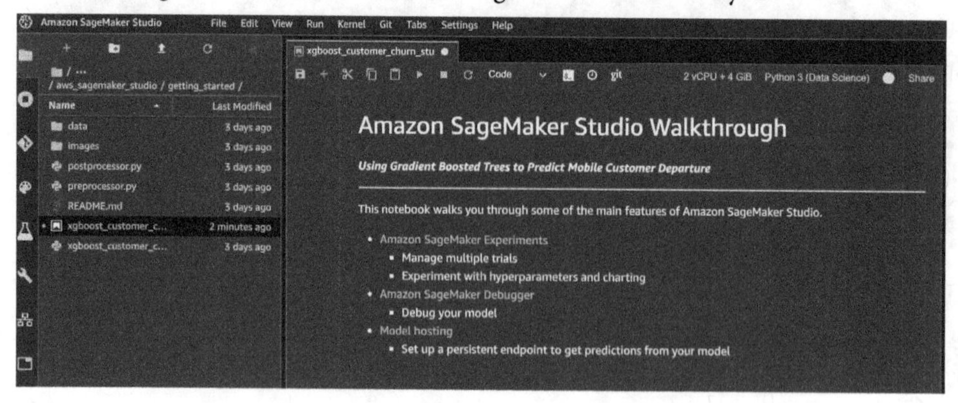

Figure 6.24 – Amazon SageMaker Studio walkthrough notebook

In this section, you have learned how to get started with AWS SageMaker and did a quick walkthrough. This is not meant to be an exhaustive summary of the SageMaker features, which is what the Amazon SageMaker Developer Guide offers. You can download it from here: `https://docs.aws.amazon.com/sagemaker/latest/dg/sagemaker-dg.pdf`.

In the next section, we will look into the automated ML features of AWS SageMaker.

AWS SageMaker Autopilot

SageMaker Autopilot, as the name suggests, is a *fully managed system that provides an automatic ML solution*. The goal, as in any automated ML solution, is to try to offload most of the redundant and time-consuming, repetitive work to the machine while humans can do higher-level cognitive tasks. In the following diagram, you can see the parts of the ML life cycle that SageMaker Autopilot covers:

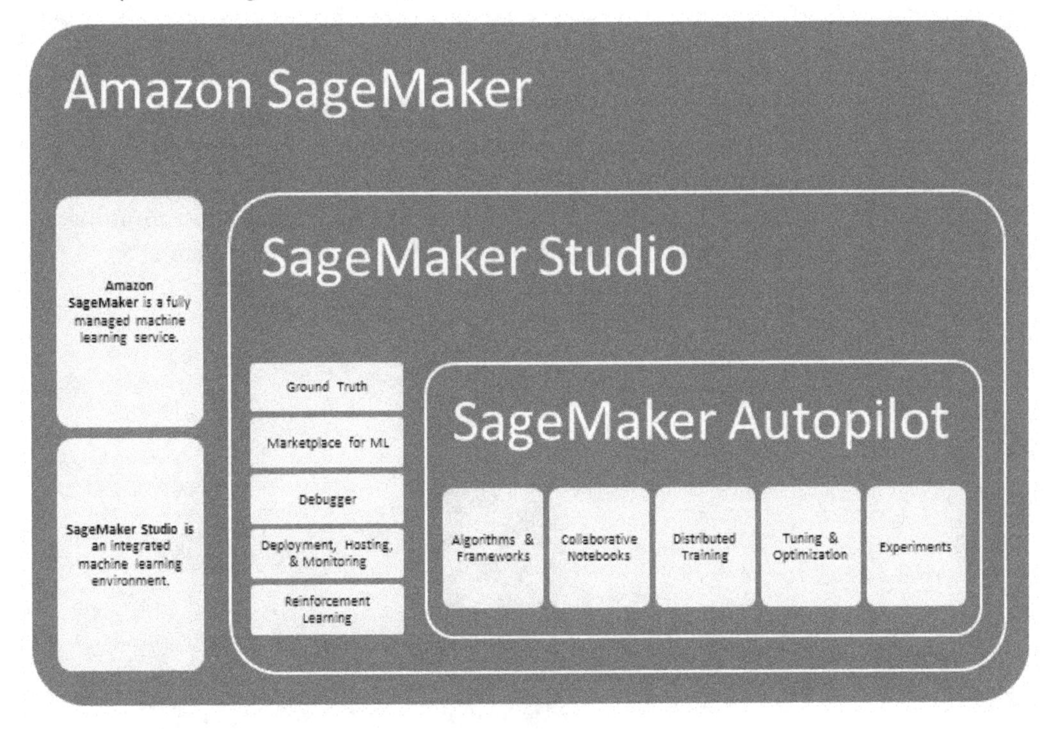

Figure 6.25 – Lifecycle of Amazon SageMaker

As part of the SageMaker ecosystem, SageMaker Autopilot is tasked with being the automated ML engine. A typical automated ML user flow is defined in the following figure, where a user analyzes the tabular data, selects the target prediction column, and then lets Autopilot do its magic of finding the correct algorithm. The secret sauce here is the underlying Bayesian optimizer as defined by *Das et al.* in their paper *Amazon SageMaker Autopilot: a white box AutoML solution at scale* (`https://www.amazon.science/publications/amazon-sagemaker-autopilot-a-white-box-automl-solution-at-scale`):

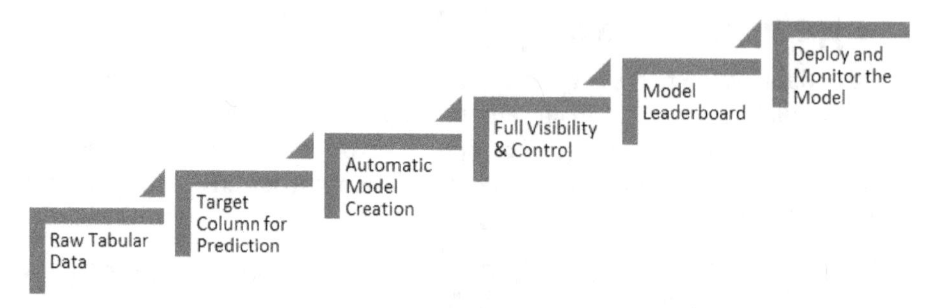

Figure 6.26 – Amazon SageMaker Autopilot life cycle

Once the automatic model creation phase completes, it provides full visibility for the model notebooks. The workflow in the following figure shows the job processing of a ML pipeline. Providing a structured dataset and the target column, AWS SageMaker Autopilot splits the data for training and validation folds, transforms the data with pipeline execution, reviews the list of models, and ranks them by the quality metric:

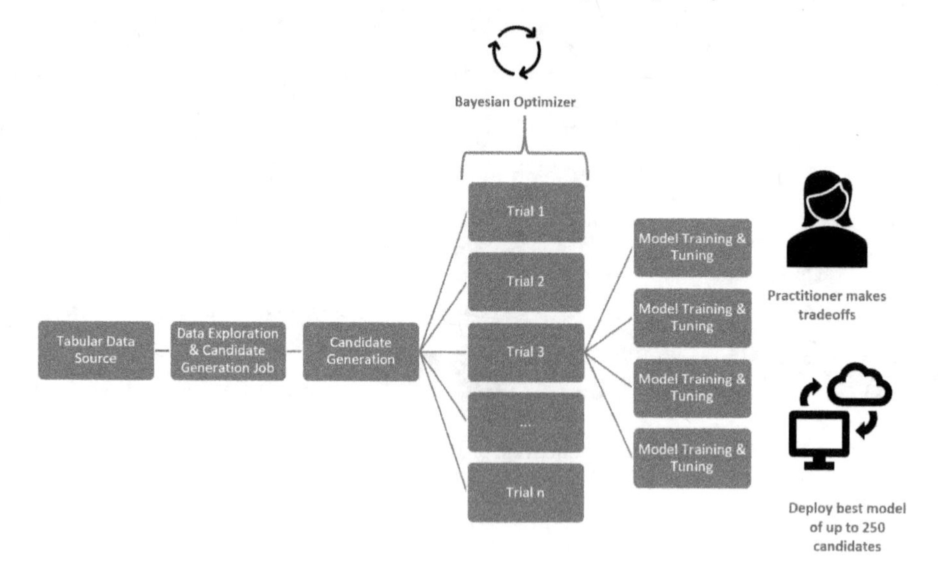

Figure 6.22 – Amazon SageMaker Autopilot – under the hood

The Bayesian optimizer and multiple trials based on the candidate notebook are at the root of this hyperparameter optimization. The following under-the-hood diagram shows how one candidate notebook spawns multiple trials, and even more model training and tuning instances. This process ends up generating a leaderboard of candidate models to be chosen and ranked on measures such as model accuracy, latency, and other trade-offs.

A trade-off can be seen in the following figure. The accuracy difference between model #1 and #2 is 2%; however, the latency, that is, the time the model took to respond, is increased by 250 ms, which is quite significant:

#	Model	Accuracy	Latency	Model Size
1	churn-xgboost-1756-013-33398f0	95%	450 ms	9.1 MB
2	churn-xgboost-1756-014-53facc2	93%	200 ms	4.8 MB
3	churn-xgboost-1756-015-58bc692	92%	200 ms	4.3 MB
4	churn-linear-1756-016-db54598	91%	50 ms	1.3 MB
5	churn-xgboost-1756-017-af8d756	91%	190 ms	4.2 MB

Figure 6.28 – Autopilot job accuracy versus latency trade-off – courtesy of Amazon re:Invent

SageMaker Autopilot publishes the final models, the training approach, that is, the pipeline showing the hyperparameters, the algorithm, and the associated measure. This demonstration helps to make these models transparent from an MLOps perspective and creates high-quality, editable ML models that scale well. These models can be published and monitored within the SageMaker ecosystem for drift, and an alternative model can be chosen and deployed at will. These capabilities are at the forefront of the AWS ML ecosystem, enabling developers to build and deploy valuable solutions for customers.

AWS JumpStart

In Dec 2020, Amazon announced SageMaker JumpStart as a capability to access pre-built model repositories also called model zoos to accelerate model development. Integrated as apart of Amazon SageMaker, JumpStart provides pre-built templates for predictive maintenance, computer vision, autonomous driving, fraud detection, credit risk prediction, OCR for extracting and analyze data from documents, churn prediction, and personalized recommendations.

JumpStart provides an excellent starting point for developers to use these pre-existing templates to JumpStart (pun intended) their development. These accelerator and starter kits are available on GitHub here. `https://github.com/awslabs/` and provide recipes and best practices to use Amazon SageMaker model development and deployment mechanisms.

Further details on using AWS JumpStart can be found here. `https://docs.aws.amazon.com/sagemaker/latest/dg/studio-jumpstart.html`

Summary

In this chapter, you learned about the AWS ML stack and how to get started with AWS SageMaker and notebook development. You also became acquainted with SageMaker Autopilot and its automated ML workflow capabilities. We provided you with an overview of the built-in algorithms, the SageMaker ML life cycle, and what algorithms and techniques are used by SageMaker automated ML. This introduction gives you the background knowledge needed for further exploration and learning of the AWS ML stack and the SageMaker automated ML life cycle.

In the next chapter, we will use some of the SageMaker Autopilot features practically to run classification, regression, and time series analysis.

Further reading

For more information on the following topics, you can refer to the given links:

- *Amazon SageMaker Autopilot: a white box AutoML solution at scale* by Piali Das et al.: `https://www.amazon.science/publications/amazon-sagemaker-autopilot-a-white-box-automl-solution-at-scale`
- *Build, train, and deploy a ML model with Amazon SageMaker*: `https://aws.amazon.com/getting-started/hands-on/build-train-deploy-machine-learning-model-sagemaker/`
- *Amazon SageMaker Studio - AutoML with Amazon SageMaker AutoPilot (part 1)*: `https://www.youtube.com/watch?v=qMEtqJPhqpA`
- *Amazon SageMaker Studio - AutoML with Amazon SageMaker AutoPilot (part 2)*: `https://www.youtube.com/watch?v=qMEtqJPhqpA&ab_channel=JulienSimon`
- *SageMaker Studio - AutoML with Amazon SageMaker AutoPilot (part 3)*: `https://www.youtube.com/watch?v=KZSTsWrDGXs&ab_channel=JulienSimon`
- *SageMaker Studio - AutoML with Amazon SageMaker AutoPilot (part 4)*: `https://www.youtube.com/watch?v=vRHyX3kDstI&ab_channel=JulienSimon`

7
Doing Automated Machine Learning with Amazon SageMaker Autopilot

"One of the holy grails of machine learning is to automate more and more of the feature engineering process."

– Pedro Domingos

"Automated machine learning, the best thing since sliced bread!"

– Anonymous

Automated Machine Learning (**AutoML**) via hyperscalers – that is, via cloud providers – has the potential to bring AI democratization to the masses. In the previous chapter, you created a **Machine Learning** (**ML**) workflow in SageMaker, and also learned about the internals of SageMaker Autopilot.

In this chapter, we will look at a couple of examples explaining how Amazon SageMaker Autopilot can be used in a visual, as well as in notebook, format.

In this chapter, we will cover the following topics:

- Creating an Amazon SageMaker Autopilot limited experiment
- Creating an AutoML experiment
- Running the SageMaker Autopilot experiment and deploying the model
- Invoking and testing the SageMaker Autopilot model
- Building and running SageMaker Autopilot experiments from the notebook

Let's get started!

Technical requirements

You will need access to an Amazon SageMaker Studio instance on your machine.

Creating an Amazon SageMaker Autopilot limited experiment

Let's gets a hands-on introduction to applying AutoML using SageMaker Autopilot. We will download and apply AutoML to an open source dataset. Let's get started!

1. From Amazon SageMaker Studio, start a data science notebook by clicking on the **Python 3** button, as shown in the following screenshot:

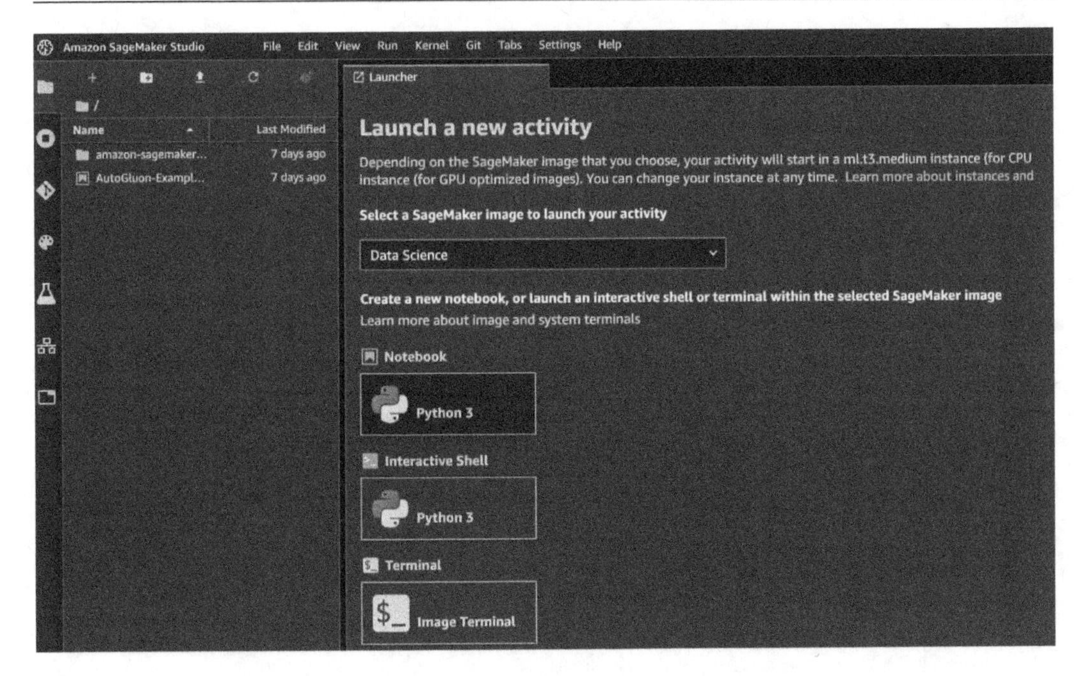

Figure 7.1 – Amazon SageMaker Launcher main screen

Download the Bank Marketing dataset from UCI by calling the following URL retrieve commands and save it in your notebook:

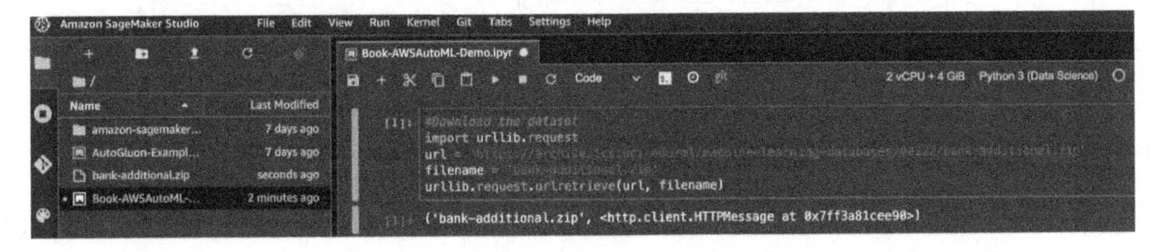

Figure 7.2 – Amazon SageMaker Studio Jupyter Notebook – downloading the dataset

This Bank Marketing dataset is from a Portuguese banking institution and has the classification goal of predicting the client's subscription to deposit (binary feature, y). The dataset is from Moro, Cortez, and Rita's paper on "*A Data-Driven Approach to Predict the Success of Bank Telemarketing. Decision Support Systems*", published by Elsevier. The dataset can be downloaded from the UCI website (`https://archive.ics.uci.edu/ml/datasets/Bank+Marketing`), as shown in the following screenshot:

Bank Marketing Data Set

Download: Data Folder, Data Set Description

Abstract: The data is related with direct marketing campaigns (phone calls) of a Portuguese banking institution. The classification goal is to predict if the client will subscribe a term deposit (variable y).

Data Set Characteristics:	Multivariate	**Number of Instances:**	45211	**Area:**	Business
Attribute Characteristics:	Real	**Number of Attributes:**	17	**Date Donated**	2012-02-14
Associated Tasks:	Classification	**Missing Values?**	N/A	**Number of Web Hits:**	1285737

Figure 7.3 – Bank Marketing dataset; "A Data-Driven Approach to Predict the Success of Bank Telemarketing. Decision Support Systems", Elsevier

The attribute information for the Bank Marketing dataset can be seen in the following screenshot:

Attribute Information:

Input variables:
bank client data:
1 - age (numeric)
2 - job : type of job (categorical: 'admin.','blue-collar','entrepreneur','housemaid','management','retired','self-employed','services','student','technician','unemployed','unknown')
3 - marital : marital status (categorical: 'divorced','married','single','unknown'; note: 'divorced' means divorced or widowed)
4 - education (categorical: 'basic.4y','basic.6y','basic.9y','high.school','illiterate','professional.course','university.degree','unknown')
5 - default: has credit in default? (categorical: 'no','yes','unknown')
6 - housing: has housing loan? (categorical: 'no','yes','unknown')
7 - loan: has personal loan? (categorical: 'no','yes','unknown')
related with the last contact of the current campaign:
8 - contact: contact communication type (categorical: 'cellular','telephone')
9 - month: last contact month of year (categorical: 'jan', 'feb', 'mar', ..., 'nov', 'dec')
10 - day_of_week: last contact day of the week (categorical: 'mon','tue','wed','thu','fri')
11 - duration: last contact duration, in seconds (numeric). Important note: this attribute highly affects the output target (e.g., if duration=0 then y='no'). Yet, the duration is not known before a call is performed. Also, after the end of the call y is obviously known. Thus, this input should only be included for benchmark purposes and should be discarded if the intention is to have a realistic predictive model.
other attributes:
12 - campaign: number of contacts performed during this campaign and for this client (numeric, includes last contact)
13 - pdays: number of days that passed by after the client was last contacted from a previous campaign (numeric; 999 means client was not previously contacted)
14 - previous: number of contacts performed before this campaign and for this client (numeric)
15 - poutcome: outcome of the previous marketing campaign (categorical: 'failure','nonexistent','success')
social and economic context attributes
16 - emp.var.rate: employment variation rate - quarterly indicator (numeric)
17 - cons.price.idx: consumer price index - monthly indicator (numeric)
18 - cons.conf.idx: consumer confidence index - monthly indicator (numeric)
19 - euribor3m: euribor 3 month rate - daily indicator (numeric)
20 - nr.employed: number of employees - quarterly indicator (numeric)

Output variable (desired target):
21 - y - has the client subscribed a term deposit? (binary: 'yes','no')

Figure 7.4 – Attributes for the Bank Marketing dataset "A Data-Driven Approach to Predict the Success of Bank Telemarketing. Decision Support Systems", Elsevier

Now that you have downloaded the dataset, you can unzip the file using the commands shown in the following screenshot:

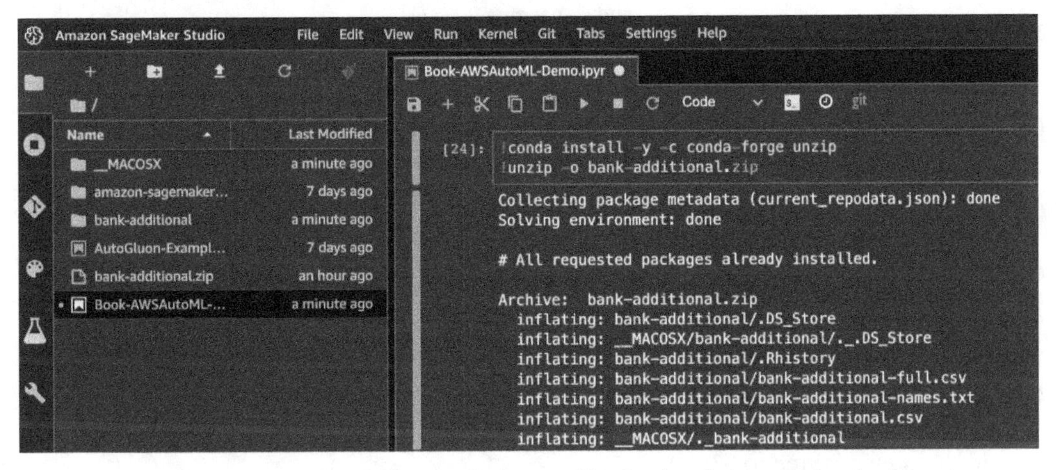

Figure 7.5 – Amazon SageMaker Studio Jupyter Notebook – decompressing the dataset

The extracted archive has the following three files inside it:

- `bank-additional-full.csv`, along with all examples (complete data), ordered by date (from May 2008 to November 2010)
- `bank-additional.csv`, with 10% of the examples (4,119) randomly selected from `bank-additional-full.csv`
- `bank-additional-names.txt`, which contains the field information described in the preceding screenshot

As shown in the following screenshot, you can view the contents of the files using pandas once you've loaded the CSV file into the pandas DataFrame:

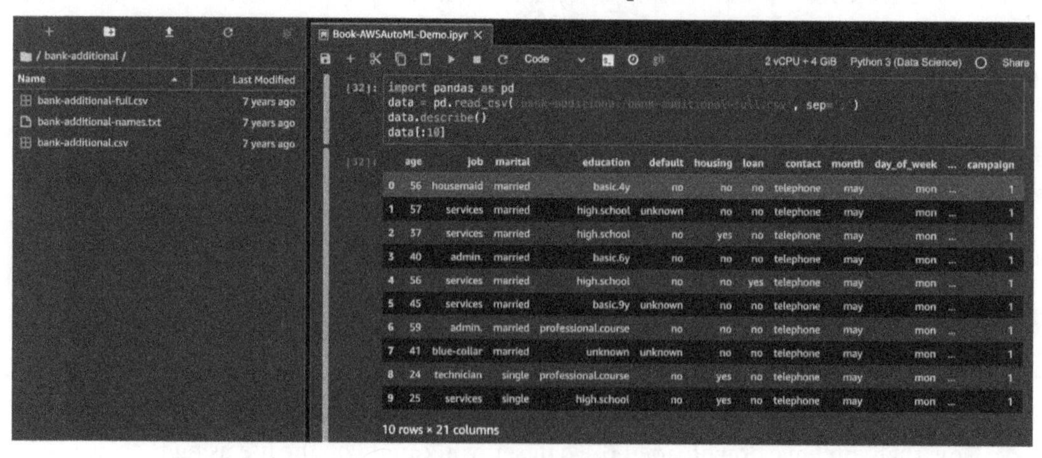

	age	job	marital	education	default	housing	loan	contact	month	day_of_week	...	campaign
0	56	housemaid	married	basic.4y	no	no	no	telephone	may	mon	...	1
1	57	services	married	high.school	unknown	no	no	telephone	may	mon	...	1
2	37	services	married	high.school	no	yes	no	telephone	may	mon	...	1
3	40	admin.	married	basic.6y	no	no	no	telephone	may	mon	...	1
4	56	services	married	high.school	no	no	yes	telephone	may	mon	...	1
5	45	services	married	basic.9y	unknown	no	no	telephone	may	mon	...	1
6	59	admin.	married	professional.course	no	no	no	telephone	may	mon	...	1
7	41	blue-collar	married	unknown	unknown	no	no	telephone	may	mon	...	1
8	24	technician	single	professional.course	no	yes	no	telephone	may	mon	...	1
9	25	services	single	high.school	no	yes	no	telephone	may	mon	...	1

Figure 7.6 – Amazon SageMaker Studio Jupyter Notebook – loading the dataset in a pandas DataFrame and visualizing it

Using NumPy, split the dataset into training and testing segments. In this case, we will use 95% of the data for training and 5% of the data for testing, as shown in the following screenshot. You will store this data in two files: one for training and another for testing.

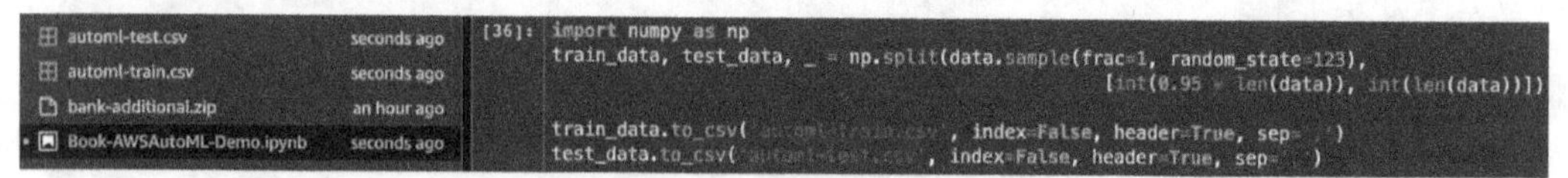

Figure 7.7 - Amazon SageMaker Studio Jupyter Notebook – splitting the dataset into training/test and saving the files in S3

Using the SageMaker API, create a session and upload the training data we created in the previous step to S3:

```
[37]: import sagemaker
      prefix = 'sagemaker/automlbook-bankds/input'
      sess   = sagemaker.Session()
      uri = sess.upload_data(path="automl-train.csv", key_prefix=prefix)
      print(uri)

      s3://sagemaker-us-east-1-385578370913/sagemaker/automlbook-bankds/input/automl-train.csv
```

Figure 7.8 – Amazon SageMaker Studio Jupyter Notebook – uploading the dataset to S3

In the previous chapter, we learned how to create an AutoML experiment using the notebook. Now, let's create an experiment via the SageMaker UI. Click on the experiment icon in the left pane and create an experiment by providing the experiment's name and S3 bucket address, as shown in the following screenshot:

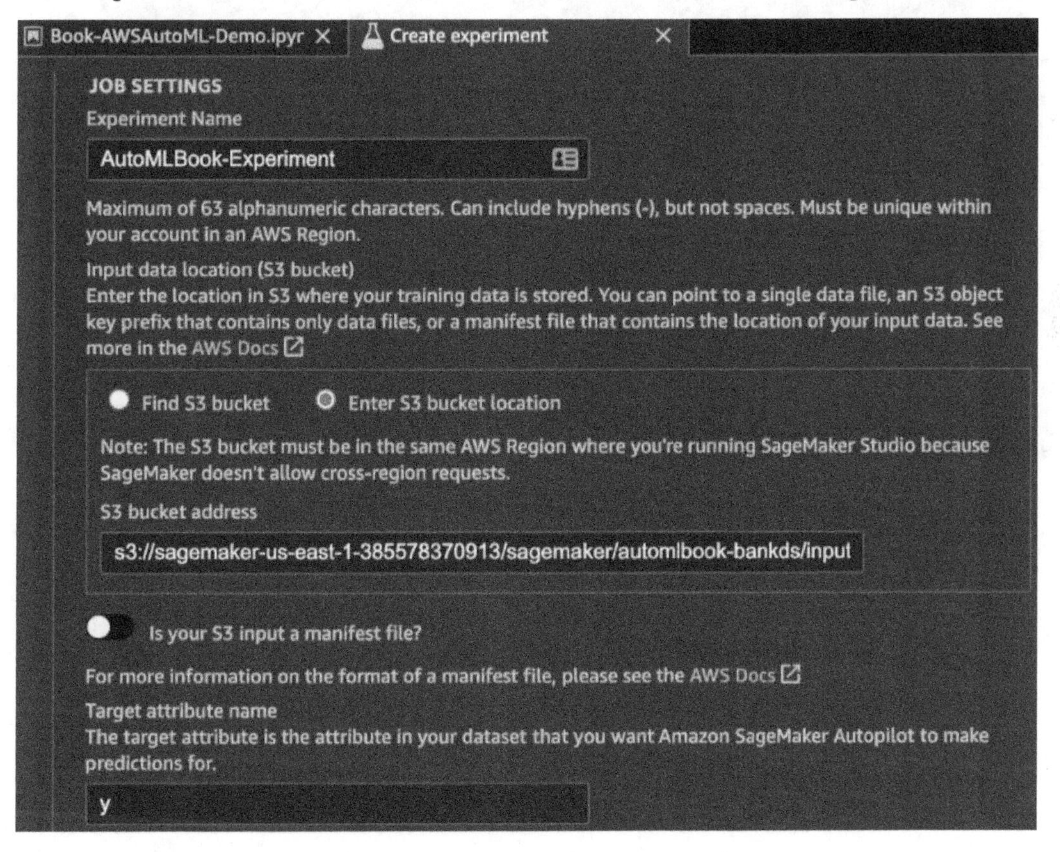

Figure 7.9 – Amazon SageMaker Studio UI – creating an experiment

2. Set the target attribute to `y`. The target attribute is described in the dataset as Output variable (desired target): `y` – has the client subscribed a term deposit? `(binary: "yes","no")`:

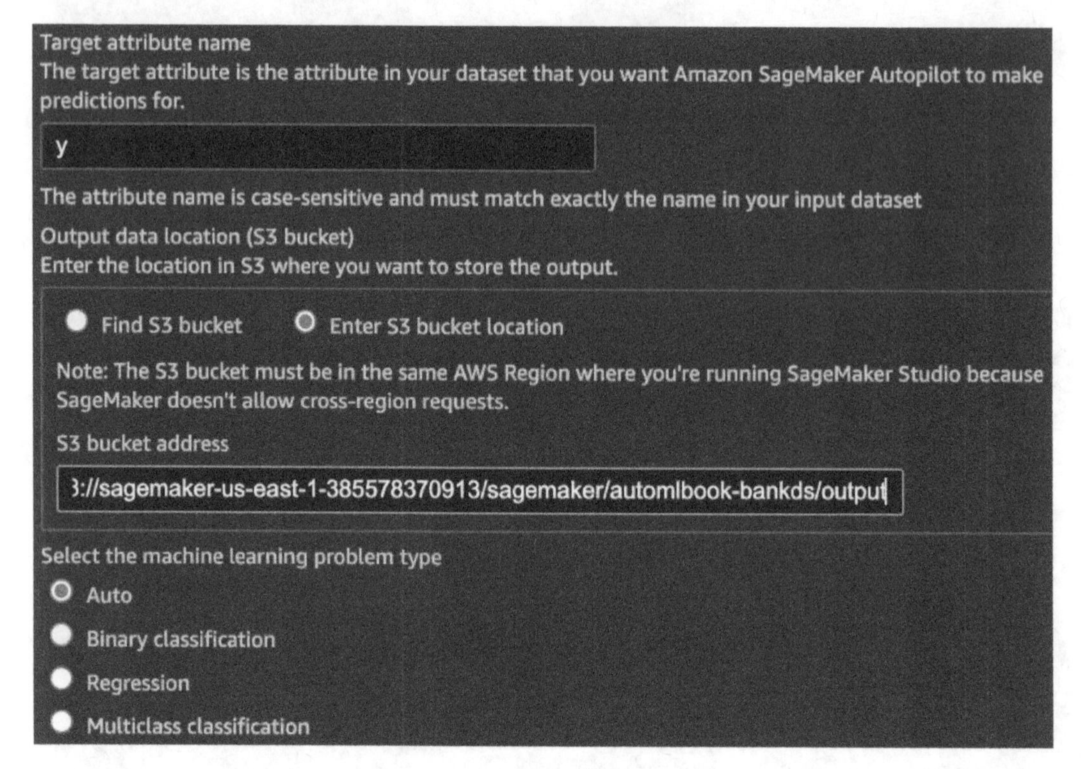

Figure 7.10 – Amazon SageMaker Studio UI – creating an experiment

As shown in the preceding screenshot, you can define the ML problem by yourself – it's binary classification in this case – or let the SageMaker AutoML engine decide this on its own. In this case, we will leave it as **Auto**, and you will see that the SageMaker will recognize this as a binary classification problem.

3. You can either run the full experiment – that is, data analysis, feature engineering, and modeling tuning – or create a notebook to view the candidate definitions. We will do both with this dataset to demonstrate the benefits of each approach:

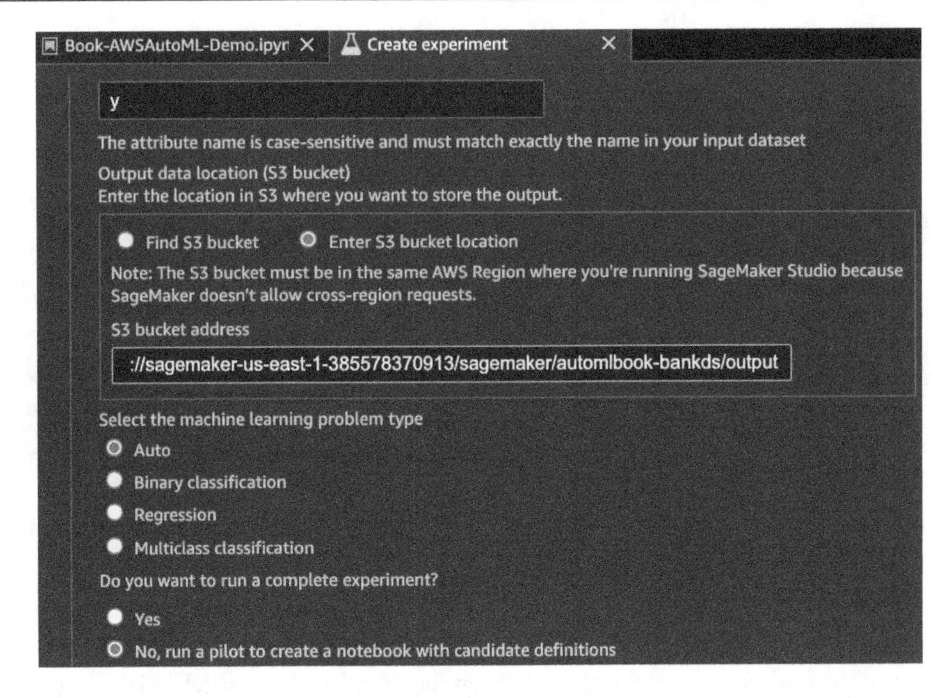

Figure 7.11 – Amazon SageMaker Studio UI – complete experiment versus pilot for candidate definitions

Lastly, you can set some advanced optional parameters, such as a custom SageMaker role, encryption key (if your S3 data is encrypted), and VPC information, if you are working with a virtual private cloud:

ADVANCED SETTINGS - *Optional*

IAM role

Amazon SageMaker Autopilot requires permissions to call other services on your behalf. We create an IAM role that provides these permissions. If you already have a role that has the AmazonSageMakerFullAccess policy attached, you can use that.

Default SageMaker role

Encryption key - *Optional*

We use the AWS managed KMS key for S3 to encrypt your data when we store them in S3. To use another KMS key, enter its ID or Amazon Resource Name (ARN).

Encrypted with AWS key

VPC - *Optional*

A virtual private cloud (VPC) is a virtual network dedicated to your AWS account. Using a VPC can help you secure your AWS resources.

No VPC

Figure 7.12 – Amazon SageMaker Studio UI – Advanced Settings

With that, we have entered all the required information and can run the experiment. Upon submitting the job, you will see the following screen, which contains two steps (analyzing data and candidate definitions generation). This is because we have chosen not to run the entire experiment; we have only chosen to generate candidate definitions:

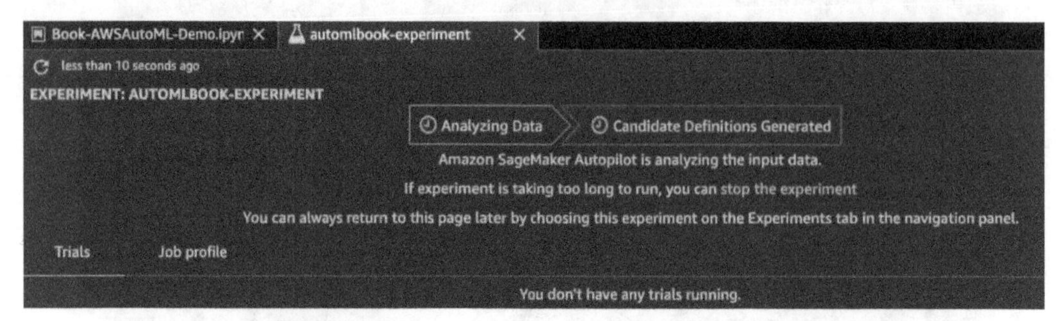

Figure 7.13 – Amazon SageMaker Studio experiment creation UI – Analyzing Data screen

4. Once this partial experiment is completed, you will see the following screen, which shows the completed job information, trials, and job profile. Since we only generated the candidates in this case, the experiment didn't take too long. The **Open candidate generation notebook** and **Open data exploration notebook** buttons can be found at the top-right of the page. Both these buttons will open the respective notebooks:

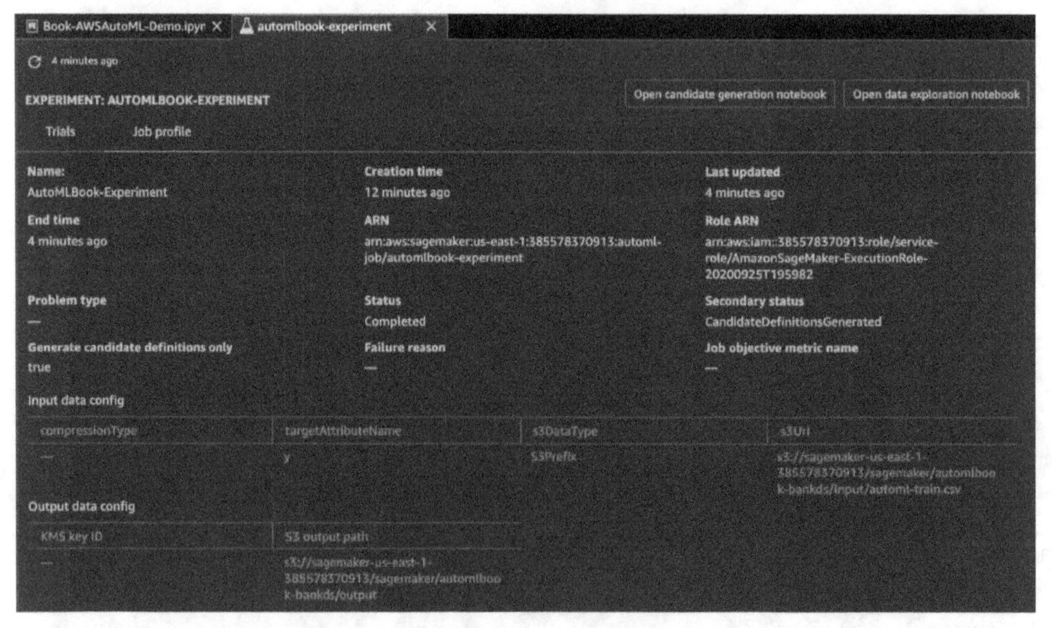

Figure 7.14 – Amazon SageMaker AutoML experiment completion view

The SageMaker Autopilot candidate definition notebook helps the data scientist take a deeper look at the dataset, its features, its classification problem, and the quality metric of the trained model. This is essentially an in-depth view of what happens behind the scenes in the SageMaker Autopilot pipeline and gives the data scientist a chance to run this manually and fine-tune or make changes as they deem necessary:

Amazon SageMaker Autopilot Candidate Definition Notebook
This notebook was automatically generated by the AutoML job AutoMLBook-Experiment. This notebook allows you to customize the candidate definitions and execute the SageMaker Autopilot workflow.
The dataset has 21 columns and the column named y is used as the target column. This is being treated as a BinaryClassification problem. The dataset also has 2 classes. This notebook will build a BinaryClassification model that maximizes the "F1" quality metric of the trained models. The "F1" metric applies for binary classification with a positive and negative class. It mixes between precision and recall, and is recommended in cases where there are more negative examples compared to positive examples.
As part of the AutoML job, the input dataset has been randomly split into two pieces, one for training and one for validation. This notebook helps you inspect and modify the data transformation approaches proposed by Amazon SageMaker Autopilot. You can interactively train the data transformation models and use them to transform the data. Finally, you can execute a multiple algorithm hyperparameter optimization (multi-algo HPO) job that helps you find the best model for your dataset by jointly optimizing the data transformations and machine learning algorithms.
Available Knobs Look for sections like this for recommended settings that you can change.
Contents
1. Sagemaker Setup
A. Downloading Generated Candidates
B. SageMaker Autopilot Job and Amazon Simple Storage Service (Amazon S3) Configuration
2. Candidate Pipelines
A. Generated Candidates
B. Selected Candidates
3. Executing the Candidate Pipelines

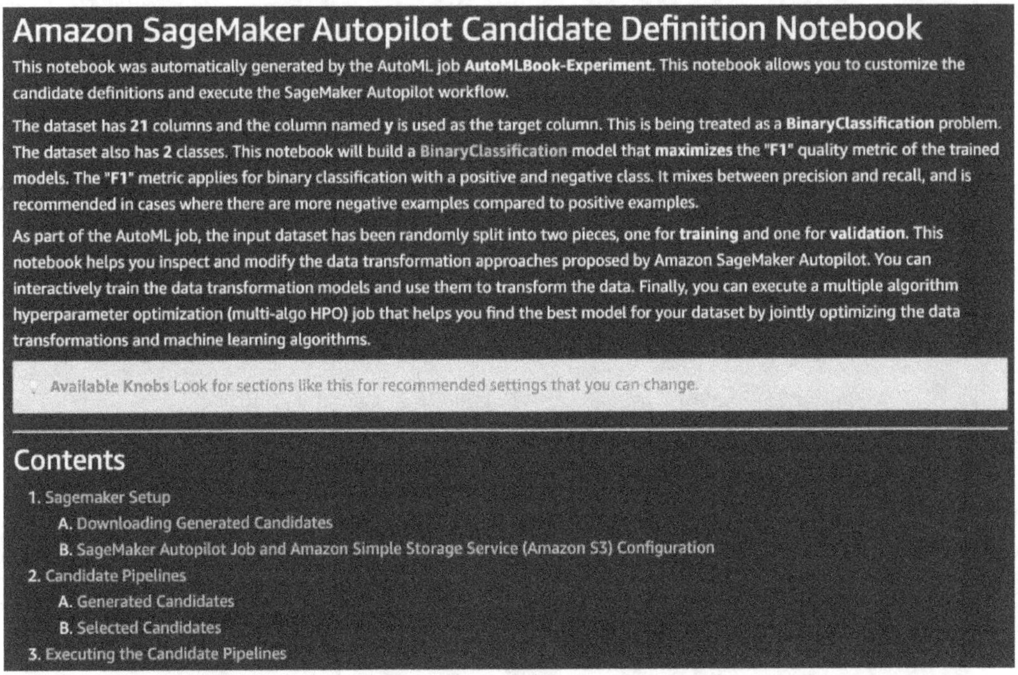

Figure 7.15 – Amazon SageMaker Autopilot candidate definition notebook

The candidate definition notebook is a fairly large file and contains a table of contents, as shown in the preceding screenshot. Similarly, the data exploration notebook provides you with insights into the dataset:

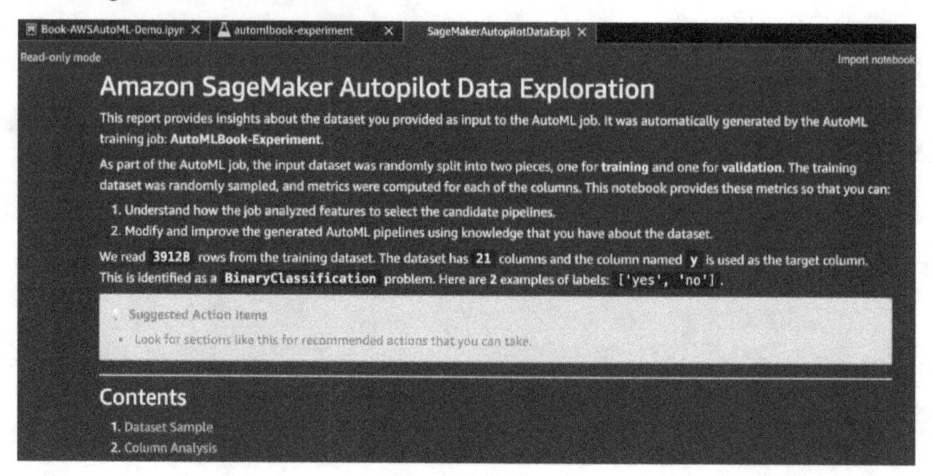

Figure 7.16 – Amazon SageMaker Autopilot data exploration notebook

These insights include what you would typically expect from a data scientist – a data scientist looks for features and their data types, range, mean, median, descriptive statistics, missing data, and more. Even if you are skeptical about the AutoML capabilities that are available in general, this is an excellent place for a data scientist to just explore the dataset and its respective candidates:

Descriptive Statistics

For each of the numerical input features, several descriptive statistics are computed from the data sample.

SageMaker Autopilot may treat numerical features as `Categorical` if the number of unique entries is sufficiently low. For `Numerical` features, we may apply numerical transformations such as normalization, log and quantile transforms, and binning to manage outlier values and difference in feature scales.

We found **10 of the 21** columns contained at least one numerical value. The table below shows the **10** columns which have the largest percentage of numerical values.

Suggested Action Items

- Investigate the origin of the data field. Are some values non-finite (e.g. infinity, nan)? Are they missing or is it an error in data input?
- Missing and extreme values may indicate a bug in the data collection process. Verify the numerical descriptions align with expectations. For example, use domain knowledge to check that the range of values for a feature meets with expectations.

	% of Numerical Values	Mean	Median	Min	Max
age	100.0%	40.0096	38.0	17.0	98.0
duration	100.0%	258.631	178.0	0.0	4918.0
campaign	100.0%	2.57031	2.0	1.0	56.0
pdays	100.0%	962.305	999.0	0.0	999.0
previous	100.0%	0.173099	0.0	0.0	7.0
emp.var.rate	100.0%	0.0813279	1.1	-3.4	1.4
cons.price.idx	100.0%	93.5751	93.837	92.201	94.767
cons.conf.idx	100.0%	-40.5078	-41.8	-50.8	-26.9
euribor3m	100.0%	3.62068	4.857	0.634	5.045
nr.employed	100.0%	5167.03	5191.0	4963.6	5228.1

Figure 7.17 – Amazon SageMaker Autopilot data exploration notebook – descriptive statistics

The Amazon SageMaker Autopilot data exploration and candidate definition notebooks provide a transparent view for users to analyze data and conduct experiments. As notebooks go, these are executable pieces of code where you can see the preprocessors, hyperparameters, algorithms, ranges of hyperparameters, and all the prescribed preprocessing steps that are used to identify the best candidates.

In the next section, we will build and run a full Autopilot experiment.

Creating an AutoML experiment

Since the Autopilot data exploration and candidate definition notebooks provide an in-depth overview of the dataset, the complete experiment actually runs these steps and give you a final, tuned model based on the steps described in these notebooks. Now, let's create a full experiment using the same UI as looked at earlier:

1. From Amazon SageMaker Studio, start a data science experiment. Click on the experiment icon in the left-hand pane and create an experiment by providing the experiment name and S3 bucket address, as shown in the following screenshot:

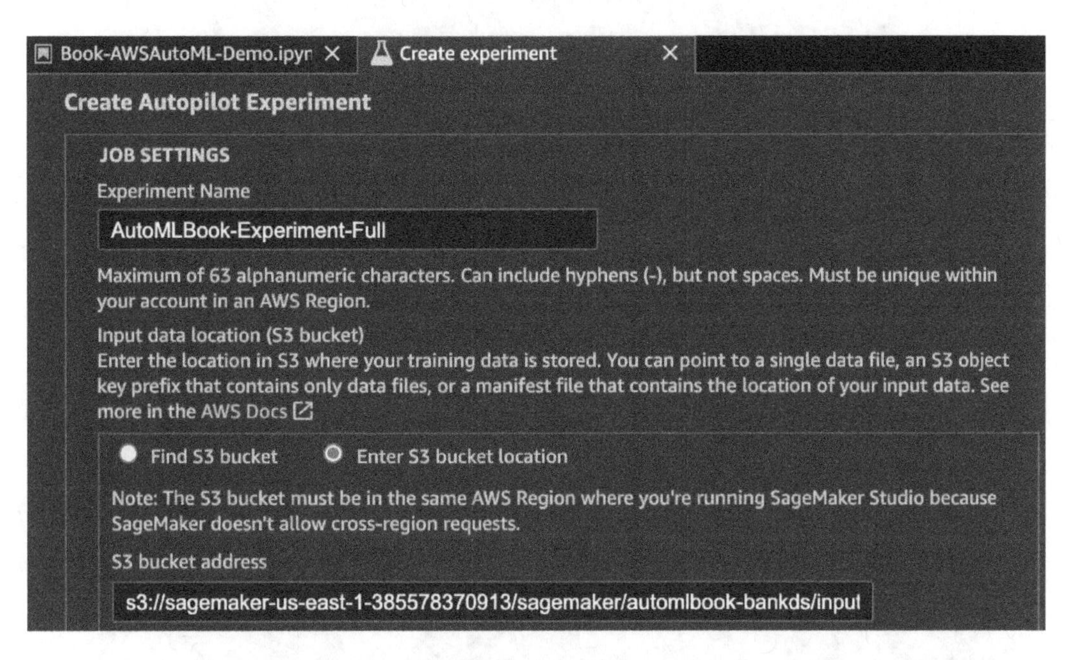

Figure 7.18 – Amazon SageMaker Autopilot – creating the experiment

In the previous *Creating an Amazon SageMaker Autopilot limited experiment section*, we did the limited run. In this section, we will use the complete experiment feature:

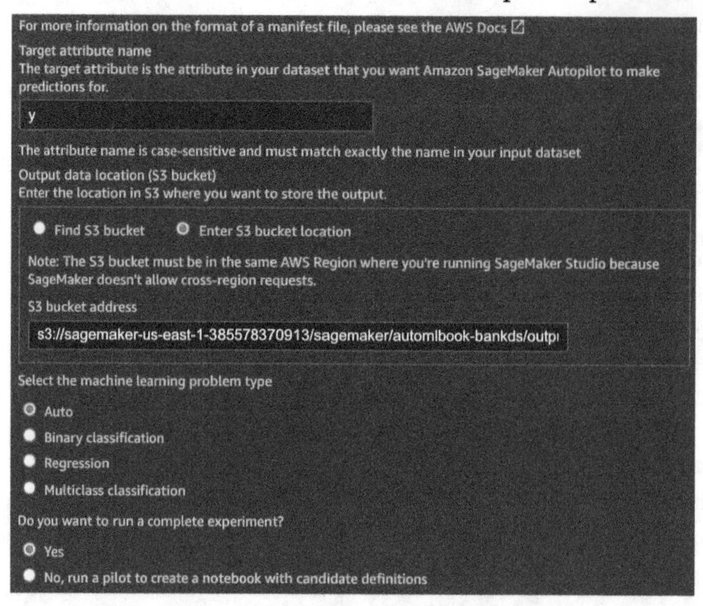

Figure 7.19 – Amazon SageMaker Autopilot – creating the complete experiment

When you start the experiment, it will behave very similar to our earlier candidate experiment, aside from the fact that this complete experiment will take longer and will build and execute the entire pipeline. You will see the following screen in the meantime while you wait for the results:

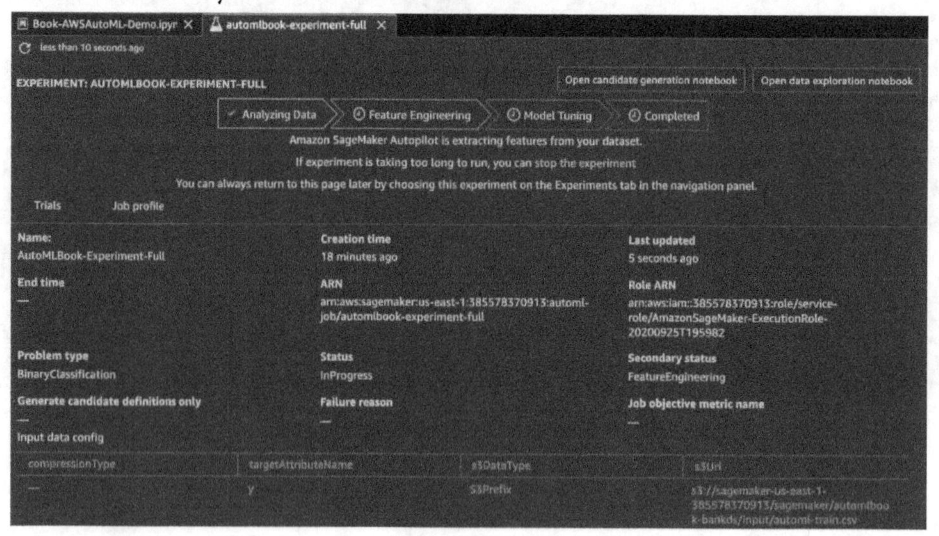

Figure 7.20 – Amazon SageMaker Autopilot – running the full experiment

While the experiment is running, you can track its progress by looking at the individual experiments and getting valuable insights from the **Trials** tab. You may also notice that the problem type here is correctly classified as binary classification:

Figure 7.21 – Amazon SageMaker Autopilot – running the full experiment

The detailed summary of the experiment shown in the following screenshot shows the inference containers that were used, the model data URI, and the environments that were utilized, along with their respective **Amazon Resource Names (ARNs)**, which uniquely identify AWS resources:

Summary

Name	Status	Creation time	Last modified
tuning-job-1-b6a568e36c7241558c-212-4c80d306	Completed	34 minutes ago	22 minutes ago

Inference containers

Image	Model Data URL	Environment - Transform mode	Environment - default invocations accept
683313688378.dkr.ecr.us-east-1.amazonaws.com/sagemaker-sklearn-automl:0.2-1-cpu-py3	s3://sagemaker-us-east-1-385578370913/sagemaker/automlbook-bankds/output-full/AutoMLBook-Experiment-Full/data-processor-models/AutoMLBook-dpp8-1-f1cfd1024b9f474ba0379f8c1ea99d224118134de50b4/output/model.tar.gz	feature-transform	application/x-recordio-protobuf
683313688378.dkr.ecr.us-east-1.amazonaws.com/sagemaker-xgboost:1.0-1-cpu-py3	s3://sagemaker-us-east-1-385578370913/sagemaker/automlbook-bankds/output-full/AutoMLBook-Experiment-Full/tuning/AutoMLBook-dpp8-xgb/tuning-job-1-b6a568e36c7241558c-212-4c80d306/output/model.tar.gz	—	text/csv
683313688378.dkr.ecr.us-east-1.amazonaws.com/sagemaker-sklearn-automl:0.2-1-cpu-py3	s3://sagemaker-us-east-1-385578370913/sagemaker/automlbook-bankds/output-full/AutoMLBook-Experiment-Full/data-processor-models/AutoMLBook-dpp8-1-f1cfd1024b9f474ba0379f8c1ea99d224118134de50b4/output/model.tar.gz	inverse-label-transform	text/csv

Figure 7.22 – Amazon SageMaker Autopilot inference container information

The **Trials** tab shows the different trials and tuning jobs that run, as well as the objective function (F1 score), which demonstrates how it improves over time:

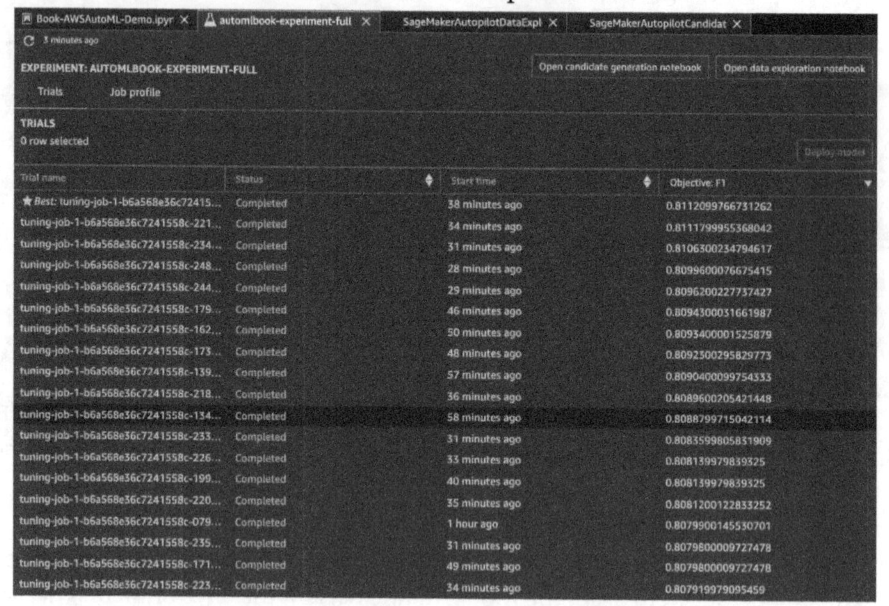

Figure 7.23 – Amazon SageMaker Autopilot experiment run trials – best model

You have seen this exact iteration in previous chapters; it is déjà vu all over again. We have seen this process unfolding in the OSS tools, but it's just different here, in that it's done in a more organized end-to-end manner. You have the entire pipeline built into one; that is, the strategy, data analysis, feature engineering, model tuning, and hyperparameter optimization processes. You can see the tuning job's details in the following screenshot:

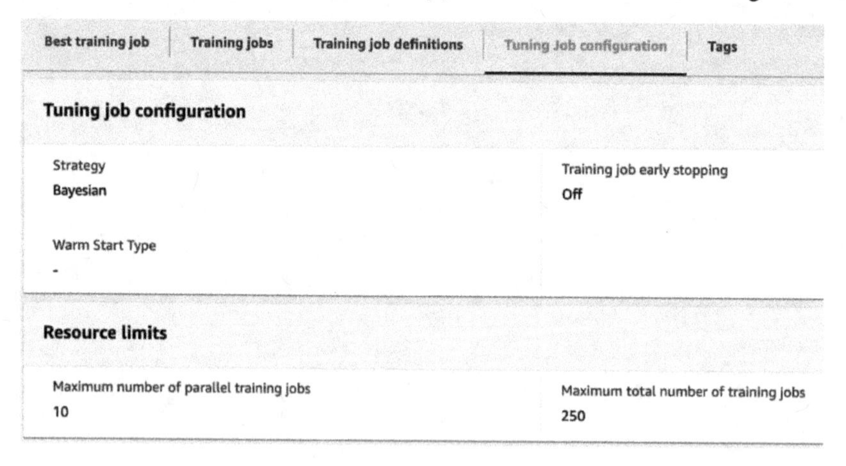

Figure 7.24 – Amazon SageMaker Autopilot tuning job details showing Bayesian strategy and resource information

Now that we've run the entire experiment and the process is completed, let's deploy the best model.

Running the SageMaker Autopilot experiment and deploying the model

Amazon SageMaker Studio makes it easy for us to build, train, and deploy machine learning models; that is, it enables the data science life cycle. To deploy the model we built in the preceding section, we will need to set certain parameters. For this, you must provide the endpoint name, instance type, how many instances (count), and if you'd like to capture the request and response information. Let's get started:

1. If you select the **Data capture** option, you will need an S3 bucket for storage, as shown in the following screenshot:

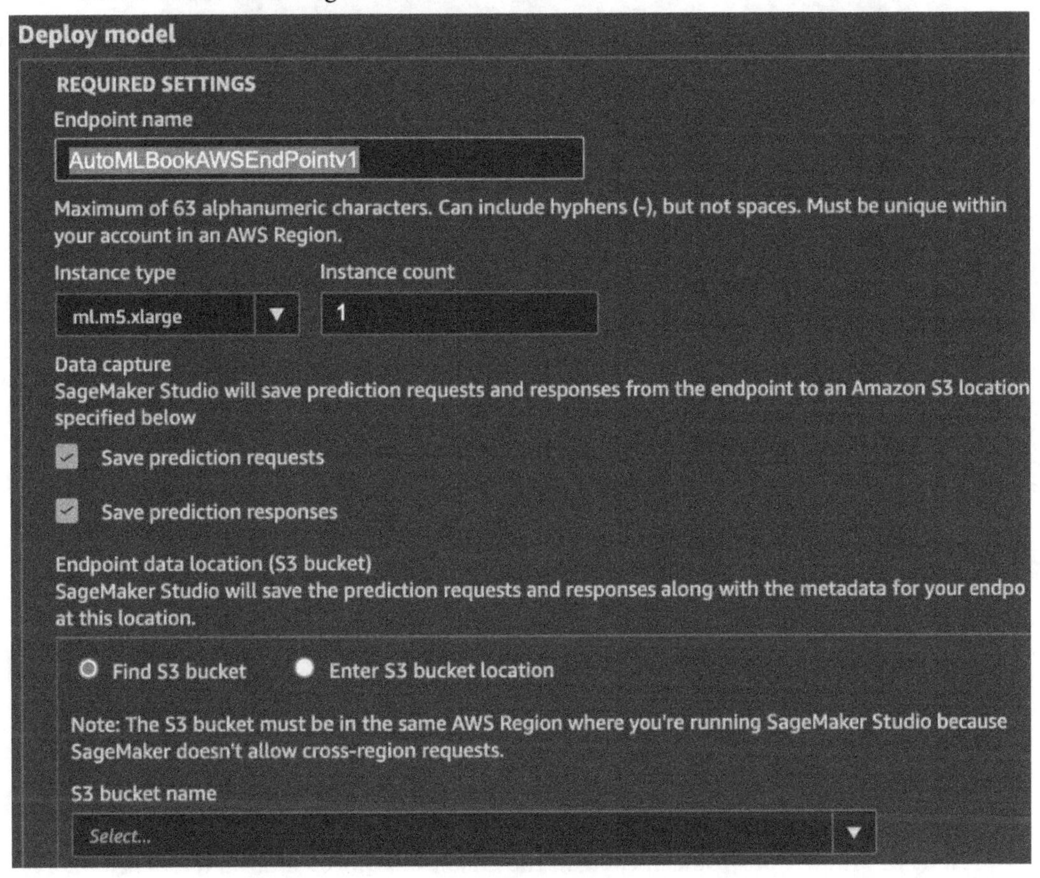

Figure 7.25 – Amazon SageMaker endpoint deployment

2. Once you've clicked on **Deploy**, you will see the following screen, which shows the progress of the new endpoint being created:

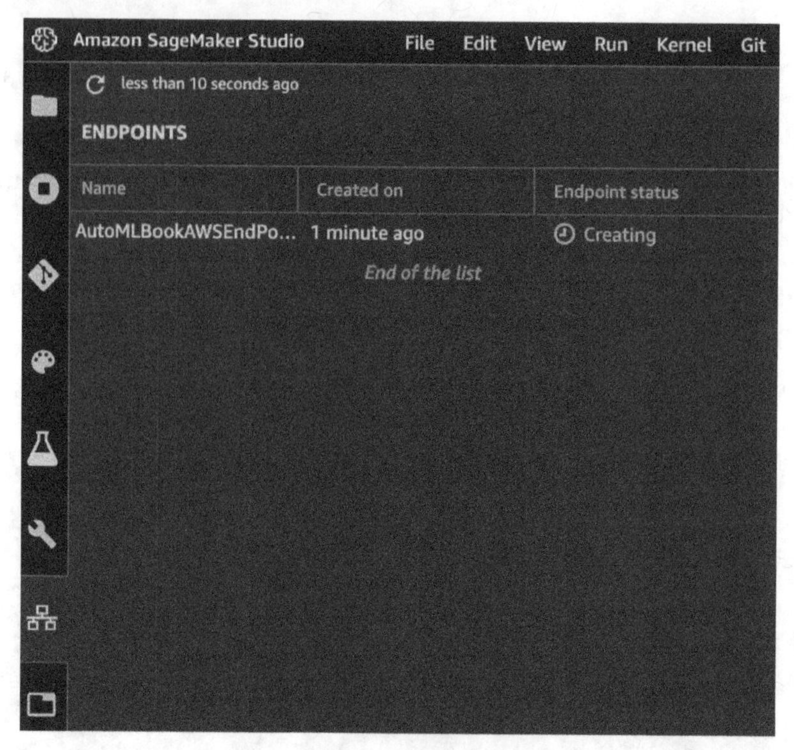

Figure 7.26 – Amazon SageMaker endpoint deployment in progress

Once the deployment is completed, you will see the following status of InService:

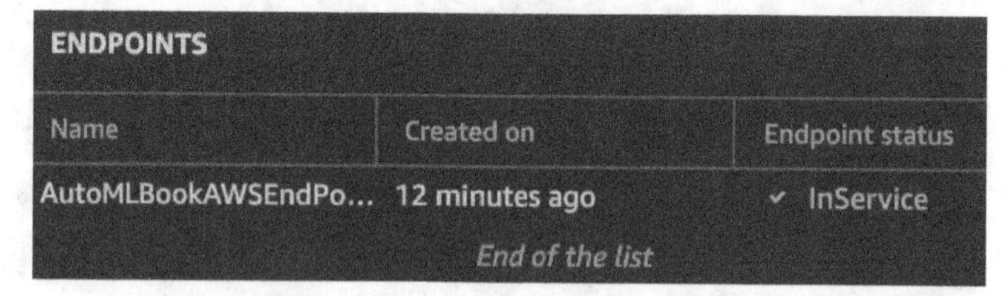

Figure 7.27 – Amazon SageMaker endpoint deployment completed

3. The model endpoint is an important resource for ensuring the quality of the model. By enabling the model monitor, you can detect data drift and monitor the quality of any models in production. This proactive detection of the model's quality helps ensure that your machine learning service does not end up providing the wrong results in production. You can click on the **Enable monitoring** button to engage the Amazon SageMaker Model Monitor:

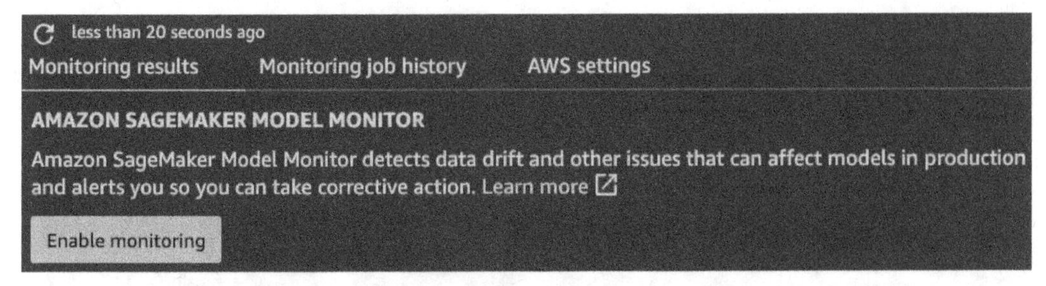

Figure 7.28 – Amazon SageMaker Autopilot Model Monitor startup screen

Model monitoring is an important area of the machine learning life cycle. As shown in the following screenshot, the Amazon SageMaker Model Monitor addresses this by capturing data, creating a baseline, scheduling monitoring jobs, and then allowing SMEs to interpret the results in the case of outliers and violations:

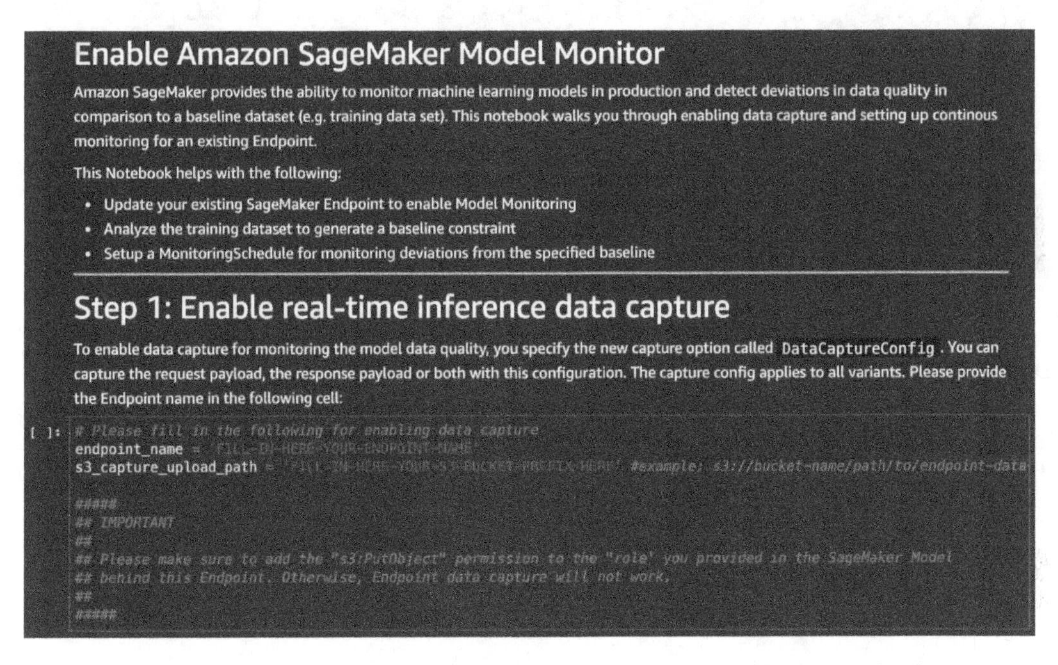

Figure 7.29 – Amazon SageMaker Autopilot Model Monitor enablement notebook

Now that we have created and the deployed the model, it is time to test it out by invoking it. This operation of invoking a machine learning model that's been exposed via a web service is typically called inferencing or evaluation.

Invoking the model

With the model built and deployed using Amazon SageMaker Autopilot, we can test it out. Remember the test data we saved earlier? Now, it's time to use it. Here, you can see that we are iterating through the `automl-test.csv` file and invoking the endpoint by passing the line of data as a request:

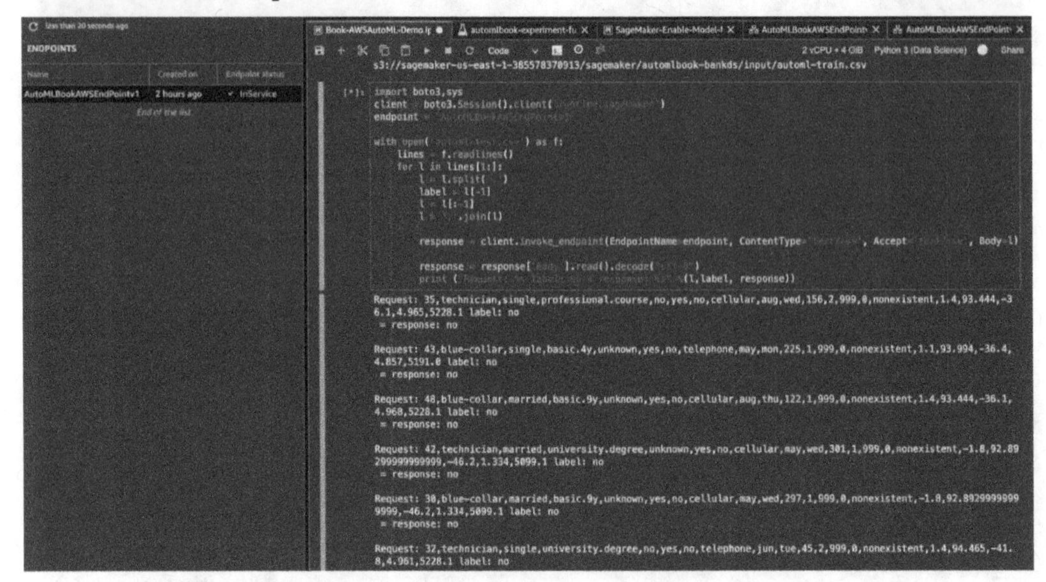

Figure 7.30 – Amazon SageMaker Autopilot – model invocation from the notebook

The request contains information about the person applying for the loan. We have removed the outcome (label) from the request, and then compared it as we wish to print the value out. You can see the request, the label, and the corresponding response from the web service in the preceding screenshot. You can use this information to calculate the accuracy of the service results; they are fairly accurate:

```
Request: 32,technician,single,university.degree,no,yes,no,telephone,jun,tue,45,2,999,0,nonexistent,1.4,94.465,-41.
8,4.961,5228.1 label: no
 = response: no

Request: 46,blue-collar,married,unknown,no,no,yes,cellular,jul,thu,36,1,999,0,nonexistent,1.4,93.91799999999999,-4
2.7,4.962,5228.1 label: no
 = response: no

Request: 29,admin.,single,university.degree,no,yes,yes,cellular,nov,fri,1222,2,999,0,nonexistent,-0.1,93.2,-42.0,4.
021,5195.8 label: yes
 = response: yes

Request: 24,blue-collar,single,basic.4y,no,yes,yes,cellular,jul,wed,132,1,999,0,nonexistent,1.4,93.91799999999999,-
42.7,4.963,5228.1 label: no
 = response: no

Request: 23,entrepreneur,married,professional.course,no,no,no,cellular,jul,tue,58,1,999,0,nonexistent,1.4,93.917999
99999999,-42.7,4.962,5228.1 label: no
 = response: no

Request: 45,management,single,basic.9y,no,yes,no,telephone,jun,thu,69,1,999,0,nonexistent,1.4,94.465,-41.8,4.961,52
28.1 label: no
 = response: no

Request: 38,admin.,married,university.degree,no,no,no,cellular,oct,wed,180,2,999,1,failure,-3.4,92.431,-26.9,0.74,5
017.5 label: no
 = response: yes

Request: 58,services,married,high.school,no,yes,no,cellular,jul,fri,72,30,999,0,nonexistent,1.4,93.91799999999999,-
42.7,4.962,5228.1 label: no
 = response: no
```

Figure 7.31 – Amazon SageMaker Autopilot – model invocation responses

Now that you have learned how to set up an AutoML experiment from the Amazon SageMaker Autopilot UI, in the next section, we will use notebooks to do the same.

Building and running SageMaker Autopilot experiments from the notebook

Customer churn is a real problem for businesses and in this example, we will use our knowledge of completing AutoML in Amazon SageMaker Autopilot to build a customer churn prediction experiment using the notebook. In this experiment, we will use a publicly available dataset of US mobile customers provided by Daniel T. Larose in his book *Discovering Knowledge in Data*. To demonstrate running the full gamut, the sample notebook executes the Autopilot experiment by performing feature engineering, building a model pipeline (along with any optimal hyperparameters), and deploying the model.

The evolution of the UI/API/CLI paradigm has helped us utilize the same interface in multiple formats; in this case, we will be utilizing the capabilities of Amazon SageMaker Autopilot directly from the notebook. Let's get started:

1. Open the `autopilot_customer_churn` notebook from the `amazon-sagemaker-examples/autopilot` folder, as shown in the following screenshot:

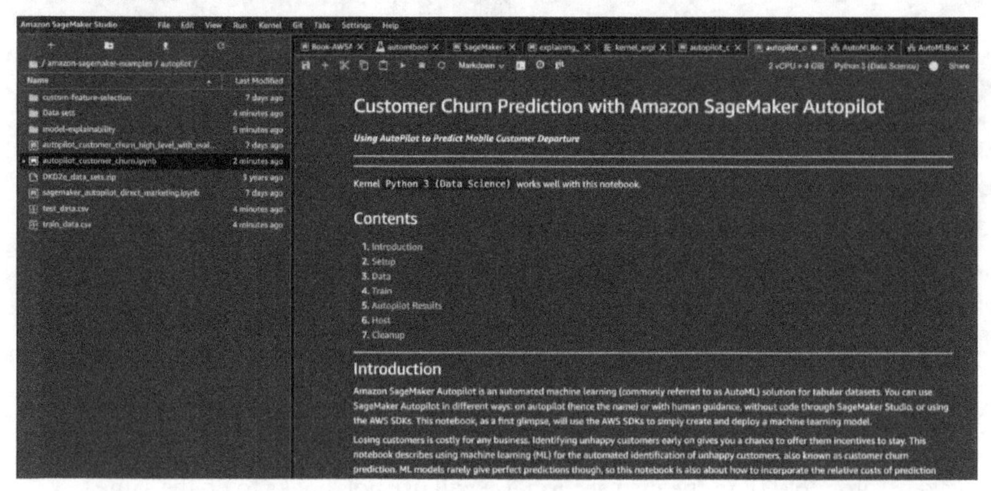

Figure 7.32 – Amazon SageMaker Autopilot – customer churn prediction Autopilot notebook

2. Run the setup by specifying the S3 bucket and the **Identity and Access Management** (**IAM**) role, as we did in the previous *Creating an AutoML experiment* section. Download the dataset, as shown in the following screenshot:

Setup

This notebook was created and tested on an ml.m4.xlarge notebook instance.

Let's start by specifying:

- The S3 bucket and prefix that you want to use for training and model data. This should be within the same region as the Notebook Instance, training, and hosting.
- The IAM role arn used to give training and hosting access to your data. See the documentation for how to create these. Note, if more than one role is required for notebook instances, training, and/or hosting, please replace the boto regexp with a the appropriate full IAM role arn string(s).

```
import sagemaker
import boto3
from sagemaker import get_execution_role

region = boto3.Session().region_name

session = sagemaker.Session()

# You can modify the following to use a bucket of your choosing
bucket = session.default_bucket()
prefix = 'sagemaker/DEMO-autopilot-churn'

role = get_execution_role()

# This is the client we will use to interact with SageMaker AutoPilot
sm = boto3.Session().client(service_name='sagemaker',region_name=region)
```

Figure 7.33 – Amazon SageMaker Autopilot – running the notebook to set up a default bucket and creating the session

3. At this point, you will need to install the prerequisites, and download the dataset, as shown in the following screenshot:

Data

Mobile operators have historical records on which customers ultimately ended up churning and which continued using the service. We can use this historical information to construct an ML model of one mobile operator's churn using a process called training. After training the model, we can pass the profile information of an arbitrary customer (the same profile information that we used to train the model) to the model, and have the model predict whether this customer is going to churn. Of course, we expect the model to make mistakes–after all, predicting the future is tricky business! But I'll also show how to deal with prediction errors.

The dataset we use is publicly available and was mentioned in the book Discovering Knowledge in Data by Daniel T. Larose. It is attributed by the author to the University of California Irvine Repository of Machine Learning Datasets. Let's download and read that dataset in now:

```
[3]: !apt-get install unzip
     !wget http://dataminingconsultant.com/DKD2e_data_sets.zip
     !unzip -o DKD2e_data_sets.zip

Reading package lists... Done
Building dependency tree
Reading state information... Done
Suggested packages:
  zip
The following NEW packages will be installed:
  unzip
0 upgraded, 1 newly installed, 0 to remove and 19 not upgraded.
Need to get 172 kB of archives.
After this operation, 580 kB of additional disk space will be used.
Get:1 http://deb.debian.org/debian buster/main amd64 unzip amd64 6.0-23+deb10u1 [172 kB]
Fetched 172 kB in 0s (10.8 MB/s)
debconf: delaying package configuration, since apt-utils is not installed
Selecting previously unselected package unzip.
(Reading database ... 16492 files and directories currently installed.)
Preparing to unpack .../unzip_6.0-23+deb10u1_amd64.deb ...
Unpacking unzip (6.0-23+deb10u1) ...
Setting up unzip (6.0-23+deb10u1) ...
Processing triggers for mime-support (3.62) ...
--2020-10-03 00:04:29--  http://dataminingconsultant.com/DKD2e_data_sets.zip
Resolving dataminingconsultant.com (dataminingconsultant.com)... 160.153.91.162
Connecting to dataminingconsultant.com (dataminingconsultant.com)|160.153.91.162|:80... connected.
```

Figure 7.34 – Amazon SageMaker Autopilot – downloading the dataset and unzipping the file

4. Once the dataset has been downloaded and uncompressed, you can add it to a pandas DataFrame and view it. It shows information about the customer, such as their calling attributes, as shown in the following screenshot:

```
[4]: churn = pd.read_csv('./Data sets/churn.txt')
     pd.set_option('display.max_columns', 500)
     churn
```

[4]:

	State	Account Length	Area Code	Phone	Int'l Plan	VMail Plan	VMail Message	Day Mins	Day Calls	Day Charge	Eve Mins	Eve Calls	Eve Charge	Night Mins	Night Calls	Night Charge	Intl Mins	Intl Calls	Intl Charge	CustS C
0	KS	128	415	382-4657	no	yes	25	265.1	110	45.07	197.4	99	16.78	244.7	91	11.01	10.0	3	2.70	
1	OH	107	415	371-7191	no	yes	26	161.6	123	27.47	195.5	103	16.62	254.4	103	11.45	13.7	3	3.70	
2	NJ	137	415	358-1921	no	no	0	243.4	114	41.38	121.2	110	10.30	162.6	104	7.32	12.2	5	3.29	
3	OH	84	408	375-9999	yes	no	0	299.4	71	50.90	61.9	88	5.26	196.9	89	8.86	6.6	7	1.78	
4	OK	75	415	330-6626	yes	no	0	166.7	113	28.34	148.3	122	12.61	186.9	121	8.41	10.1	3	2.73	
...	...	...	...	...	...	...	...	...	...	...	...	...	...	...	...	...	...	...	...	
3328	AZ	192	415	414-4276	no	yes	36	156.2	77	26.55	215.5	126	18.32	279.1	83	12.56	9.9	6	2.67	
3329	WV	68	415	370-3271	no	no	0	231.1	57	39.29	153.4	55	13.04	191.3	123	8.61	9.6	4	2.59	
3330	RI	28	510	328-8230	no	no	0	180.8	109	30.74	288.8	58	24.55	191.9	91	8.64	14.1	6	3.81	
3331	CT	184	510	364-6381	yes	no	0	213.8	105	36.35	159.6	84	13.57	139.2	137	6.26	5.0	10	1.35	
3332	TN	74	415	400-4344	no	yes	25	234.4	113	39.85	265.9	82	22.60	241.4	77	10.86	13.7	4	3.70	

3333 rows × 21 columns

By modern standards, it's a relatively small dataset, with only 3,333 records, where each record uses 21 attributes to describe the profile of a customer of an unknown US mobile operator. The attributes are:

Figure 7.35 – Amazon SageMaker notebook showing the dataset's information

5. We can now sample the dataset as test and training buckets, and then upload these files to S3 for future use. Once they've been uploaded, you will get the S3 buckets' names, as shown in the following screenshot:

Reserve some data for calling inference on the model

Divide the data into training and testing splits. The training split is used by SageMaker Autopilot. The testing split is reserved to perform inference using the suggested model.

```
[5]: train_data = churn.sample(frac=0.8,random_state=200)
     test_data = churn.drop(train_data.index)
     test_data_no_target = test_data.drop(columns=['Churn?'])
```

Now we'll upload these files to S3.

```
[6]: train_file = 'train_data.csv';
     train_data.to_csv(train_file, index=False, header=True)
     train_data_s3_path = session.upload_data(path=train_file, key_prefix=prefix + "/train")
     print('Train data uploaded to: ' + train_data_s3_path)

     test_file = 'test_data.csv';
     test_data_no_target.to_csv(test_file, index=False, header=False)
     test_data_s3_path = session.upload_data(path=test_file, key_prefix=prefix + "/test")
     print('Test data uploaded to: ' + test_data_s3_path)

     Train data uploaded to: s3://sagemaker-us-east-1-385578370913/sagemaker/DEMO-autopilot-churn/train/train_data.csv
     Test data uploaded to: s3://sagemaker-us-east-1-385578370913/sagemaker/DEMO-autopilot-churn/test/test_data.csv
```

Figure 7.36 – Amazon SageMaker Autopilot – sample dataset for test and training, and uploading the files to the S3

So far, everything we have done is traditional notebook work. Now, we will set up the Autopilot job.

6. Let's define the configuration, as shown in the following screenshot:

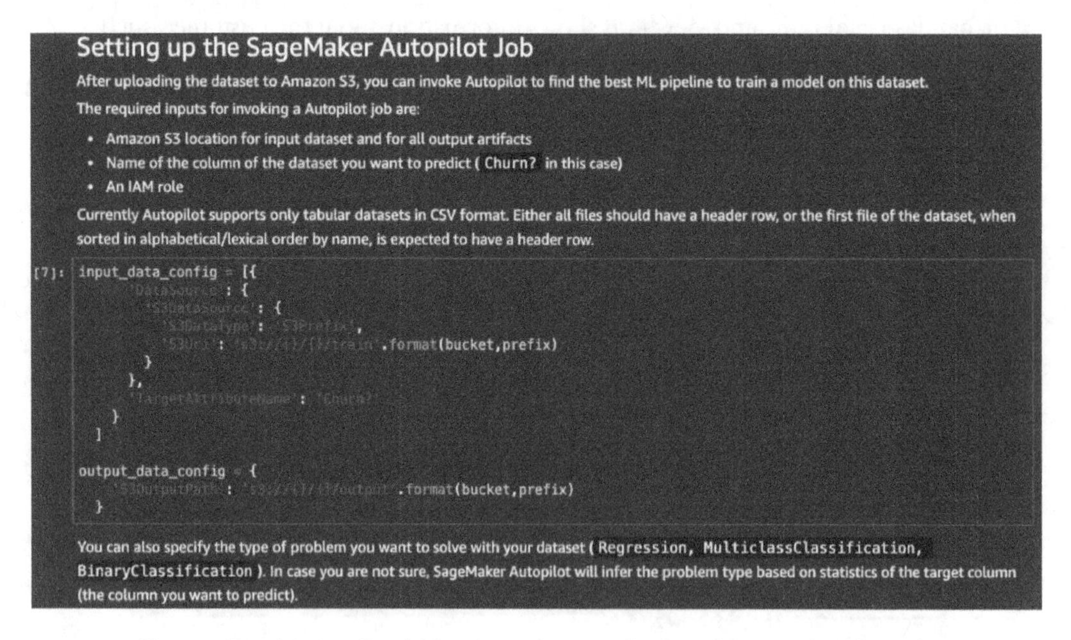
Setting up the SageMaker Autopilot Job
After uploading the dataset to Amazon S3, you can invoke Autopilot to find the best ML pipeline to train a model on this dataset.
The required inputs for invoking a Autopilot job are:
• Amazon S3 location for input dataset and for all output artifacts
• Name of the column of the dataset you want to predict (Churn? in this case)
• An IAM role
Currently Autopilot supports only tabular datasets in CSV format. Either all files should have a header row, or the first file of the dataset, when sorted in alphabetical/lexical order by name, is expected to have a header row.
[7]: input_data_config = [{
'DataSource': {
'S3DataSource': {
'S3DataType': 'S3Prefix',
'S3Uri': 's3://{}/{}/train'.format(bucket,prefix)
}
},
'TargetAttributeName': 'Churn?'
}
]
output_data_config = {
'S3OutputPath': 's3://{}/{}/output'.format(bucket,prefix)
}
You can also specify the type of problem you want to solve with your dataset (Regression, MulticlassClassification, BinaryClassification). In case you are not sure, SageMaker Autopilot will infer the problem type based on statistics of the target column (the column you want to predict).

Figure 7.37 – Amazon SageMaker Autopilot – configuring the Autopilot job config

7. Now, let's launch the SageMaker Autopilot job by invoking the `create_auto_ml_job` API call, like so:

Launching the SageMaker Autopilot Job

You can now launch the Autopilot job by calling the `create_auto_ml_job` API. We limit the number of candidates to 20 so that the job finishes in a few minutes.

```
[8]: from time import gmtime, strftime, sleep
     timestamp_suffix = strftime('%d-%H-%M-%S', gmtime())

     auto_ml_job_name = 'automl-churn-' + timestamp_suffix
     print('AutoMLJobName: ' + auto_ml_job_name)

     sm.create_auto_ml_job(AutoMLJobName=auto_ml_job_name,
                           InputDataConfig=input_data_config,
                           OutputDataConfig=output_data_config,
                           AutoMLJobConfig={'CompletionCriteria':
                                            {'MaxCandidates': 20}
                                           },
                           RoleArn=role)

     AutoMLJobName: automl-churn-03-00-04-31
[8]: {'AutoMLJobArn': 'arn:aws:sagemaker:us-east-1:385578370913:automl-job/automl-churn-03-00-04-31',
      'ResponseMetadata': {'RequestId': '50a2c4c1-f90c-4f28-a669-560c3d8f4254',
       'HTTPStatusCode': 200,
       'HTTPHeaders': {'x-amzn-requestid': '50a2c4c1-f90c-4f28-a669-560c3d8f4254',
        'content-type': 'application/x-amz-json-1.1',
        'content-length': '95',
        'date': 'Sat, 03 Oct 2020 00:04:32 GMT'},
       'RetryAttempts': 0}}
```

Figure 7.38 – Amazon SageMaker Autopilot – configuring the Autopilot job

The job runs with multiple trials, including the components of each experiment, as shown in the following screenshot:

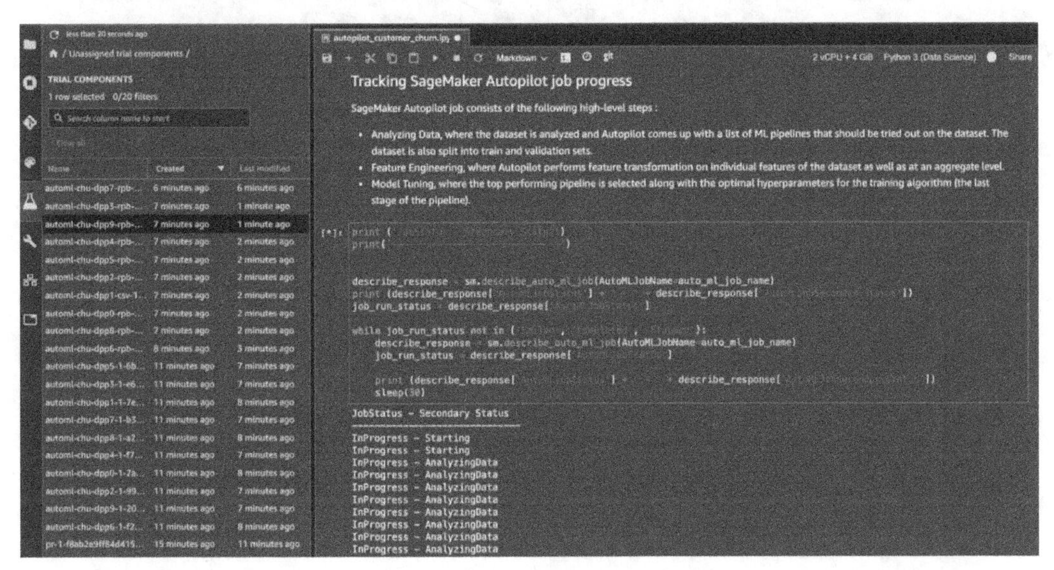

Figure 7.39 – Amazon SageMaker Autopilot – trial components in the Autopilot job notebook

While tracking the Amazon SageMaker Autopilot job's progress, you can print its status, along with any delays, as shown in the following screenshot. However, to view the details of individual trial runs in a meaningful manner visually, you can use the user interface:

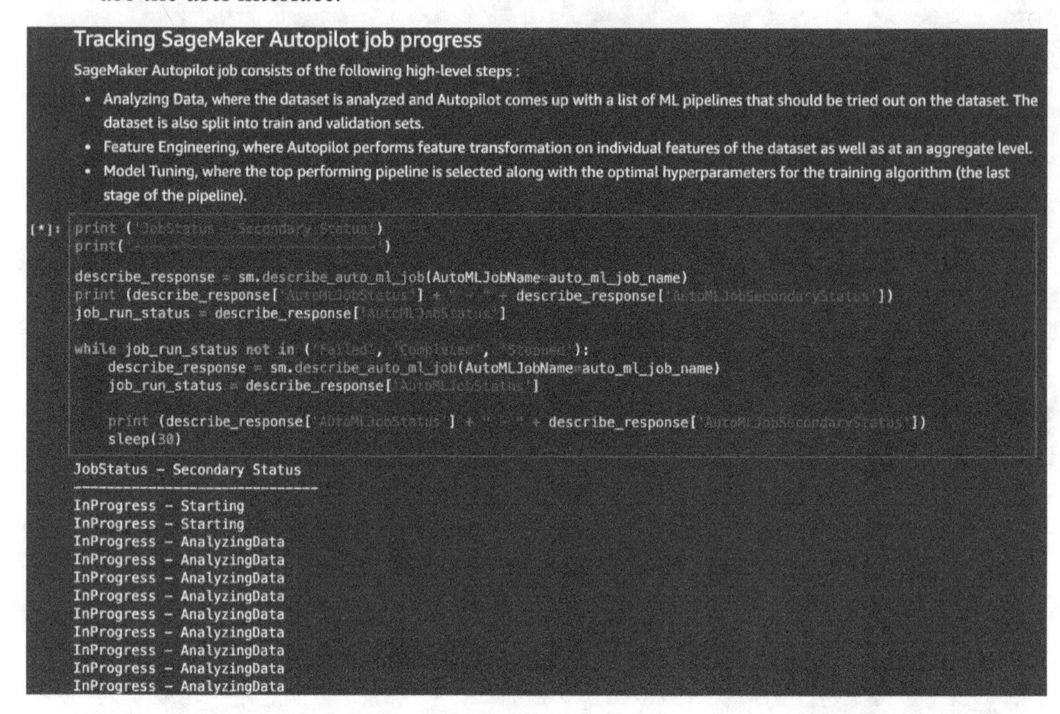

Figure 7.40 – Amazon SageMaker Autopilot – trial components in the Autopilot job notebook

8. Once the feature engineering and model tuning jobs in the trials are complete, you can run `describe_auto_ml_job` to get the best candidate information. Then, you can traverse the `best_candidate` object to get information about the underlying score and metric, as shown in the following screenshot:

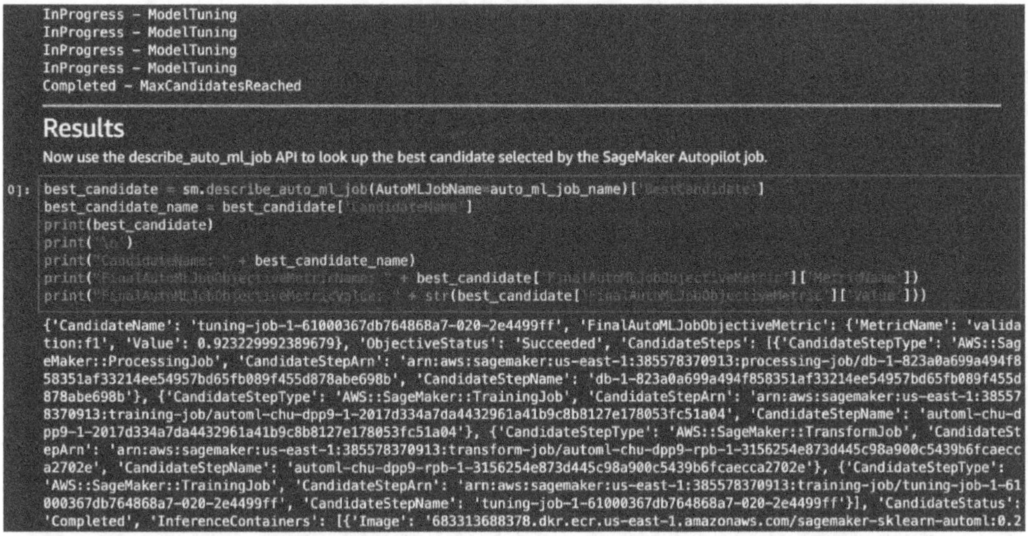

```
InProgress - ModelTuning
InProgress - ModelTuning
InProgress - ModelTuning
InProgress - ModelTuning
Completed - MaxCandidatesReached
```

Results

Now use the describe_auto_ml_job API to look up the best candidate selected by the SageMaker Autopilot job.

```
0]: best_candidate = sm.describe_auto_ml_job(AutoMLJobName=auto_ml_job_name)['BestCandidate']
    best_candidate_name = best_candidate['CandidateName']
    print(best_candidate)
    print('\n')
    print("CandidateName: " + best_candidate_name)
    print("FinalAutoMLJobObjectiveMetricName: " + best_candidate['FinalAutoMLJobObjectiveMetric']['MetricName'])
    print("FinalAutoMLJobObjectiveMetricValue: " + str(best_candidate['FinalAutoMLJobObjectiveMetric']['Value']))
```


```
{'CandidateName': 'tuning-job-1-61000367db764868a7-020-2e4499ff', 'FinalAutoMLJobObjectiveMetric': {'MetricName': 'valida
tion:f1', 'Value': 0.923229992389679}, 'ObjectiveStatus': 'Succeeded', 'CandidateSteps': [{'CandidateStepType': 'AWS::Sag
eMaker::ProcessingJob', 'CandidateStepArn': 'arn:aws:sagemaker:us-east-1:385578370913:processing-job/db-1-823a0a699a494f8
58351af33214ee54957bd65fb089f455d878abe698b', 'CandidateStepName': 'db-1-823a0a699a494f858351af33214ee54957bd65fb089f455d
878abe698b'}, {'CandidateStepType': 'AWS::SageMaker::TrainingJob', 'CandidateStepArn': 'arn:aws:sagemaker:us-east-1:38557
8370913:training-job/automl-chu-dpp9-1-2017d334a7da4432961a41b9c8b8127e178053fc51a04', 'CandidateStepName': 'automl-chu-d
pp9-1-2017d334a7da4432961a41b9c8b8127e178053fc51a04'}, {'CandidateStepType': 'AWS::SageMaker::TransformJob', 'CandidateSt
epArn': 'arn:aws:sagemaker:us-east-1:385578370913:transform-job/automl-chu-dpp9-rpb-1-3156254e873d445c98a900c5439b6fcaecc
a2702e', 'CandidateStepName': 'automl-chu-dpp9-rpb-1-3156254e873d445c98a900c5439b6fcaecca2702e'}, {'CandidateStepType':
'AWS::SageMaker::TrainingJob', 'CandidateStepArn': 'arn:aws:sagemaker:us-east-1:385578370913:training-job/tuning-job-1-61
000367db764868a7-020-2e4499ff', 'CandidateStepName': 'tuning-job-1-61000367db764868a7-020-2e4499ff'}], 'CandidateStatus':
'Completed', 'InferenceContainers': [{'Image': '683313688378.dkr.ecr.us-east-1.amazonaws.com/sagemaker-sklearn-automl:0.2
```


Figure 7.41 – Amazon SageMaker Autopilot – trial components in the Autopilot job notebook

Once the job is completed, you will see the candidate model, the final metric (the F1 score in this case), and any associated values:

```
CandidateName: tuning-job-1-61000367db764868a7-020-2e4499ff
FinalAutoMLJobObjectiveMetricName: validation:f1
FinalAutoMLJobObjectiveMetricValue: 0.923229992389679
```

Due to some randomness in the algorithms involved, different runs will provide slightly different results, but accuracy will be around or above 93%, which is a good result.

Figure 7.42 – Amazon SageMaker Autopilot job results

We will deploy and invoke the best candidate model, which has a 93% F1 score, in the next section.

Hosting and invoking the model

Similar to how we invoked the model we built using the Experiment UI earlier, we will now host and invoke the model we built in the notebook. The difference is that in the first instance, we were low-code, while here we are building it using code:

1. To host the service, you will need to create a model object, endpoint configuration, and eventually an endpoint. Previously, this was done using the UI, but here, we will use the Amazon SageMaker Python instance to accomplish the same. This can be seen in the following screenshot:

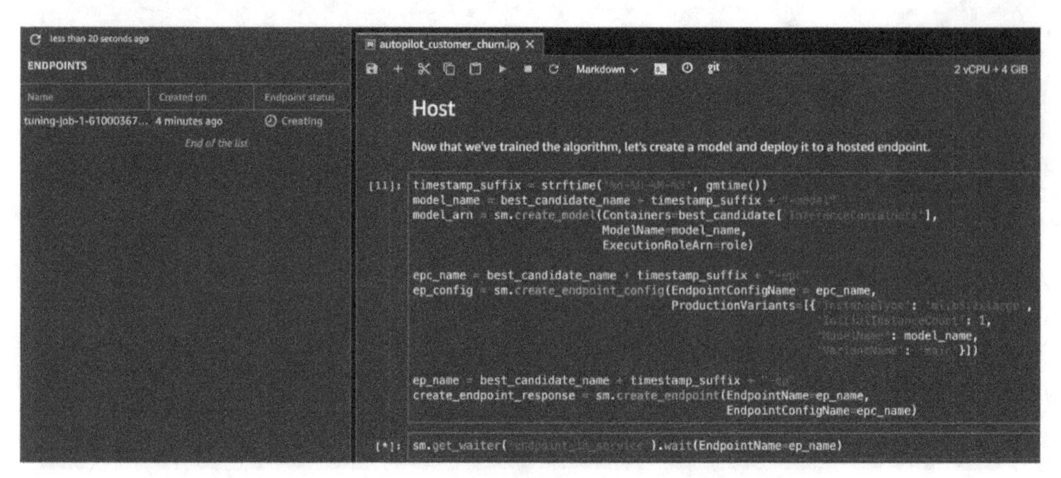

Figure 7.43 – Amazon SageMaker notebook – hosting the model

The `get_waiter` method is part of Boto3, which is the Python SDK for AWS. Like other waiters, it polls until a successful state is reached. An error is typically returned after 60 failed checks. You can read about the methods by looking at the API documentation for it, which can be found here: `https://boto3.amazonaws.com/v1/documentation/api/latest/reference/services/sagemaker.html#SageMaker.Client.create_endpoint`.

Now that the endpoint has been created and the model has been hosted, we can invoke the service. To evaluate the model, you will need to create a predictor instance and pass it the endpoint's information, along with the parameters for prediction. Instead of calling the endpoint line by line, we can perform bulk predictions by passing in the entire test data CSV file and comparing the results against the ground truth. You can see the accuracy numbers in the following screenshot:

```
[12]: sm.get_waiter('endpoint_in_service').wait(EndpointName=ep_name)
```

Evaluate

Now that we have a hosted endpoint running, we can make real-time predictions from our model very easily, simply by making an http POST request. But first, we'll need to setup serializers and deserializers for passing our `test_data` NumPy arrays to the model behind the endpoint.

```
[13]: from io import StringIO
      from sagemaker.predictor import RealTimePredictor
      from sagemaker.content_types import CONTENT_TYPE_CSV

      predictor = RealTimePredictor(
          endpoint=ep_name,
          sagemaker_session=session,
          content_type=CONTENT_TYPE_CSV,
          accept=CONTENT_TYPE_CSV)

      # Remove the target column from the test data
      test_data_inference = test_data.drop('Churn?', axis=1)

      # Obtain predictions from SageMaker endpoint
      prediction = predictor.predict(test_data_inference.to_csv(sep=',', header=False, index=False)).decode('utf-8')

      # Load prediction in pandas and compare to ground truth
      prediction_df = pd.read_csv(StringIO(prediction), header=None)
      accuracy = (test_data.reset_index()['Churn?'] == prediction_df[0]).sum() / len(test_data_inference)
      print('Accuracy: {}'.format(accuracy))

      Accuracy: 0.9685157421289355
```

Figure 7.44 – Amazon SageMaker model evaluation for accuracy

2. Once you have finished testing the endpoint, we must clean up. In cloud environments, you must clean up after yourself, so make this a priority checklist item. If you don't do this, you won't like the billing statement from leaving a server running. Virtual or not, it all adds up.

   When you're cleaning up a UI, turn off and delete the compute instances and the endpoints. Since we are doing a manual cleanup, you must delete the endpoint, endpoint config, and the model:

Cleanup

The Autopilot job creates many underlying artifacts such as dataset splits, preprocessing scripts, or preprocessed data, etc. This code, when un-commented, deletes them. This operation deletes all the generated models and the auto-generated notebooks as well.

```
[14]: #s3 = boto3.resource('s3')
      #s3_bucket = s3.Bucket(bucket)

      #job_outputs_prefix = '{}/output/{}'.format(prefix, auto_ml_job_name)
      #s3_bucket.objects.filter(Prefix=job_outputs_prefix).delete()
```

Finally, we delete the endpoint and associated resources.

```
[15]: sm.delete_endpoint(EndpointName=ep_name)
      sm.delete_endpoint_config(EndpointConfigName=epc_name)
      sm.delete_model(ModelName=model_name)

[15]: {'ResponseMetadata': {'RequestId': '7ebbee1b-301d-49f3-bdc7-8149fe5c0b34',
        'HTTPStatusCode': 200,
        'HTTPHeaders': {'x-amzn-requestid': '7ebbee1b-301d-49f3-bdc7-8149fe5c0b34',
         'content-type': 'application/x-amz-json-1.1',
         'content-length': '0',
         'date': 'Sat, 03 Oct 2020 00:42:43 GMT'},
        'RetryAttempts': 0}}
```

Figure 7.45 – Amazon SageMaker Autopilot cleanup with resulting response codes

Even though these examples have shown you how AWS AutoML enables you to perform feature engineering, model tuning, and hyperparameter optimization, you don't have to limit yourself to the algorithms provided by AWS. You can bring your own data processing code to SageMaker Autopilot, as shown at `https://github.com/aws/amazon-sagemaker-examples/blob/master/autopilot/custom-feature-selection/Feature_selection_autopilot.ipynb`.

Summary

Building AutoML systems to democratize AI from scratch is a considerable effort. Therefore, cloud hyperscalers act as enablers and accelerators to jumpstart this journey. In this chapter, you learned how to use Amazon SageMaker Autopilot, both via notebooks and via the experimentation user interface. You were also exposed to the larger AWS machine learning ecosystem and SageMaker's capabilities.

In the next chapter, we will study another major cloud computing platform, Google Cloud Platform, and the AutoML offerings provided it. Happy coding!

Further reading

For more information on the topics that were covered in this chapter, please refer to the following links and resources:

- *Mastering Machine Learning on AWS: Advanced machine learning in Python using SageMaker, Apache Spark, and TensorFlow*, by Dr. Saket S.R. Mengle , Maximo Gurmendez, Packt Publishing: `https://www.amazon.com/Mastering-Machine-Learning-AWS-TensorFlow/dp/1789349796`
- *Learn Amazon SageMaker: A guide to building, training, and deploying machine learning models for developers and data scientists*, by Julien Simon and Francesco Pochetti, Packt Publishing: `https://www.amazon.com/Learn-Amazon-SageMaker-developers-scientists/dp/180020891X`

8
Machine Learning with Google Cloud Platform

"I have always been convinced that the only way to get artificial intelligence to work is to do the computation in a way similar to the human brain. That is the goal I have been pursuing. We are making progress, though we still have lots to learn about how the brain actually works."

– Geoffrey Hinton

In the previous chapter, you were introduced to the major hyperscalers, the **Amazon Web Services** (**AWS**) platform, AWS SageMaker, and its automated **machine learning** (**ML**) capabilities with AWS SageMaker Autopilot.

Gartner named Google a leader in its 2020 Magic Quadrant for Cloud Infrastructure and Platform Services report. The Google Cloud computing services provide a suite of computing technology, tools, and services to power enterprises using the same infrastructure that powers Google's own products and services. In this chapter, we will do a review of the Google Cloud computing services and their **artificial intelligence** (**AI**) and ML offerings, especially the automated ML capabilities of cloud AutoML Tables.

In this chapter, we will cover the following topics:

- Getting started with the Google Cloud Platform services
- AI and ML with Google Cloud Platform
- Google Cloud AI Platform and AI Hub
- Getting started with Google Cloud AI Platform
- Automated ML with Google Cloud

Getting started with the Google Cloud Platform services

Like other hyperscalers and cloud computing platforms, the Google Cloud computing services also provide a large variety of general-purpose compute, analytics, storage, and security services. Besides the well-known App Engine, Cloud Run, Compute Engine, Cloud Functions, Storage, Security, Networking, and IoT offerings, there are more than 100 products listed on the **Google Cloud Platform** (**GCP**) products page. You can visit the GCP console by typing in `console.cloud.google.com`, as seen in the following figure:

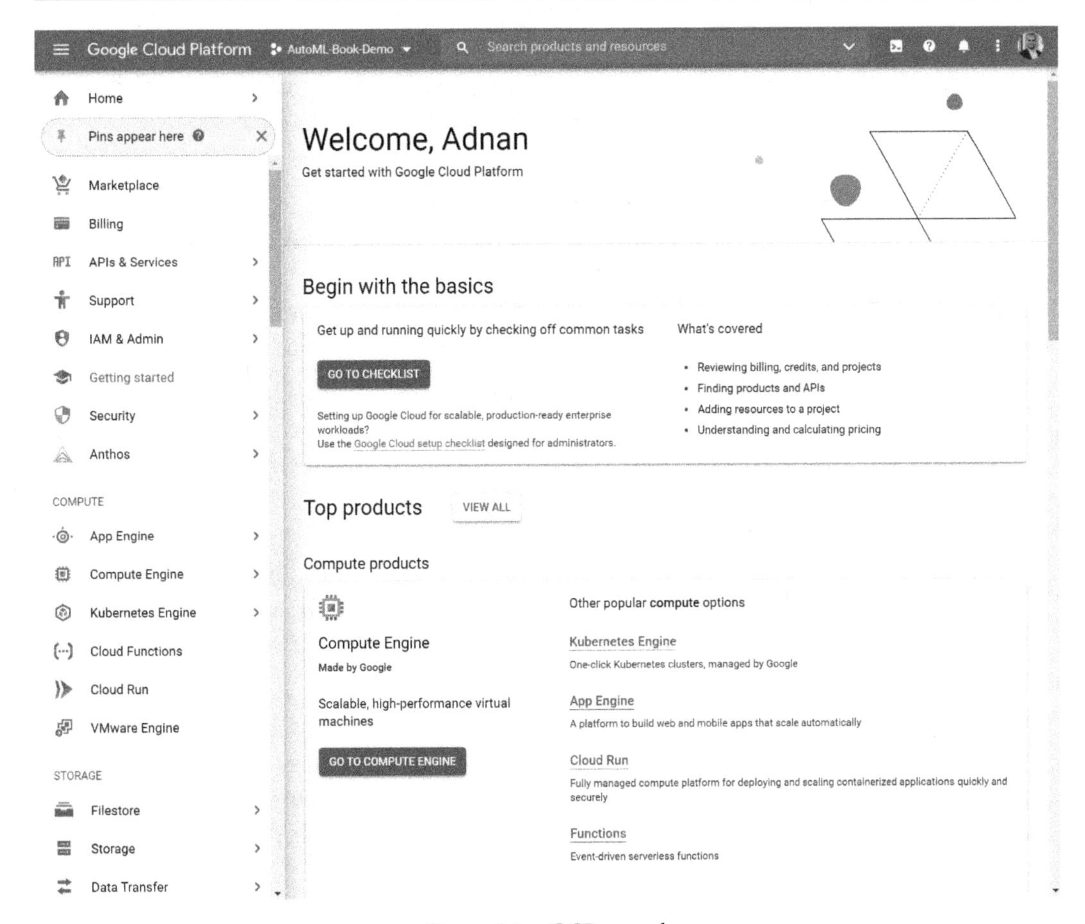

Figure 8.1 – GCP console

The key products and offerings are divided into categories such as compute, storage, database, networking, operations, big data, AI, and general development tools. The GCP Services Summary (`https://cloud.google.com/terms/services`) offers the most up-to-date list of all the services and offerings from GCP. The scope and breadth of each of these services is enormous and beyond the scope of this book. However, by way of an introduction, here is a brief overview:

- **Compute**: In this category, GCP offers services such as App Engine, Compute Engine, Kubernetes Engine, Cloud Functions, Cloud Run, and VMware Engine. These services cover a wide spectrum of computational capabilities, with different modalities such as CPUs, GPUs, or Cloud TPUs.

- **Storage**: With Cloud Storage, Cloud Filestore, Persistent Disk, Cloud Storage for Firebase, and advance data transfer capabilities, GCP achieves feature parity with other cloud storage providers. These storage repositories can be accessed from Compute Engine as needed.
- **Databases**: In the database realm, GCP offerings include Cloud Bigtable, Datastore, Firestore (a NoSQL document database), Memorystore (for Redis and Memcached), Cloud Spanner (a relational database), and Cloud SQL. With a large marketplace offering, you can migrate and run Microsoft SQL Server on Compute Engine via Cloud SQL. The GCP Cloud SQL offering helps to migrate, maintain, manage, and administer relational databases such as MySQL, PostgreSQL, and SQL Server on GCP.
- **Networking**: The GCP networking services are on par with any other hyperscaler. Their popular networking services include Cloud Load Balancing, Cloud DNS, Cloud VPN (a virtual private cloud), Cloud **CDN** (short for **content delivery network**), Cloud Router, Cloud Armor (a policy framework), Cloud **NAT** (short for **network address translation**), Service Directory, Traffic Director, and Virtual Private Cloud. The networking services offered by GCP provide hybrid connectivity, network security, and intelligence services.
- **Operations**: In the operations area, GCP is no slouch. Whether it is monitoring, debugging, error reporting, logging, profiling, or tracing, tools such as Cloud Debugger, Cloud Logging, Cloud Monitoring, Cloud Profiler, and Cloud Trace provide dashboards and alerts for uptime checks, ensuring your systems are running reliably.
- **Developer tools**: These tools include Artifact Registry (for managing containers), Cloud SDK, Container Registry, Cloud Build (for CI/CD with GitHub and Bitbucket to create artifacts such as Docker containers or Java archives), Cloud Source Repositories, Firebase Test Lab (mobile app testing), and Test Lab.

  The other tools are Cloud Build for build management, Cloud Tasks, Container Registry, Artifact Registry, Cloud Scheduler, Deployment Manager, API Gateway, Endpoints, Identity Platform, Source Repositories, Workflows, and Private Catalog.
- **Data analytics**: GCP's offering includes managed data analysis services such as BigQuery, and managed workflow orchestration services including Cloud Composer, Cloud Data Fusion (integration service), Dataflow (data-processing pipelines), Datalab (**exploratory data analysis (EDA)** and visualization), Dataproc (managed Spark and Hadoop), Pub/Sub (asynchronous messaging), Data Catalog (metadata management), and Cloud Life Sciences for working with life sciences data.

- **API management services**, including full life cycle API management, are provided with Apigee. Other tools include Cloud Endpoints and Anthos offerings for hybrid and multi-cloud management. **Google Kubernetes Engine** (**GKE**) is the open source container scheduler, while Connect and Hub are used to manage features and services on customer-registered clusters.
- **Migration and data transfer**: Migration tools include the BigQuery Data Transfer Service, which helps transfer data from HTTP/S-reachable locations including **Amazon Simple Storage Service** (**Amazon S3**) and Google Cloud products. Transfer Appliance is a solution that uses hardware and software to transfer data into GCP.
- **Security and identity**: The services offered here include Space, Access Transparency, Assured Workloads for Government, Binary Authorization, Certificate Authority Service, Cloud Asset Inventory, Cloud Data Loss Prevention, **Cloud External Key Manager** (**Cloud EKM**), **Cloud Hardware Security Module** (**Cloud HSM**), Cloud Key Management Service, Event Threat Detection, Security Command Center, VPC Service Controls, Secret Manager, and Web Security Scanner for vulnerability scans.
- **Identity and access**: The services and tools offered include Access Approval, Access Context Manager, Cloud Identity Services, Firebase Authentication, Google Cloud Identity-Aware Proxy, **Identity and Access Management** (**IAM**), Identity Platform, Managed Service for Microsoft **Active Directory** (**AD**), and Resource Manager API to programmatically manage GCP. There are also user protection services such as `*reCAPTCHA` and the Web Risk APIs.
- **Serverless computing**: In the realm of serverless computing, GCP offers Cloud Run (stateless containers), Cloud Functions, Cloud Functions for Firebase, Cloud Scheduler, and Cloud Tasks for distributed task management. There is also **Internet of Things** (**IoT**) Core offered as a fully managed service.
- **Management tools**: These tools include the Cloud Console app (native mobile app), Cloud Deployment Manager, Cloud Shell, and Recommenders (for recommending and predicting usage). There are also service infrastructure components built as foundational platforms, including the Service Management API, Service Consumer Management APIs, and Service Control APIs.

There are also a variety of partner solutions and vertical services available as part of the GCP offering. Vertical services in healthcare and life sciences include Cloud Healthcare, while for the media and gaming industry, GCP offers Game Servers. GCP premium software and partner solutions include Redis Enterprise, Apache Kafka on Confluent, DataStax Astra, Elasticsearch Service, MongoDB Atlas, and Cloud Volumes.

This was a brief enumeration of the GCP offerings in the different categories of information systems. In the next section, we will discuss the AI and ML services offered by GCP.

AI and ML with GCP

Among the early AI-first companies, Google has the first-mover advantage in building and maintaining advanced AI platforms, accelerators, and AI building blocks. Google Cloud AI Platform is a highly comprehensive, cloud-based cognitive computing offering where you can build models and notebooks, perform data labeling, create ML jobs and pipelines, and access AI Hub:

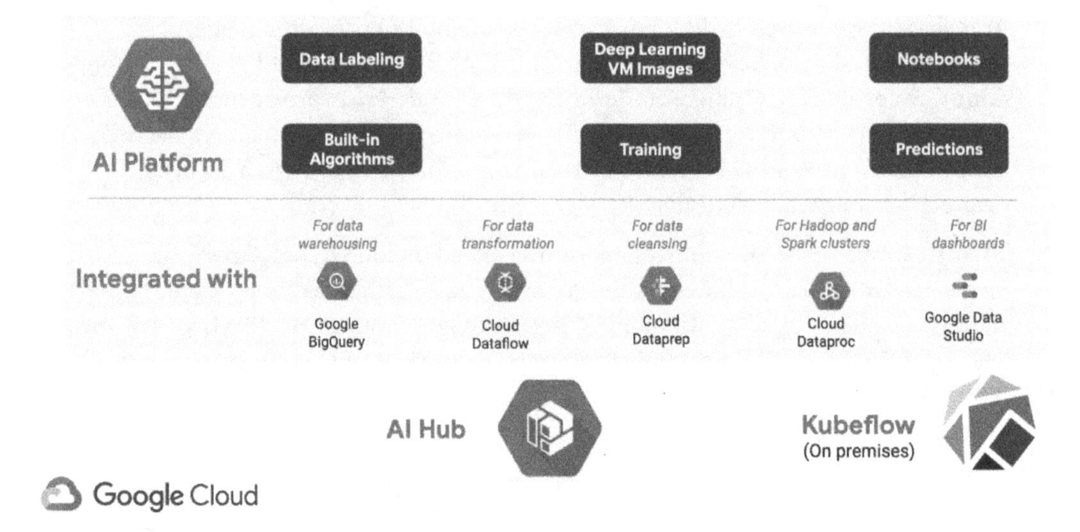

Figure 8.2 – Google Cloud AI Platform capabilities

The AI platforms and accelerators include capabilities such as building ML pipelines, deployment, and so on. It contains data labeling capabilities, platform notebooks, neural architecture search, training, and prediction functionalities.

Automated ML is also one of the key building blocks, comprising AutoML Natural Language, AutoML Tables, AutoML Translation, AutoML Video, AutoML Vision, and Recommendations AI. Other AI offerings include the Cloud Natural Language API, Cloud Translation, Cloud Vision, Dialogflow (offered in both the Essentials and Customer Experience Editions), Document AI, the Media Translation API, speech-to-text and text-to-speech services, and the Video Intelligence API.

AI Platform can be accessed at `https://console.cloud.google.com/ai-platform/dashboard` as shown in the following figure:

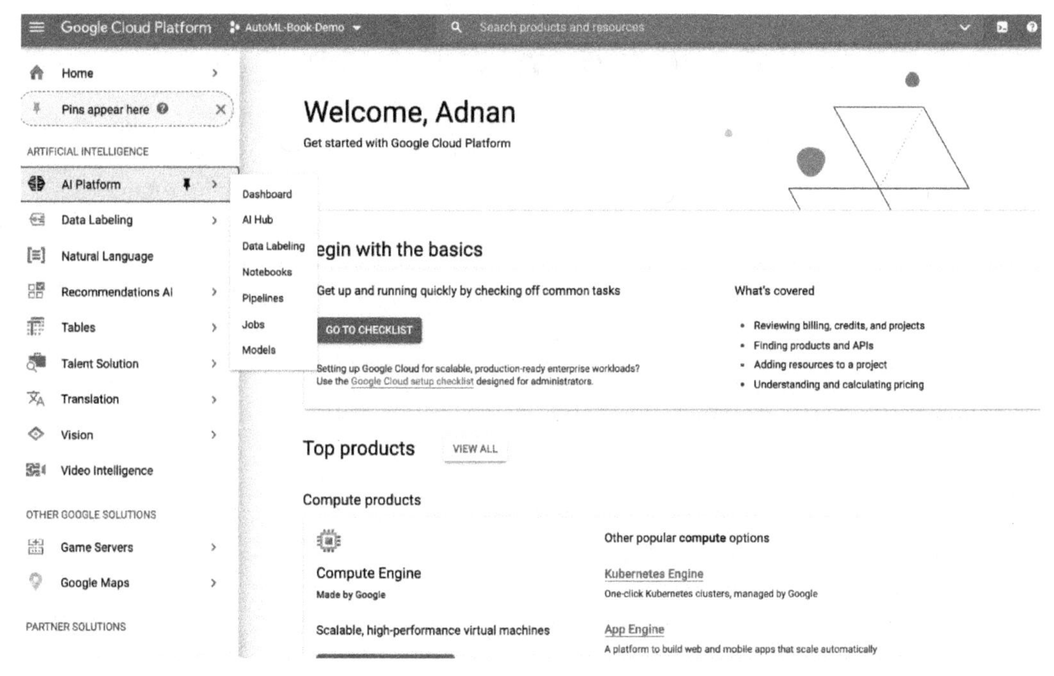

Figure 8.3 – AI Platform's main screen

AI Platform, being a one-stop shop for developers, serves as a portal from where you can navigate to other areas for data labeling, natural language processing, recommendations, translations, and other functionalities. The key areas for automated ML focus on vision, language, conversation, and structured data:

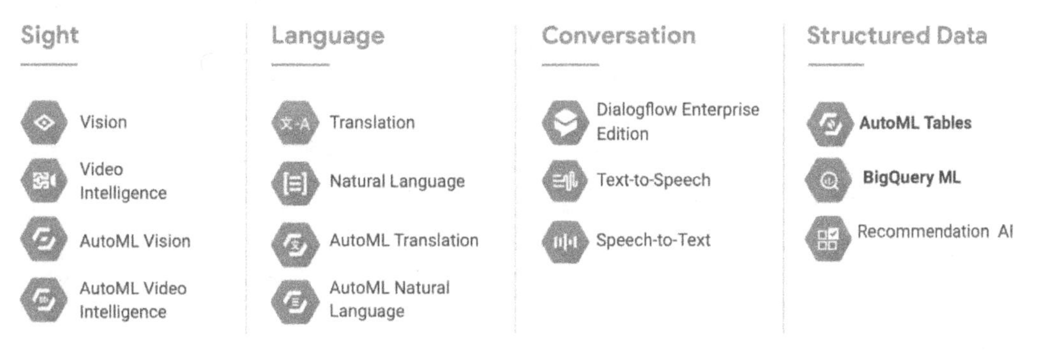

Figure 8.4 – Components of AI Platform including AutoML

For the purposes of this book, we will focus on AutoML capabilities within structured data, specifically AutoML Tables.

Google Cloud AI Platform and AI Hub

A part of the larger AI Platform offering, Google Cloud AI Hub is the one-stop shop for all things AI – it even says so on the home page. AI Hub is in beta at the time of writing. However, that shouldn't stop you from trying out its amazing one-click deployment capabilities. AI Hub and AI Platform can be confusing; the difference is in how GCP frames the problem. AI Hub focuses on enterprise-grade sharing capabilities to enable private collaboration and hosting, while AI Platform is a larger ecosystem of all things AI, including notebooks, jobs, and platforms. This is not to say that these capabilities don't overlap, and the GCP marketing team will probably come up with a cohesive strategy one day – but until then, the duality continues.

The following screenshot shows the AI Hub home page. You navigate to this page by clicking the AI Hub link on the AI Platform page at `https://console.cloud.google.com/ai-platform/dashboard`, or directly by typing `https://aihub.cloud.google.com/` in your browser:

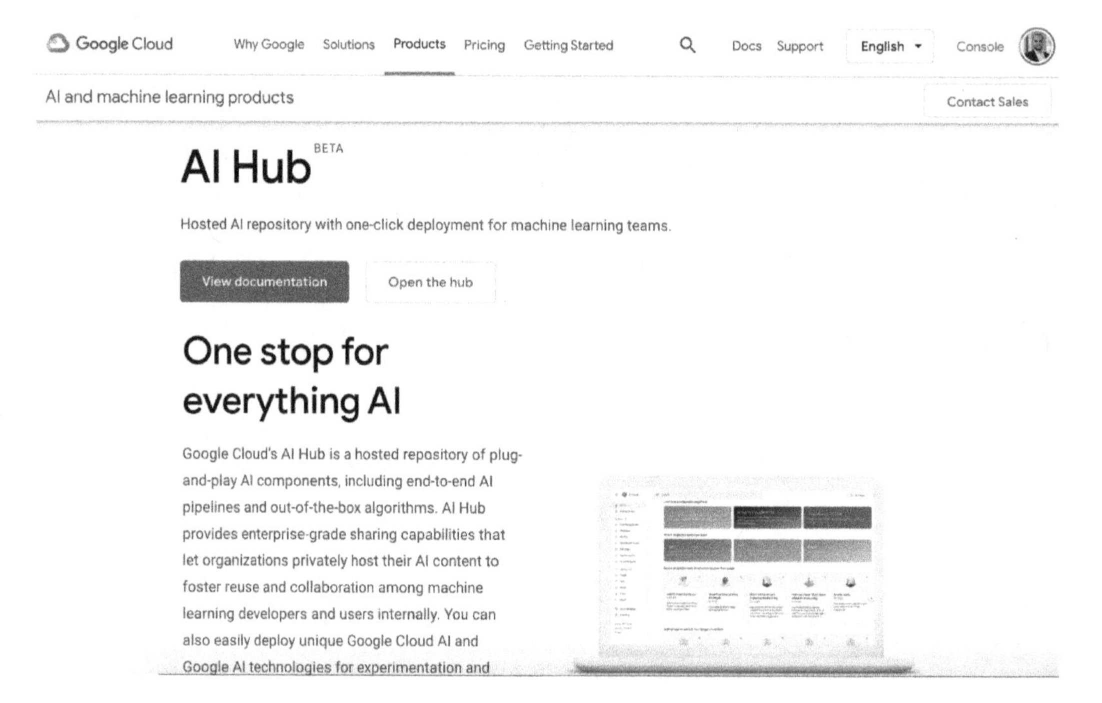

Figure 8.5 – AI Hub's main screen

The AI Hub home page frames its problem statement in the simplest form. It provides you with starter kits, the latest news around the ML use cases, and cutting-edge ML technologies and tutorials. Here, you can build Kubeflow pipelines and ML containers, start **virtual machine** (**VM**) images, use trained models, and explore and share the recipes built by others, all in one place:

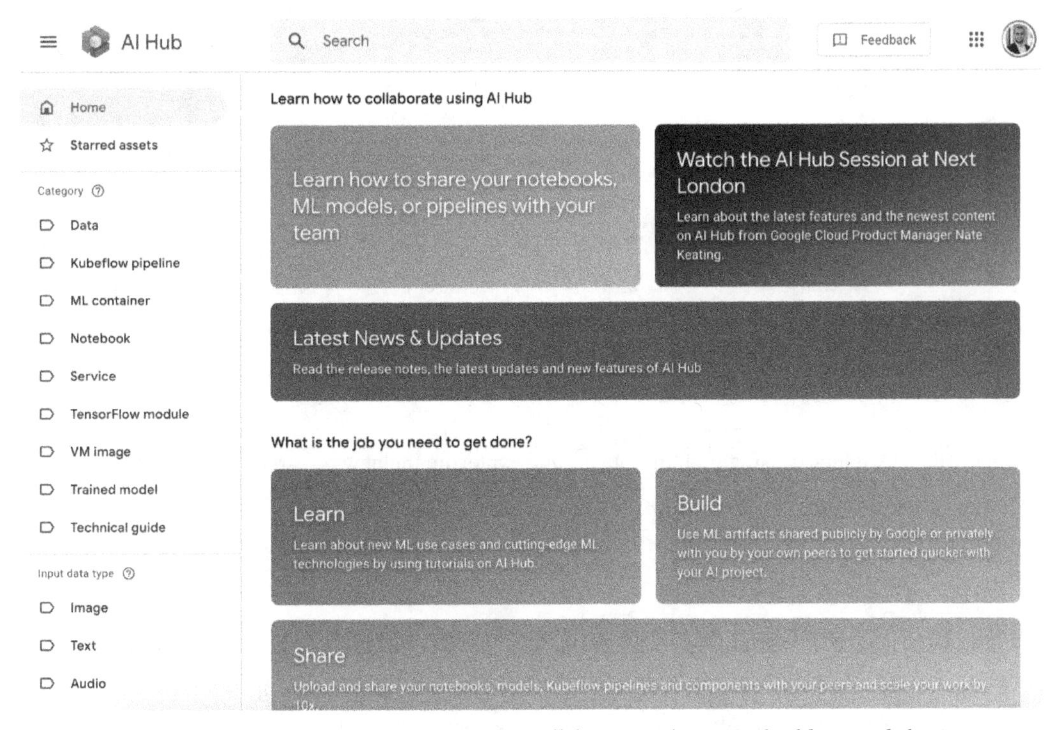

Figure 8.6 – AI Hub's main screen, covering collaboration, learning, building, and sharing

Google Cloud AI Hub aims to provide a comprehensive understanding of the most powerful algorithms in the world and how to optimize them with the cutting-edge AI research from Google's DeepMind. You might recall Google's DeepMind from AlphaGo, where the DeepMind team were trying to teach an AI to "learn," eventually being able to beat a human Go player for the first time. The same company has provided cutting-edge research and usable models around time series forecasting, generative adversarial networks, computer vision, text analytics, and automated ML, all of which can be utilized as part of AI Hub.

You can see some of the offerings listed in AI Hub in the following screenshot:

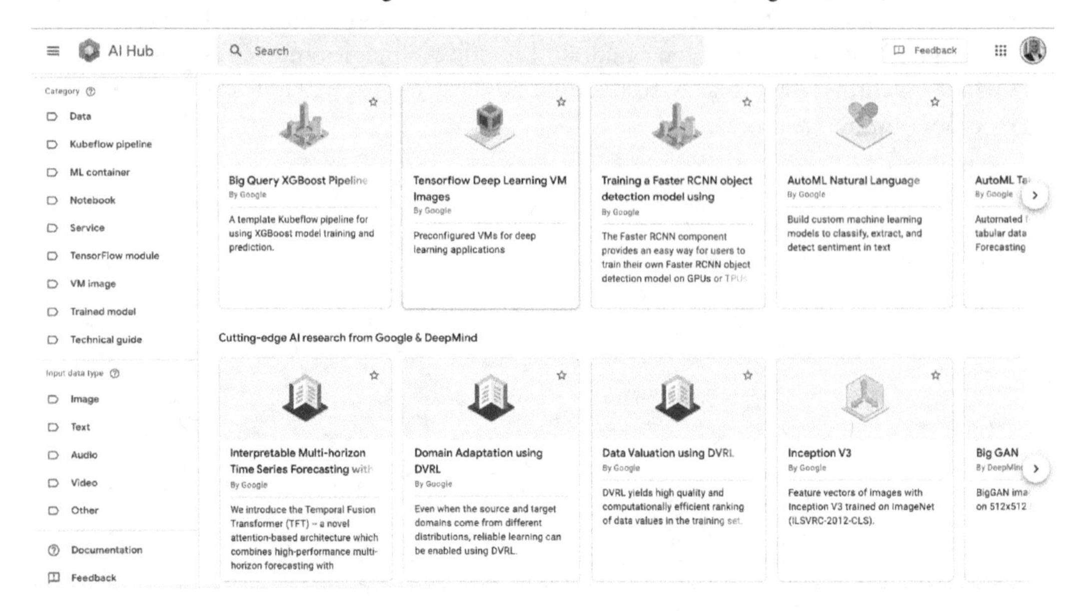

Figure 8.7 – AI Hub on using prebuilt models and exploring the new research from DeepMind

With this preliminary introduction to AI Hub and AI Platform, let's dive into making a simple notebook, a tool you are quite familiar with, using the Cloud AI Platform.

Getting started with Google Cloud AI Platform

There are a few things you can do with Google Cloud AI Platform, including creating notebooks and Kubeflow pipelines, or starting up a pre-installed VM. Kubeflow, an ML framework for Kubernetes, is a simple-to-learn workflow management system with excellent performance when building ML pipelines. This can be seen on the **Get started with Cloud AI Platform** home page shown in the following figure:

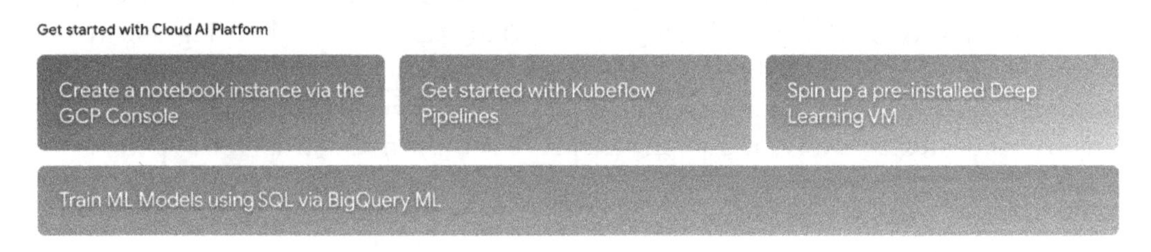

Figure 8.8 – Get started with Cloud AI Platform

In order to start with AI Platform notebooks, navigate to `http://cloud.google.com/ai-platform-notebooks` where you will see the screen shown in *Figure 8.9*.

This is the starting point for building an AI Platform notebook. Click on the **Go to console** button to navigate to the Platform console as shown in the following figure:

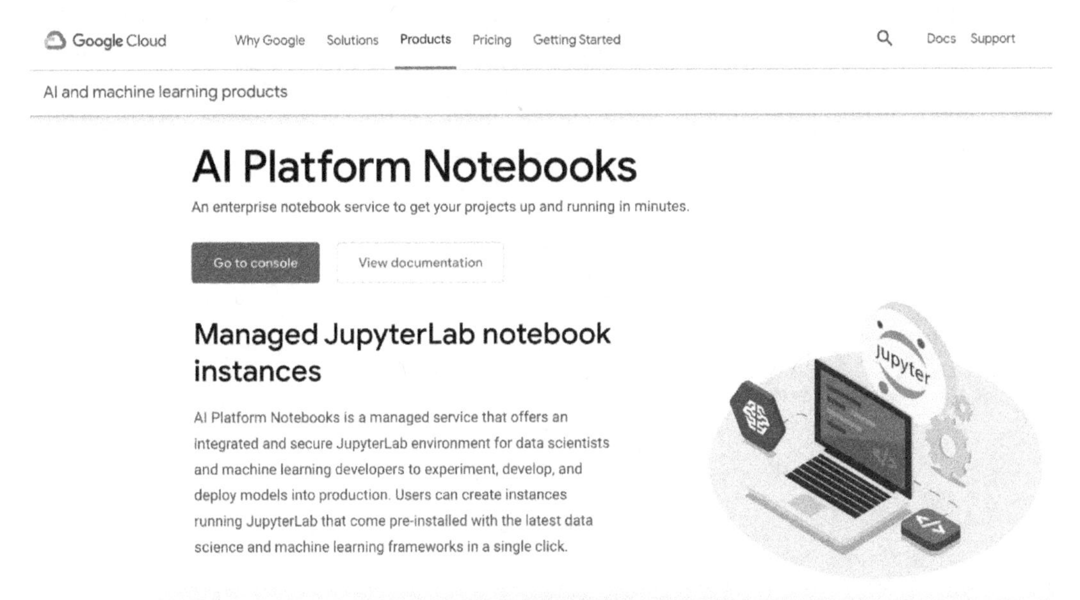

Figure 8.9 – AI Platform Notebooks page

Alternatively, you can click on the **Notebooks** link in the left pane of the AI Platform home page, as shown in the following figure:

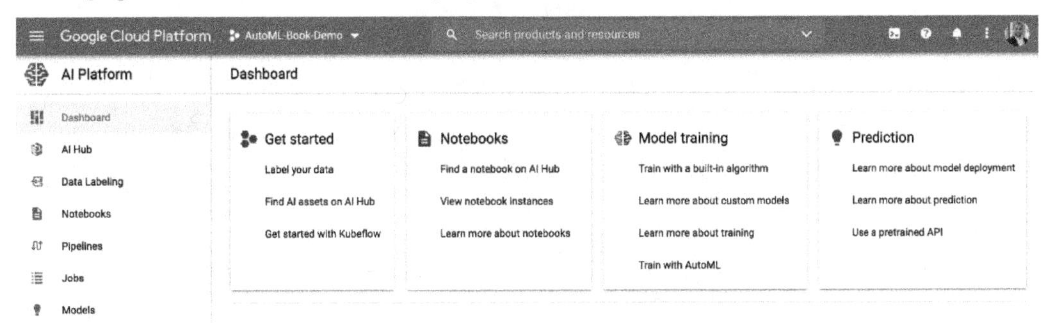

Figure 8.10 – Getting started with Google Cloud AI Platform

Either of these actions will take you to the **Notebook instances** page, as shown in the following figure.

Now, to create a notebook, you would create a new instance that can then be customized based on your needs. You can choose a specific language (Python 2 or 3) and framework (TensorFlow or PyTorch, among other flavors), as shown in the following figure. For the purposes of this demo, we will create a simple Python notebook. Select the **Python 2 and 3** option from the dropdown shown in the following figure and click **Next**:

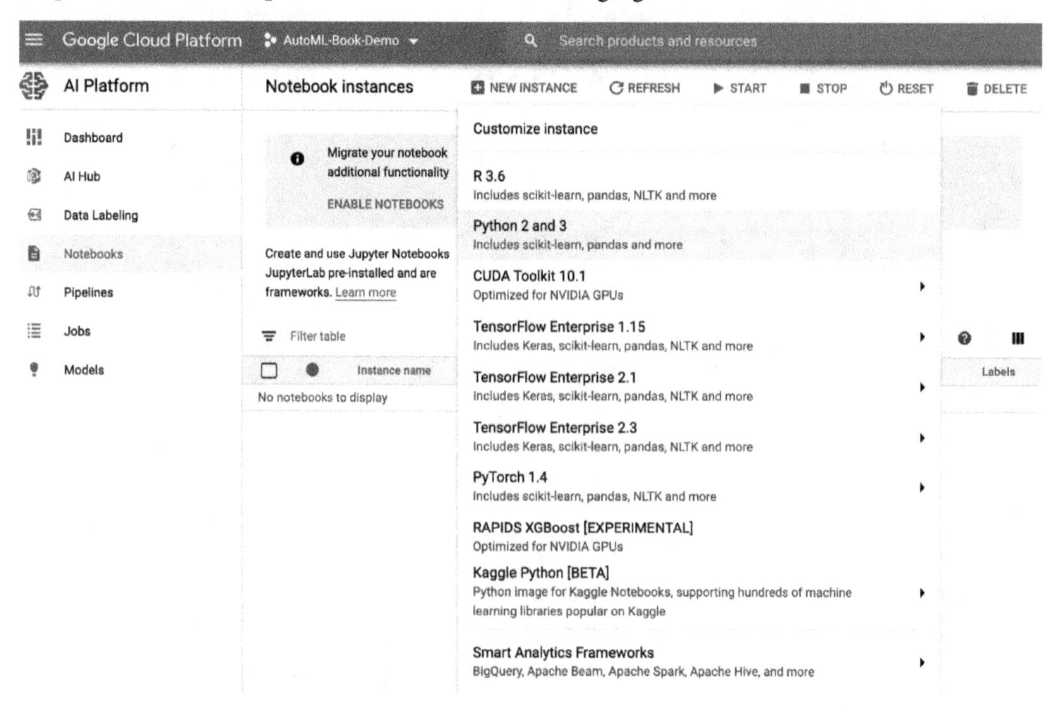

Figure 8.11 – Creating a notebook instance on AI Platform

Now you will be asked to choose the parameters of your notebook instance. This includes the region and zone where you would like to deploy the notebook. Typically, you would choose the one closest to you. Next are the operating system and environment options. In our case, we will go with **Debian Linux Distro and Intel processor**, which as it turns out, works really well at printing a Hello World message. You will see the environment information onscreen as follows:

New notebook instance

Instance name
automl-book-python-20201008-202233

63-char limit with lowercase letters, digits, or '-' only. Must start with a letter. Cannot end with a '-'.

Region *
us-east1 (South Carolina)

Zone *
us-east1-b

Instance Configuration

Environment	Intel® optimized Base (with Intel® MKL)
Machine type	4 vCPUs, 15 GB RAM
Boot disk	100 GB Standard persistent disk
Subnetwork	default(10.142.0.0/20)
External IP	Ephemeral(Automatic)
Extensions	SELECT EXTENSIONS None selected
Permission	Compute Engine default service account
Estimated cost	$102.69 monthly, $0.141 hourly

ADVANCED OPTIONS CANCEL CREATE

Figure 8.12 – Creating an AI Platform notebook instance – setting up the environment

Once you have selected the environment variables, you can see the estimated costs of how much you may end up spending to run this notebook instance. Speaking of money, you should be careful about leaving idle resources running in the cloud – it could be bad for your financial wellbeing.

Click **CREATE** as shown in *Figure 8.12* to move forward and GCP will instantiate a notebook for you with the specified parameters. You will be able to see all your notebook instances on the main **Notebook instances** page, as shown in the following figure:

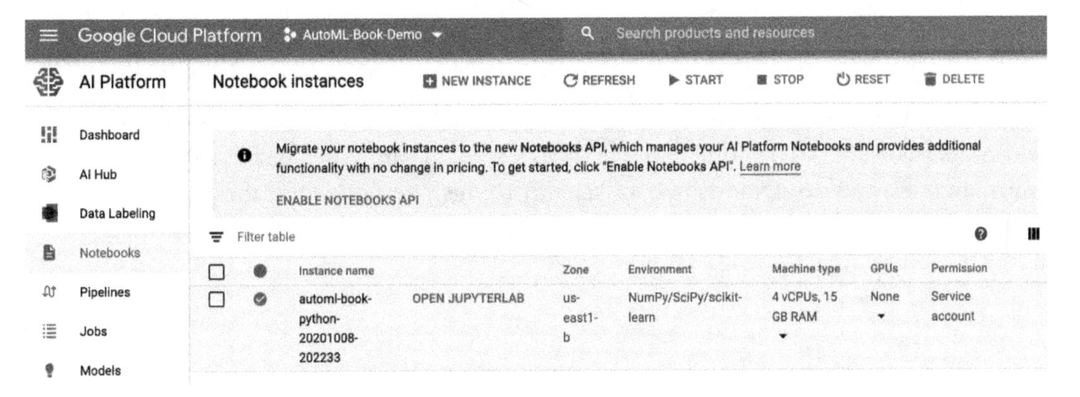

Figure 8.13 – AI Platform notebook instances enumeration

The notebook and its associated compute resources are ready. Now click on the **OPEN JUPYTERLAB** link shown in the preceding screenshot to start the familiar Jupyter environment, which would appear as follows. Click on the **Python 3** icon shown under the **Notebook** heading in the following screenshot:

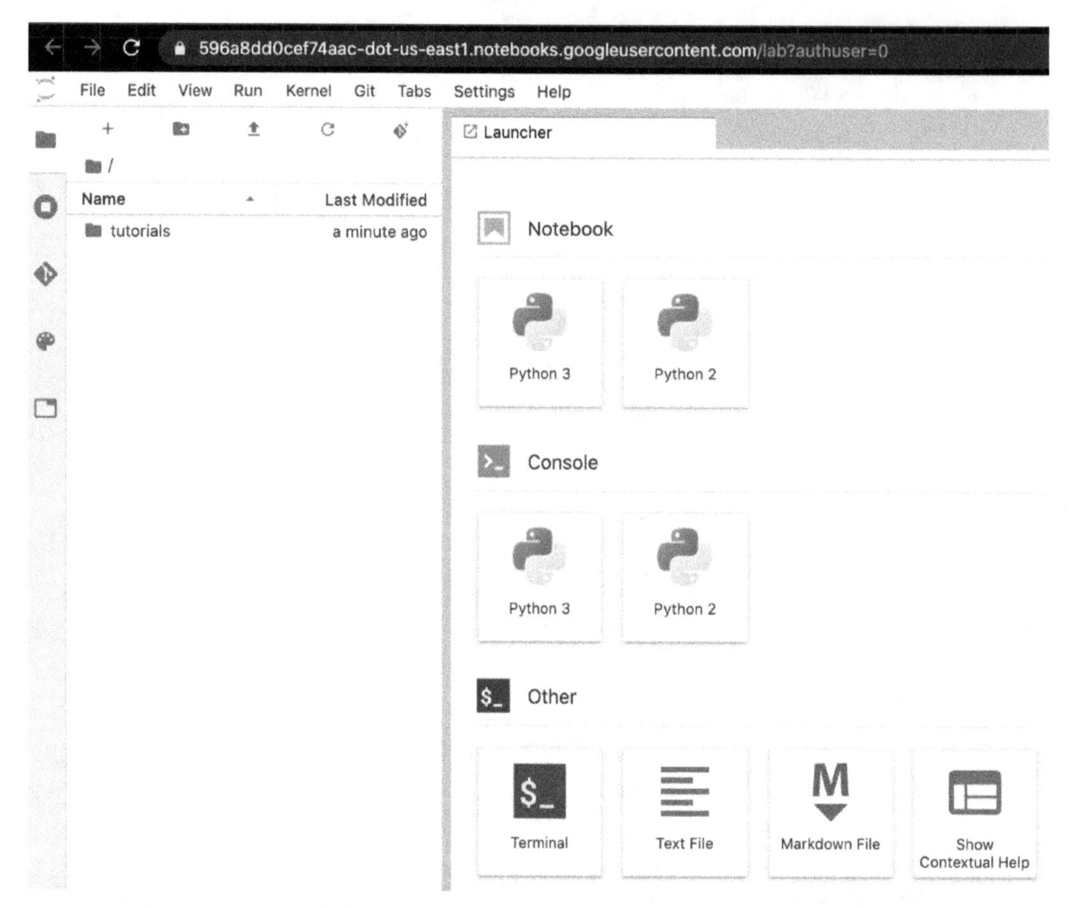

Figure 8.14 – AI Platform notebook options – the Jupyter Notebook environment

Once you select the Python 3 notebook, you will be taken to a new Jupyter notebook – an environment you should be quite familiar with. Here you can write your Python code, import libraries, and do all kinds of data-sciency things. We settled, for the purposes of this demo of course, for keeping it simple and printing **Hello World**. You can see this demonstration in the following figure:

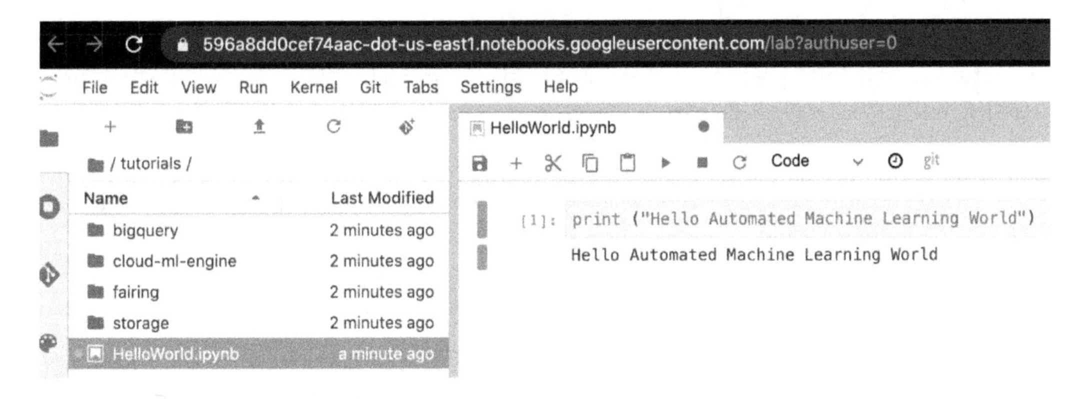

Figure 8.15 – AI Platform notebook instance – running a simple Jupyter notebook

This concludes our basic introduction on how to get started with AI Hub and running a simple notebook, which is the first step into the amazing GCP world. AI Platform is full of amazing tools, and we will be exploring the automated ML part of it in the next section.

Automated ML with Google Cloud

Automated ML is one of the key building blocks of Google Cloud AI Platform. The suite of automated ML products includes AutoML Natural Language, AutoML Tables, AutoML Translation, AutoML Video, and AutoML Vision, as shown in the following diagram:

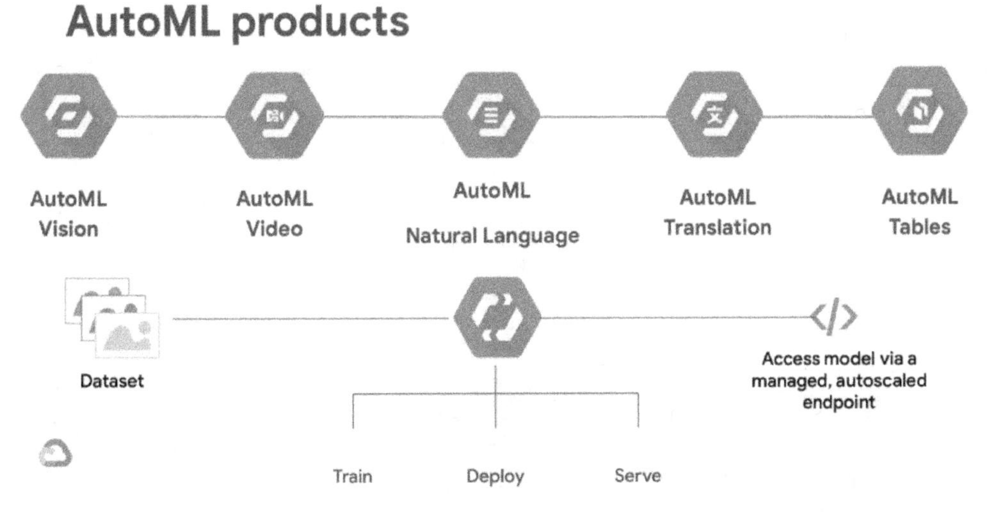

Figure 8.16 – AutoML products offered as part of Google Cloud AI Platform

The underlying components of Google Cloud's automated ML offerings involve neural architecture search and hyperparameter optimization approaches. However, by abstracting out all the intricacies, it is made easy for consumers to use.

Google Cloud AutoML Vision is a computer vision-based capability that helps train ML models on custom labels. You can also perform the same on Edge devices using the AutoML Vision Edge service.

The AutoML Video Intelligence range of products provides classification and object tracking capabilities. Currently in PreGA (beta), you can use these services to train your models to identify specific objects, shots, and segments within videos according to your own custom label definitions. These results can be extrapolated over the rest of the video to discover, detect, and track similar labels.

AutoML Natural Language is an unstructured text analysis service that helps you build, manage, and deploy models dealing with text files and documents. Natural language processing is an area of great interest among industry professionals and researchers alike, and if you plan to do tasks such as single - or multi-label classification entity extraction, or sentiment analysis using customized labels, AutoML makes it very easy.

AutoML Translation is where the automated ML approach meets **Google Neural Machine Translation** (**Google NMT**). Google Cloud AutoML Translation allows you to upload your own datasets and augment the translations. With the calculation of BLEU, base-BLEU, and BLEU gains, AutoML Translation provides a sophisticated environment for custom model development and testing.

AutoML Tables truly manifests what we have read so far about automated ML – taking the power of battle-tested, planet-scale neural algorithms from Google and unleashing it on unstructured data. The Google Cloud AutoML Tables workflow is shown in the following figure:

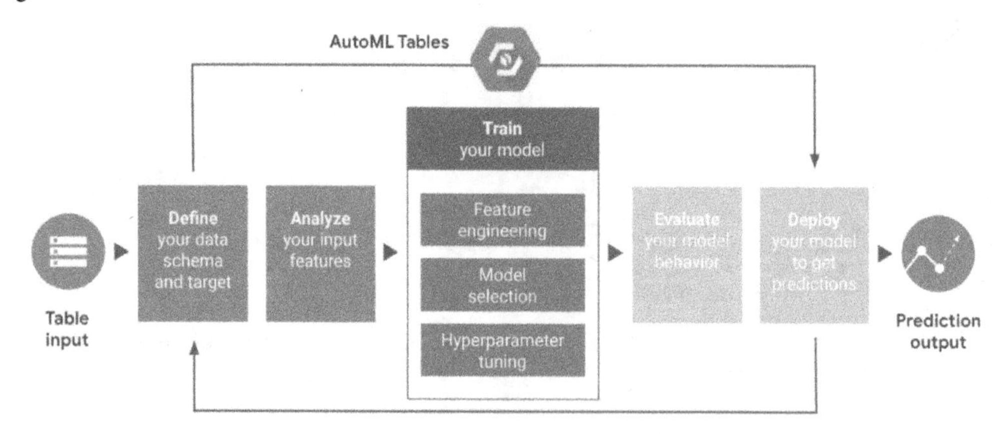

Figure 8.17 – Google Cloud AutoML Tables workflow

Getting the structured data (table input), AutoML Tables jumps into action by analyzing the input features (feature engineering), selects the model (neural architecture search), performs hyperparameter tuning, and repeatedly evaluates the model behavior to ensure consistency, accuracy, and reliability. Google AutoML Tables is widely used in a variety of scenarios, from maximizing revenue to optimizing financial portfolios and understanding customer churn. AutoML Tables fulfills the promise of building and deploying state-of-the-art ML models on structured data, and serves as the crown jewel of the self-service section of the Google Cloud AutoML suite.

Summary

In this chapter, you learned how to get started with Google Cloud AI Platform and learned about AI Hub, how to build a notebook instance, and how to run a simple program. You also learned about the different flavors of automated ML offered by GCP, including AutoML Natural Language, AutoML Tables, AutoML Translation, AutoML Video, and AutoML Vision. If the breadth of GCP offerings, capabilities, and services have left you overwhelmed, you are in good company.

In the next chapter, we will do a deep dive into Google Cloud AutoML Tables. We will build models and explain how the automated ML functionality works with AutoML Tables, that is, how you can take unstructured data and perform automated ML tasks of analyzing the input features (feature engineering), selecting the model (neural architecture search), and doing hyperparameter tuning. We will deploy these models on GCP and test them via web services to demonstrate the operationalization of these capabilities. Stay tuned.

Further reading

For more information on the following topics, you can visit the given links:

- *Using Machine Learning to Explore Neural Network Architecture*:

  `https://ai.googleblog.com/2017/05/using-machine-learning-to-explore.html`

- **Cloud AutoML** – Train high-quality custom machine learning models with minimal effort and machine learning expertise:

  `https://cloud.google.com/automl`

9
Automated Machine Learning with GCP

"The first rule of any technology used in a business is that automation applied to an efficient operation will magnify the efficiency. The second is that automation applied to an inefficient operation will magnify the inefficiency."

-Bill Gates

This has been a long yet rewarding journey of learning about major hyperscalers and how they implement automated machine learning in their respective platforms. In the previous chapter, you learned how to get started with Google Cloud AI Platform, learned about AI Hub, and learned how to build a notebook instance in GCP. You also learned about the different flavors of automated machine learning offered by GCP, including AutoML Natural Language, AutoML Tables, AutoML Translation, AutoML Video, and AutoML Vision.

Continuing with the breadth of GCP offerings, capabilities, and services, we will now do a deep dive into Cloud AutoML Tables. We will build models and explain how automated machine learning works with AutoML Tables; that is, how you can take unstructured data and perform automated machine learning tasks by analyzing the input features (feature engineering), selecting the model (neural architecture search), and performing hyperparameter tuning. We will deploy these models to GCP and test them via web services to demonstrate their operationalization.

In this chapter, we will cover the following topics:

- Getting started with Google Cloud AutoML Tables
- Creating an AutoML Tables experiment
- Understanding AutoML Tables model deployment
- AutoML Tables with BigQuery public datasets
- Automated machine learning for price prediction

Getting started with Google Cloud AutoML Tables

AutoML Tables helps harness the insights in your structured data. In any large enterprise, there are multiple modalities of data, including structured, unstructured, and semi-structured data. For most organizations dealing with databases and transactions, there is indeed a lot of structured data out there. This data is quite suitable for advances analytics, and GCP's AutoML Tables is just the tool to help you automatically build and deploy machine learning models based on structured data.

AutoML Tables enables machine learning engineers and data scientists to automatically build and deploy state-of-the-art machine learning models on structured data faster than anyone could manually do. It automates modeling on a wide range of data types, from numbers and classes to strings, timestamps, lists, and nested fields. Google Cloud AutoML tables make this happen with minimal code. In this chapter, we will learn how to take an exported CSV file, click a few buttons, wait a while, and get a very highly tuned model on the other end.

Google's automated machine learning team has worked hard so that the tool works for a wide variety of data types. Google's AutoML Tables explores that vast space of possible models and hyperparameters so that it can try and optimize things on your behalf. As we explore the examples in this chapter, you will see that the first step is to import your training data via the **Import** tab, give it a name, and select the source – either a table from BigQuery, a file on your machine, or a file on Google Cloud Storage. This first step takes a bit of time as the system analyzes the columns of your dataset. Once it's done this, you'll get to edit the auto-generated schema and select the column for predicting on. Here, you can also update the column type, as well as whether it is nullable.

You can also view datasets that may have a lot of columns to get a nice overview of their data. You can click on the different column names to see some statistics about your columns. After analyzing the data, we can start the training process. This is where AutoML really shines because all you have to do is click **Train**.

There are some options that you can set, including a maximum budget of training hours. This enables you to experiment with your data if you want and limit that training time before committing to a full, longer training run. You'll notice that the training times shown are somewhat on the long side of things. This is because it's not only doing model tuning but also selecting what model to use in the first place. So, as a result, there's a lot of things happening during training. But we don't have to do anything here.

Once the training is completed, we must evaluate and deploy the model. You will be able to see how the training did, as well as the metrics about the model's performance. Finally, we will deploy the model to get predictions. There's even an editor in the browser that will make requests to your endpoint, so you don't need to set up a local environment to make these calls to try and test it out.

Now that you have learned how AutoML Tables works, let's explore it in practice.

Creating an AutoML Tables experiment

AutoML Tables automatically builds and deploys state-of-the-art machine learning models on structured data. Let's start with our first experiment to put this knowledge into practice:

1. Access the Google Cloud AI Platform home page by visiting this link: `https://console.cloud.google.com/home/`. Click on the **Datasets** link in the left pane; you will see the following screen:

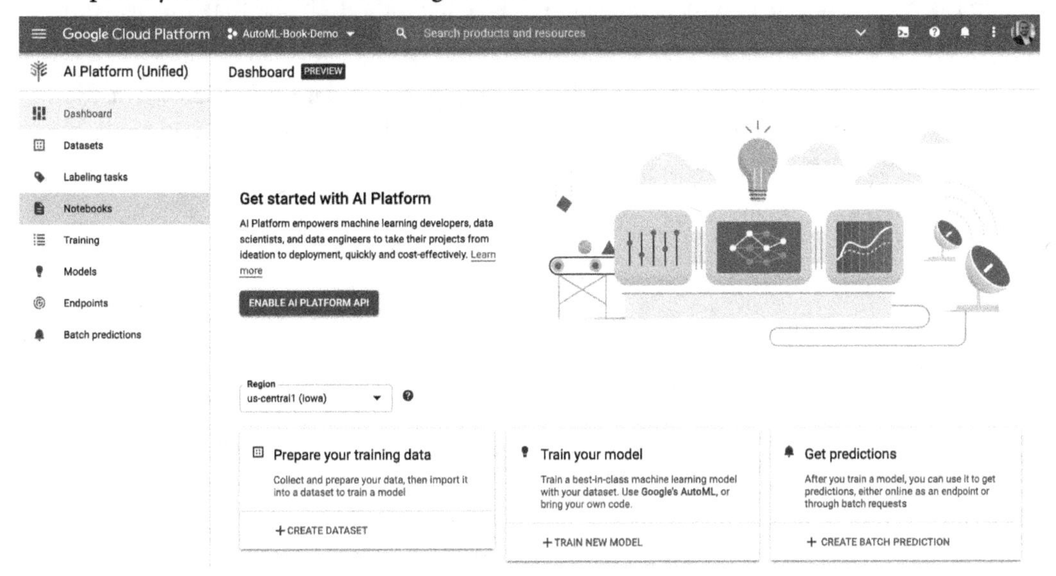

Figure 9.1 – Google Cloud AI Platform home page

2. On the **Google AutoML Tables** main screen, start the process by creating a new dataset. Click on the **NEW DATASET** button to create a new dataset and name it `IrisAutoML`. Then, click on **CREATE DATASET**:

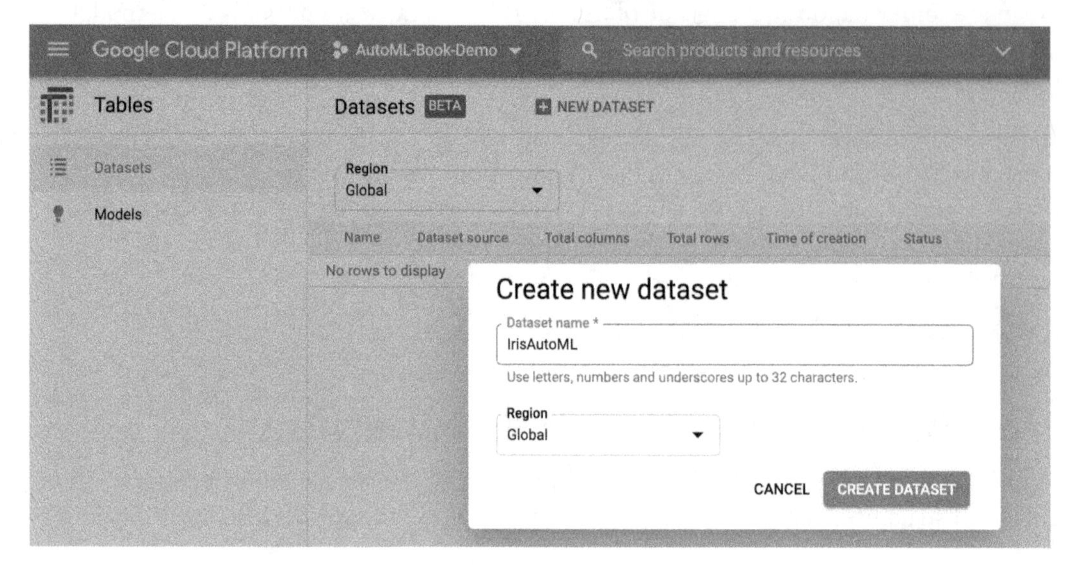

Figure 9.2 – AutoML Tables – Create new dataset screen

3. For this experiment, we will start with the Iris dataset. You can download the CSV file from `https://www.kaggle.com/uciml/iris` since we will be using it in the next step. The dataset is too small to be used for automated machine learning though, but you will see how this unfolds soon.

4. Now, you need to import the data (CSV) file into **Google AutoML Tables**. The file needs to be uploaded to a storage bucket. Select the file from your machine and click on **BROWSE** to create a storage destination on GCP, as shown in the following screenshot:

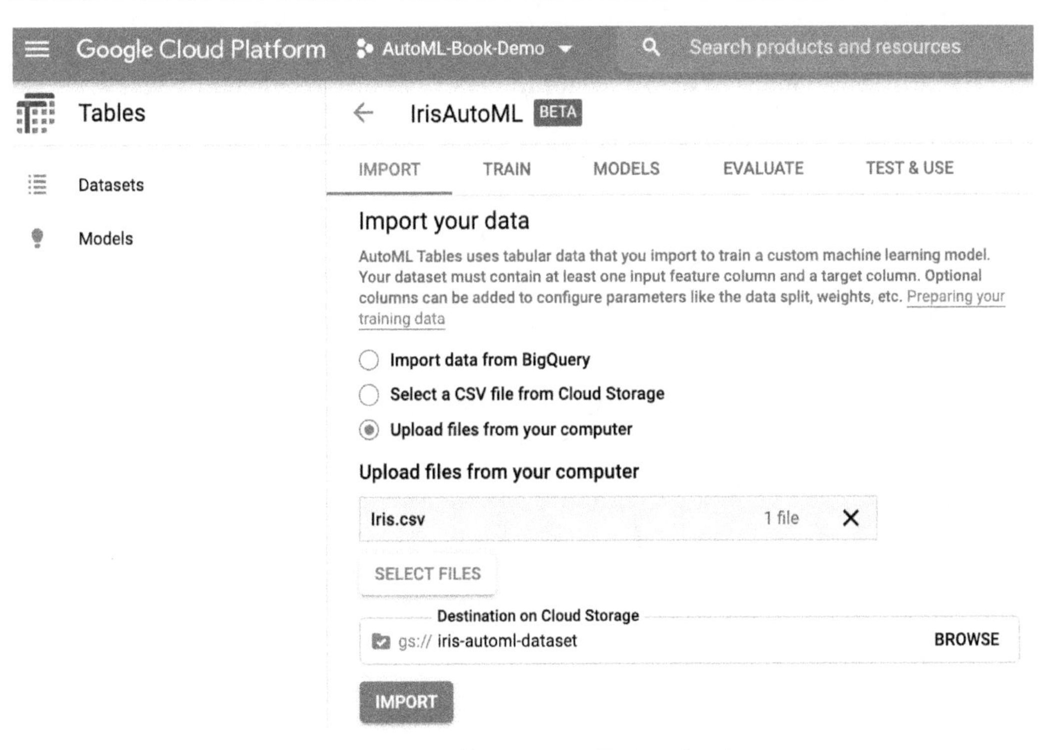

Figure 9.3 – AutoML Tables – import file from local computer

To create a storage bucket, you need to follow these steps.

1. First, provide a name for your storage bucket:

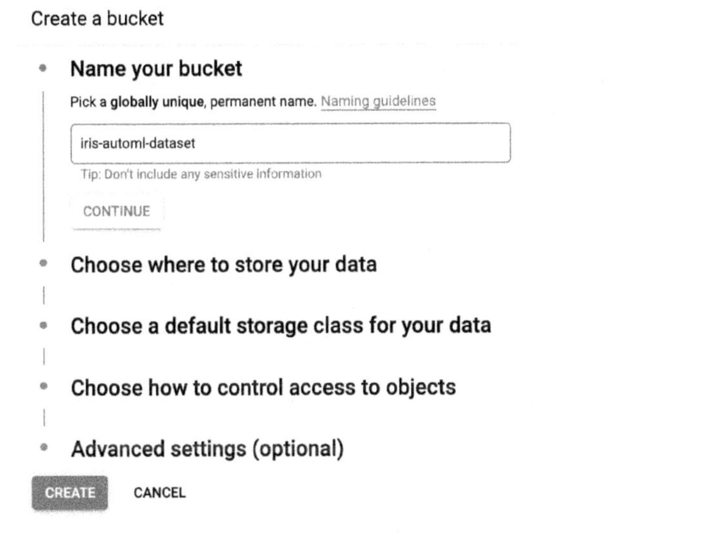

Figure 9.4 – AutoML Tables – creating a storage bucket on GCP

2. Then, choose where you would like to store the data. The options are region (single region), dual region with **high availability** (**HA**), or multi-region for highest availability across multiple locations. For this example, we will choose **us-central1** as a single region, but you can choose another region if it suits you better geographically:

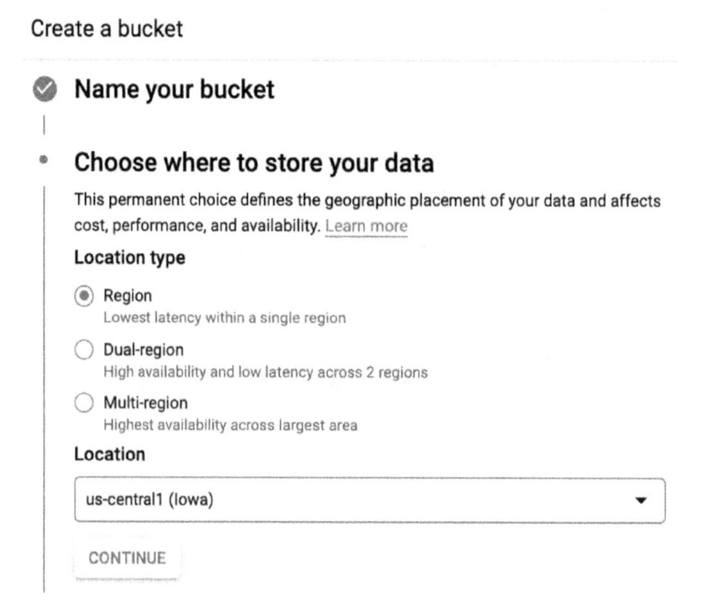

Figure 9.5 – AutoML Tables – choosing a location to create a storage bucket

3. Next, choose the default storage class for the data. The storage classes you can choose from are **Standard**, **Nearline** (backup), **Coldline** (disaster recovery), and **Archive** (for archival use). Choose the **Standard** class for the purpose of this implementation, as seen in the following screenshot:

Create a bucket

Name your bucket

Choose where to store your data

Choose a default storage class for your data

A storage class sets costs for storage, retrieval, and operations. Pick a default storage class based on how long you plan to store your data and how often it will be accessed. Learn more

Standard
Best for short-term storage and frequently accessed data

Nearline
Best for backups and data accessed less than once a month

Coldline
Best for disaster recovery and data accessed less than once a quarter

Archive
Best for long-term digital preservation of data accessed less than once a year

CONTINUE

Figure 9.6 – AutoML Tables – choosing a storage class for your data on GCP

4. Finally, the encryption settings need to be configured. Here, you can provide your own key or use the default setting of using Google's managed key. Click on **CREATE** to complete the process of making the bucket:

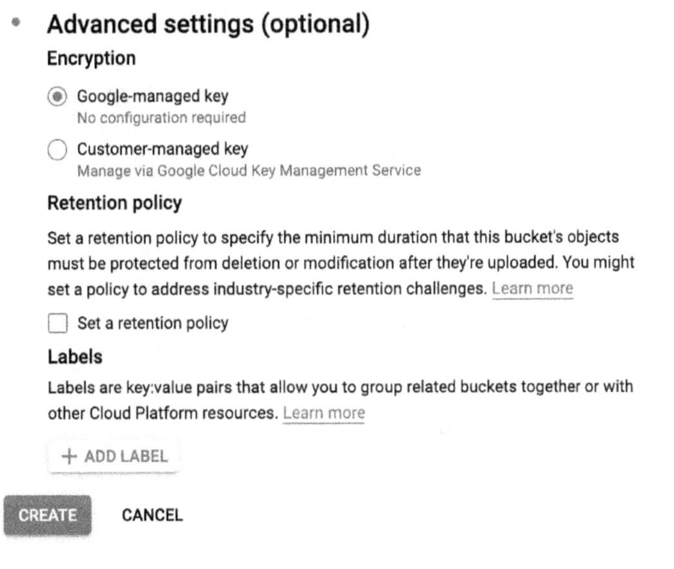

Advanced settings (optional)
Encryption
Google-managed key
No configuration required
Customer-managed key
Manage via Google Cloud Key Management Service
Retention policy
Set a retention policy to specify the minimum duration that this bucket's objects must be protected from deletion or modification after they're uploaded. You might set a policy to address industry-specific retention challenges. Learn more
Set a retention policy
Labels
Labels are key:value pairs that allow you to group related buckets together or with other Cloud Platform resources. Learn more
+ ADD LABEL
CREATE
CANCEL

Figure 9.7 – AutoML Tables – choosing encryption settings

This will trigger the bucket being created and the data being imported. You will see the following screen:

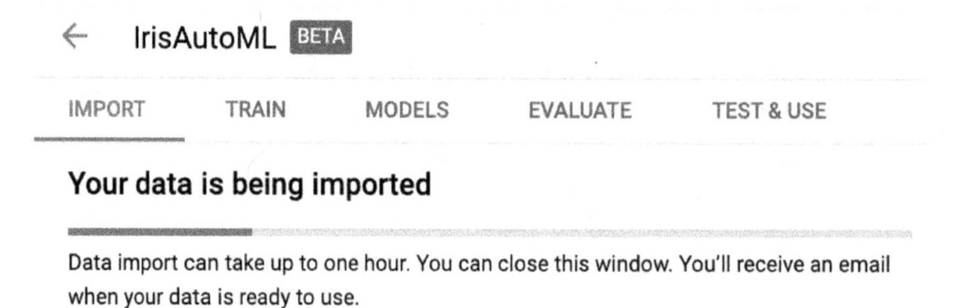

Figure 9.8 – AutoML Tables – data being imported into the GCP bucket

This is where we will learn an important lesson: not all data is suitable for automated machine learning. Once the import process is complete, you will see the following error message:

Error details

Operation ID: projects/262569142203/locations/us-central1/operations/TBL993971155893223424

Error Messages: Too few rows: 150. Minimum number is: 1000

Figure 9.9 – AutoML Tables – too few rows error

Even though you can do this experiment with other tools, cloud automated machine learning platforms set a minimum bar to ensure the quality of their algorithms is not compromised. This example provides us with a critical lesson that not all problems are automated machine learning-worthy.

Let's repeat the same experiment with a larger dataset – the loan risk dataset – which contains 21 fields and 1,000 instances. You can download it from BigML (`http://bml.io/W2SpyF`, BigML, Inc. Corvallis, Oregon, USA, 2011). This dataset was created by Dr. Hans Hofmann at Institut für Statistik und Ökonometrie, Universität Hamburg, and contains fields such as checking status, duration, credit history, purpose, credit amount, and savings status. These can be used to create a model that will predict the level of the risk of a loan application.

Let's carry out the aforementioned steps for the loan risk dataset by creating a bucket:

1. Download the dataset from `http://bml.io/W2SpyF`. Click on the **CREATE DATASET** button and import the loan risk dataset:

Create new dataset

Dataset name *

AutoMLCredit

Use letters, numbers and underscores up to 32 characters.

Region

Global

CANCEL CREATE DATASET

Figure 9.10 – AutoML Tables – choosing a new dataset bucket

2. As the import process starts, upload the csv file extracted from the dataset that we downloaded in step 5 file and point it to the destination cloud storage by clicking the **SELECT FILES** button:

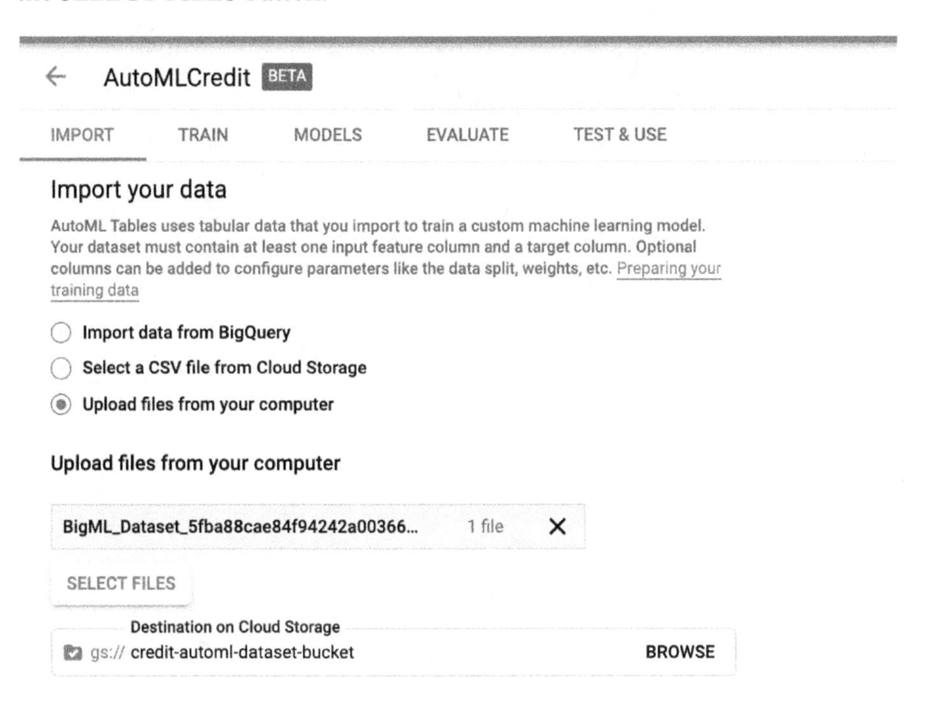

Figure 9.11 – AutoML Tables – choosing storage for data

Since the loan dataset meets the required size limitations, it is imported successfully, and you will see the following training screen. This is where you can edit the auto-generated schema and select the column for predicting on. You can also update the column type, as well as whether it is nullable.

This screen gives you a detailed overview of the dataset. You can click on the different column names to see some statistics about your columns:

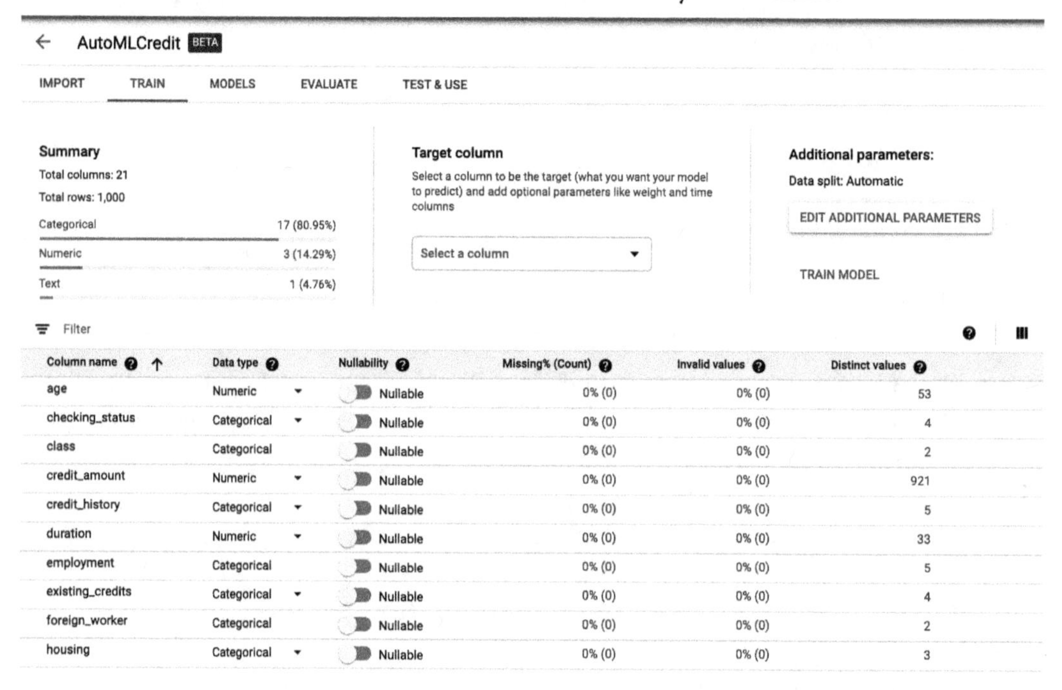

Figure 9.12 – AutoML Tables – training screen

3. Now, select the target column; that is, the column to predict, which is the class. The class is a categorical field with two possible values: good credit or bad credit. This determines whether this specific individual qualifies for credit. Once you've selected the class, click on **TRAIN MODEL**:

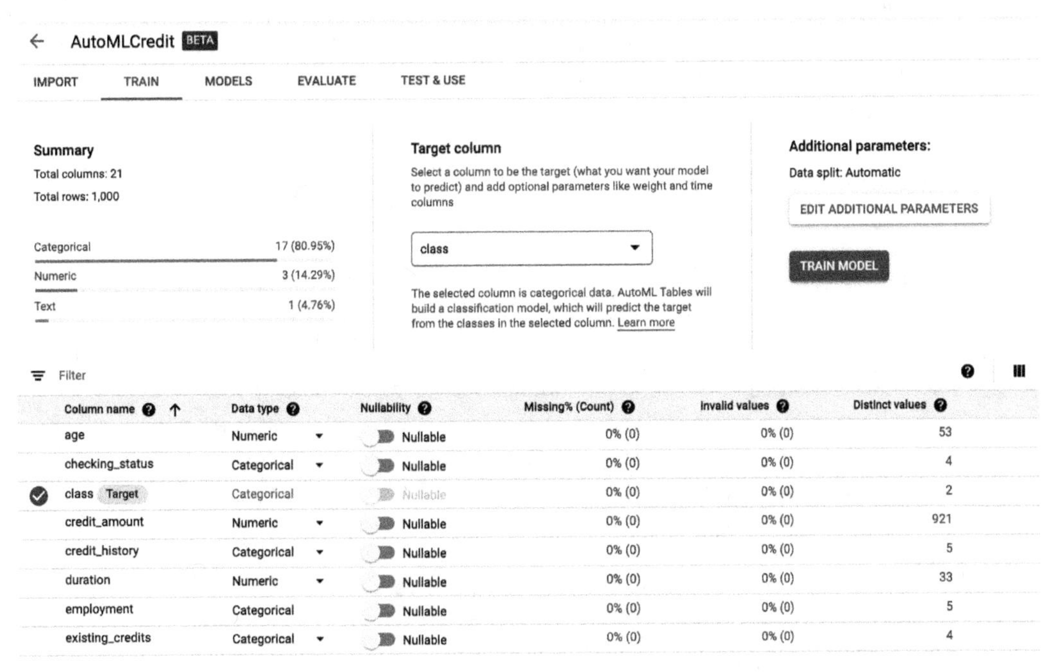

Figure 9.13 – AutoML Tables – selecting the target column for training

Upon clicking on the **TRAIN MODEL** button, you will see a fluid menu on the right. The menu can be seen in the following screenshot. This is where you can set experimental parameters and is where AutoML really shines because all you have to do is click **Train**. There are some options that you can set, including a maximum budget for training hours. This enables you to experiment with your data if you want and limit that training time before committing to a full, longer training run. You'll notice that the training times shown are somewhat on the long side of things. This is because it's not only doing model tuning but also selecting what model to use in the first place:

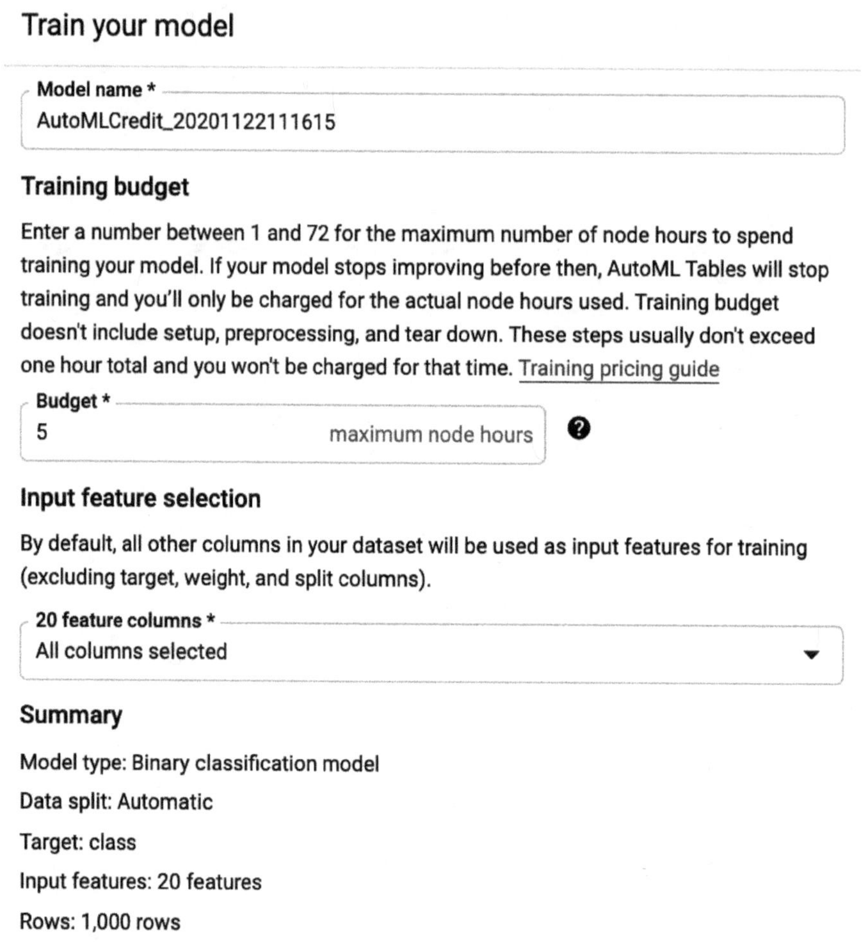

Train your model

Model name *

AutoMLCredit_20201122111615

Training budget

Enter a number between 1 and 72 for the maximum number of node hours to spend training your model. If your model stops improving before then, AutoML Tables will stop training and you'll only be charged for the actual node hours used. Training budget doesn't include setup, preprocessing, and tear down. These steps usually don't exceed one hour total and you won't be charged for that time. Training pricing guide

Budget *

5 maximum node hours

Input feature selection

By default, all other columns in your dataset will be used as input features for training (excluding target, weight, and split columns).

20 feature columns *

All columns selected

Summary

Model type: Binary classification model

Data split: Automatic

Target: class

Input features: 20 features

Rows: 1,000 rows

Figure 9.14 – AutoML Tables – choosing a storage class for your data on GCP

How long should you train your model for? The suggested training time from GCP is defined at `https://cloud.google.com/automl-tables/docs/train`, as shown in the following screenshot:

Rows	Suggested training time
Less than 100,000	1-3 hours
100,000 - 1,000,000	1-6 hours
1,000,000 - 10,000,000	1-12 hours
More than 10,000,000	3 - 24 hours

Figure 9.15 – AutoML Tables – suggested training times

You can find the respective pricing guide at `https://cloud.google.com/automl-tables/pricing`.

You can also review the advanced options, which is where you can see the optimization objectives for the experiment. Since this is a classification experiment, the objectives listed include **AUC ROC**, **Log loss**, **AUC PR**, **Precision**, and **Recall**. The **Early stopping** toggle ensures that upon detecting that no more improvements can be made, it stops the process. Otherwise, AutoML Tables will continue training until the budget is met.

Click on **TRAIN MODEL** to start this operation:

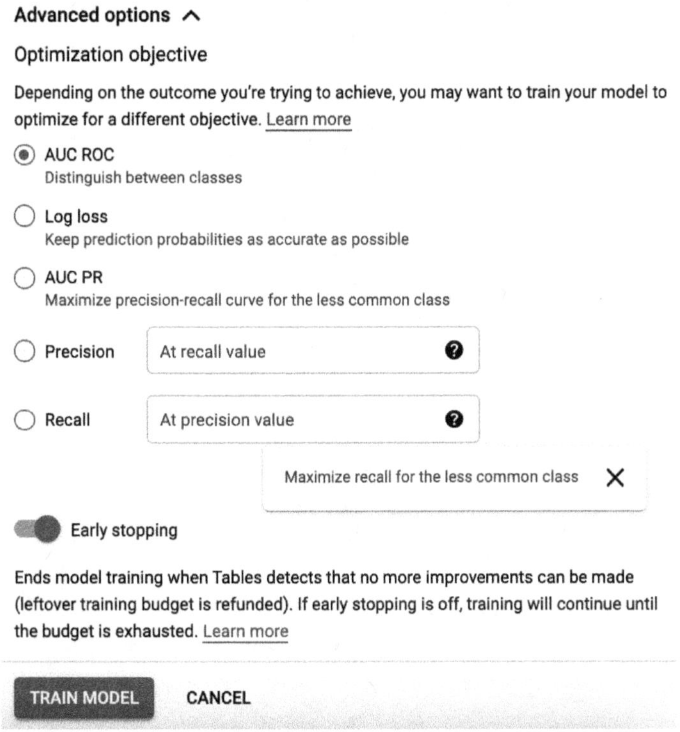

Advanced options ^

Optimization objective

Depending on the outcome you're trying to achieve, you may want to train your model to optimize for a different objective. Learn more

(◉) AUC ROC
Distinguish between classes

( ) Log loss
Keep prediction probabilities as accurate as possible

( ) AUC PR
Maximize precision-recall curve for the less common class

( ) Precision | At recall value

( ) Recall | At precision value

Maximize recall for the less common class ✕

Early stopping

Ends model training when Tables detects that no more improvements can be made (leftover training budget is refunded). If early stopping is off, training will continue until the budget is exhausted. Learn more

TRAIN MODEL CANCEL

Figure 9.16 – AutoML Tables – advanced options for training

Upon starting the experiment, you will see the following screen. It's initiated when the infrastructure is set up, as well as when the model is eventually trained:

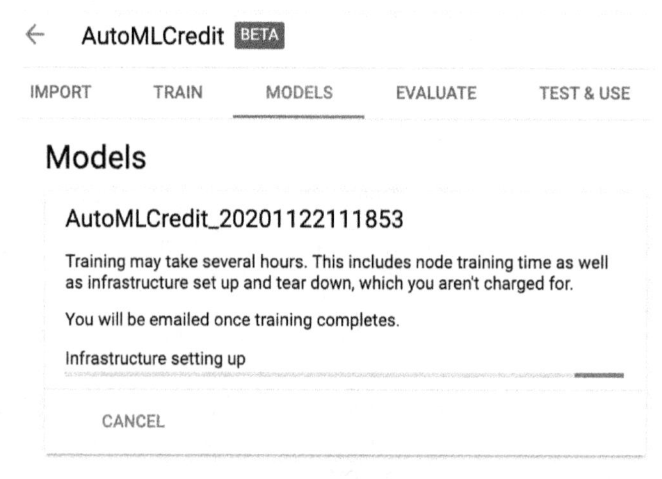

Figure 9.17 – AutoML Tables – starting the AutoML Tables experiment

Once the training is completed, we must evaluate and deploy the model. At this stage, you will be able to see how the training did, as well as any metrics about the model's performance. Finally, you can deploy the model to get predictions about credit worthiness:

Figure 9.18 – AutoML Tables – choosing a storage class for your data on GCP

The accuracy is measured as a percentage, while the area under the **precision recall** (**PR**) curve ranges from `0` to `1`. Training the model with different training costs (duration) will get you a higher value, which indicates a higher-quality model:

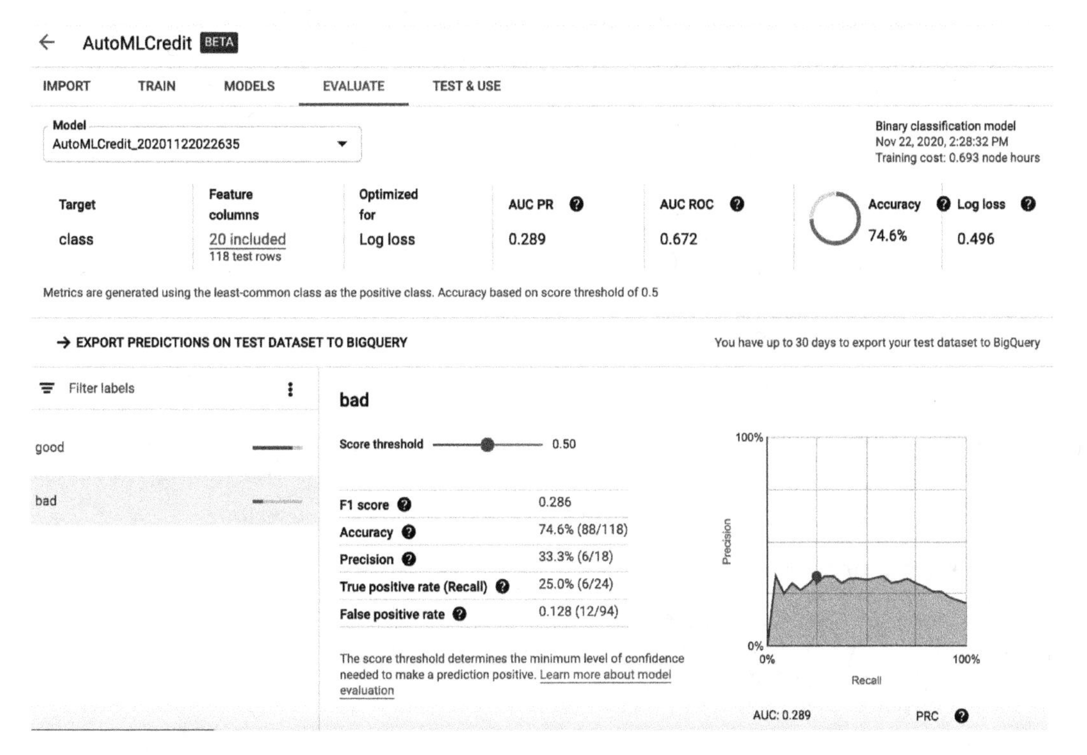

Figure 9.19 – AutoML Tables – details of the trained model, including its F1 score, accuracy, and precision

The following page also shows the confusion matrix, which shows the quality of the model on the data; that is, how many data-points were predicted correctly, compared to how many were incorrectly predicted:

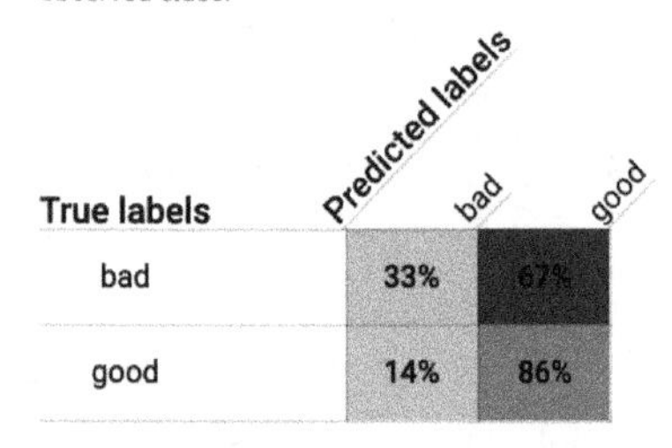

Confusion matrix
A confusion matrix helps you understand where misclassifications occur (which classes get "confused" with each other). Each row is a predicted class and each column is an observed class. The cells of the table indicate how often each classification prediction coincides with each observed class.
Predicted labels
bad
good
True labels
bad
33%
67%
good
14%
86%

Figure 9.20 – AutoML Tables – confusion matrix

Feature importance – that is, which feature has the biggest impact on the resulting model – is also shown. In this case, you can observe that the checking status, the duration of credit, and the purpose seem to have the most impact on the credit decision:

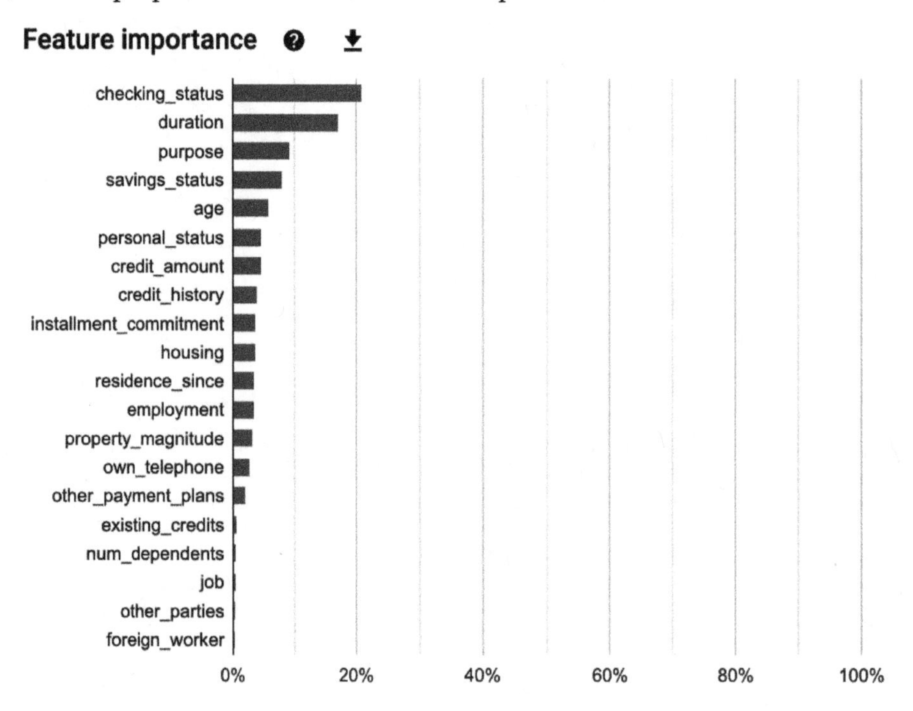

Figure 9.21 – AutoML Tables – feature importance table

Now that we've trained the model, let's proceed with its deployment.

Understanding AutoML Tables model deployment

In order to deploy the model that we trained in the previous section, perform the following steps:

1. We must click on the **TEST & USE** tab to deploy the model. There are multiple ways of testing the trained model: you can either test it as a batch prediction (file-based), as an online prediction (API), or export it in a Docker container. The option at the top of the page lets you toggle between online predictions via the REST API and batch predictions. This allows you to upload a CSV file or point to a BigQuery table and get prediction results for that entire file or table. Considering the amount of time it takes to use, AutoML Tables enables you to achieve a much higher level of model performance than you could reach manually. We will be doing online API-based prediction in this section:

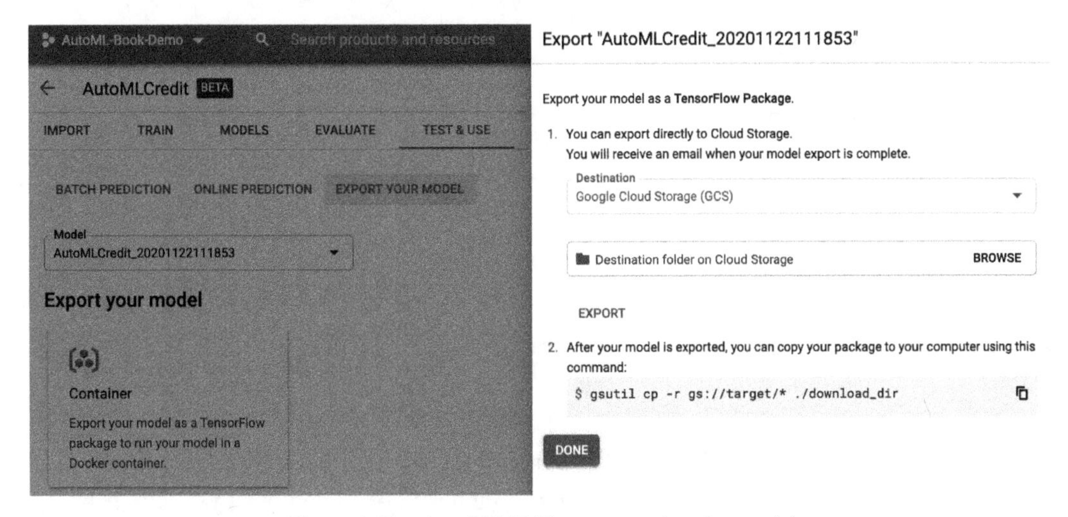

Figure 9.22 – AutoML Tables – exporting the model

2. Click on the **ONLINE PREDICTION** tab. You will see the following screen. Here, you can call the API right from the console:

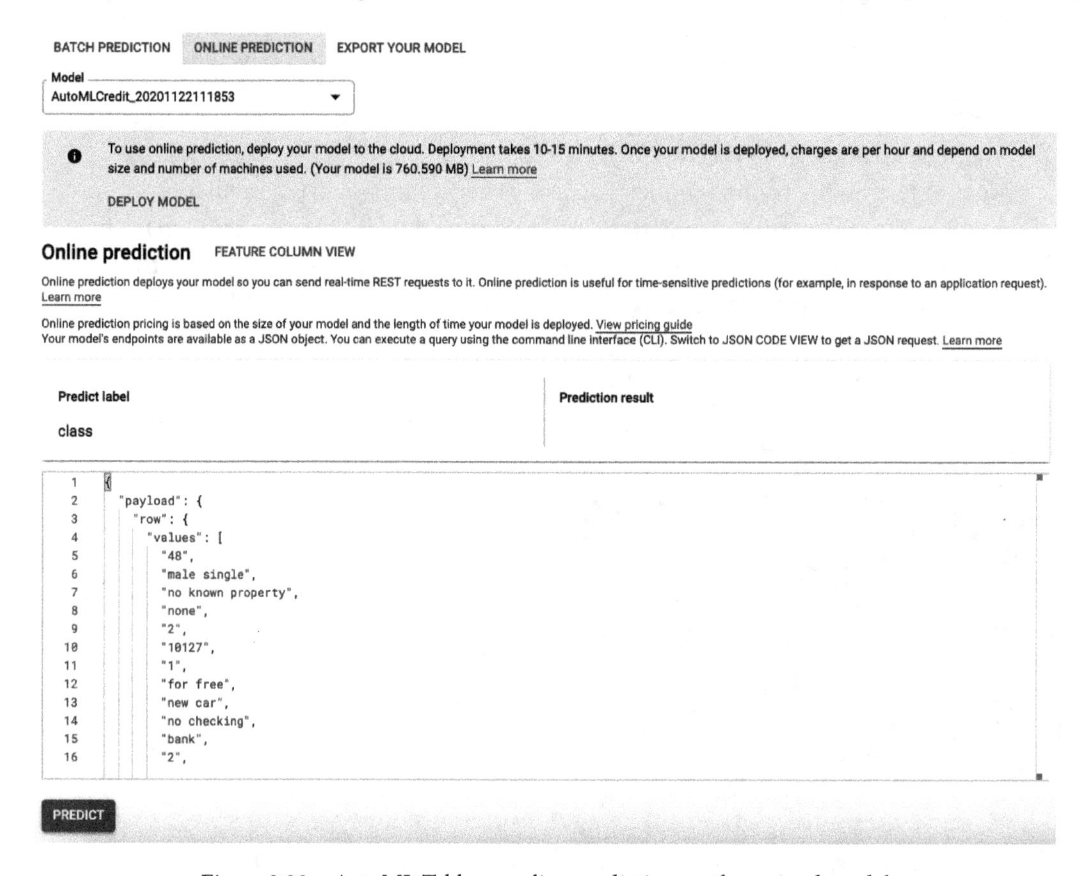

Figure 9.23 – AutoML Tables – online prediction on the trained model

3. However, if you just click on **PREDICT**, it will give you the error shown in the following screenshot. Why? Because the model hasn't been deployed yet, which means there is no endpoint to call:

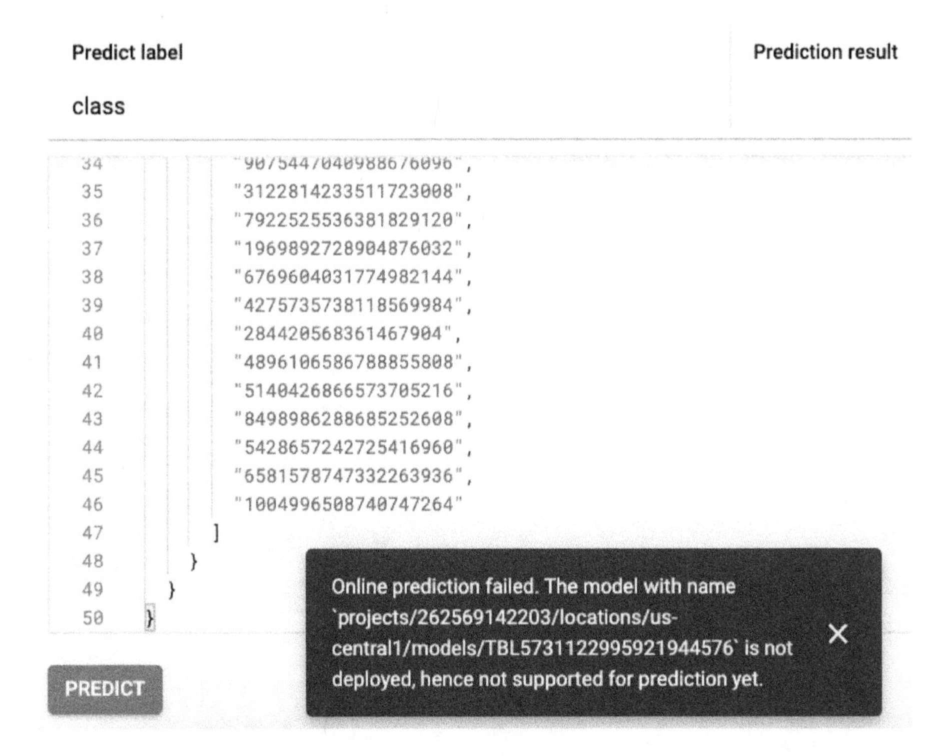

Figure 9.24 – AutoML Tables – online prediction on the trained model's error

4. Click on the **Deploy model** button. You will see the following popup, confirming the deployment details. Now, click on **DEPLOY**:

Figure 9.25 – AutoML Tables – Deploying the trained model popup

This starts the process of model deployment. Once completed, you will see the following screen, which states that the model has been successfully deployed and is available for requests, along with the size it takes up. You should remember that the model is now running on a server, so you will be spending compute and storage costs associated with running the model. This is what the prior warning was about.

5. At this point, you can go ahead and click on the **PREDICT** button:

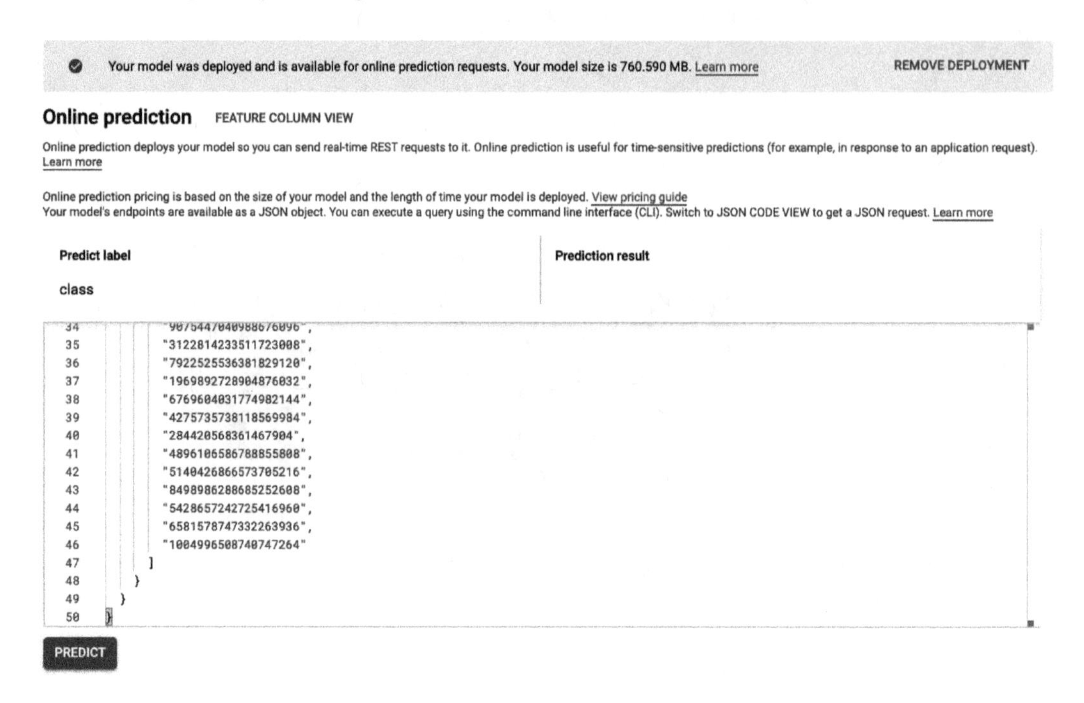

Figure 9.26 – AutoML Tables – calling the prediction API

It will pass the JSON request to the API and invoke the prediction function. This function will return the response, along with the prediction confidence score, as shown in the following screenshot:

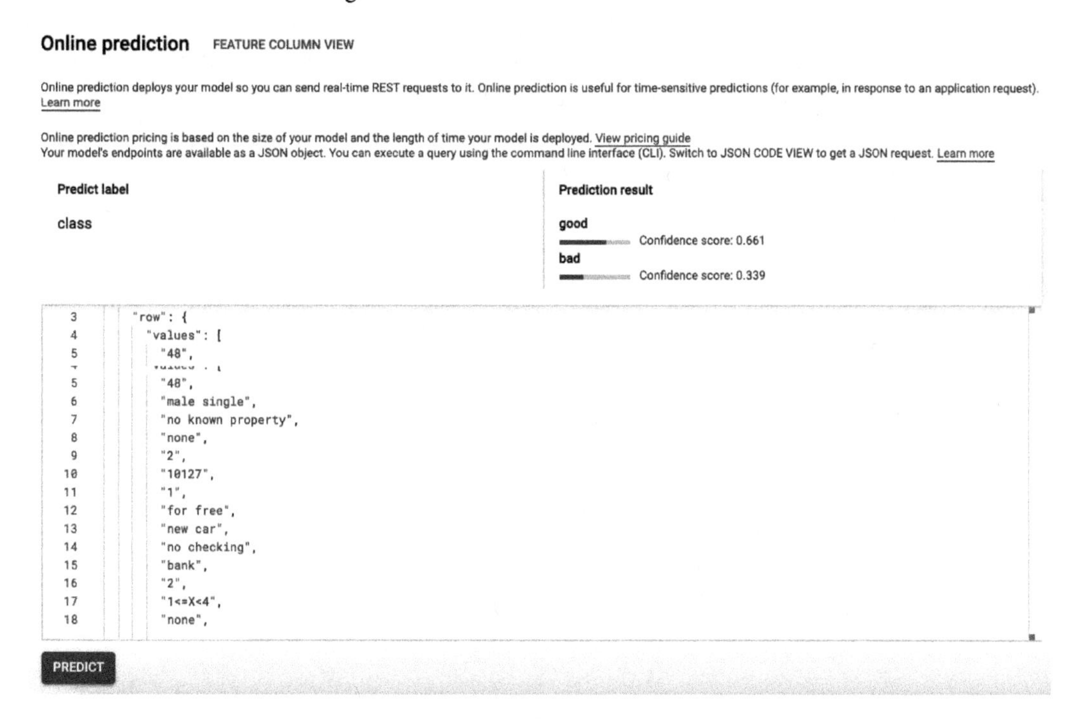

Figure 9.27 – AutoML Tables – response from the online prediction API

The preceding screenshot shows the results of the model. It has a good credit response since it has a `0.661` confidence score. At this point, you can switch to the feature column view and edit some of the parameters. We intuitively known that age and the duration of the credit have a significant impact on our credit results. Lowering the age from `48` to `18` and increasing the credit term to `60` in the editable form turns this good credit decision into a bad one.

6. Let's change these values and invoke the API again. You will see that the results have changed to bad ones, as seen in the following screenshot:

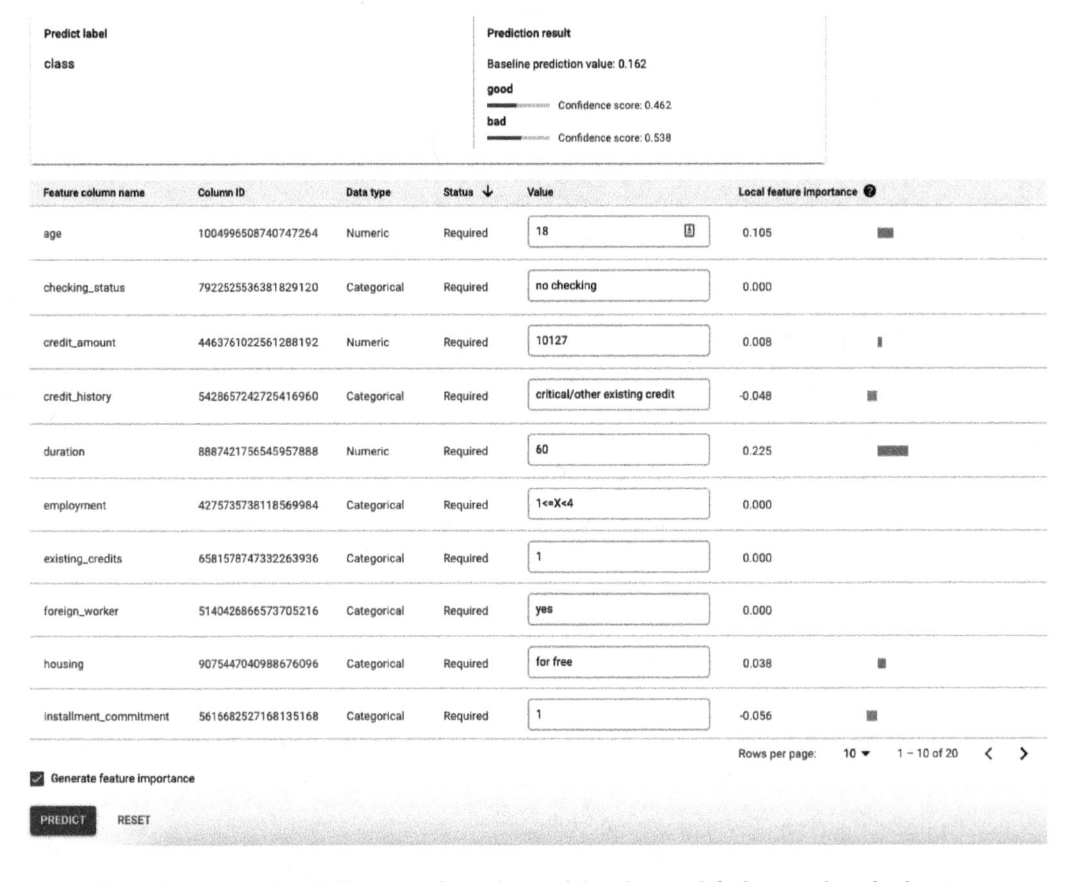

Figure 9.28 – AutoML Tables – invoking the model with a modified age and credit duration

The preceding experiments have shown you how to train, deploy, and test a model. Now, let's explore how to use BigQuery-based public datasets with AutoML Tables.

AutoML Tables with BigQuery public datasets

Data has been called the new oil of the digital economy. To extend this analogy, automated machine learning is the engine that uses data to provide advanced analytics without custom manual plumbing each time, but I digress. Real-world data for performing machine learning experiments comes from various organizations, though counterparts are needed to perform experiments and try out hypotheses. Such a data repository is the Google BigQuery cloud data warehouse – specifically, its large collection of public datasets. In this example, we will use BigQuery, one of the three methods specified in the data ingestion process for AutoML Tables, for our experiment.

Like the loan dataset we used earlier, the adult income dataset is a public dataset derived from the 1994 United States Census Bureau and uses demographic information to predict the income of two classes: above or below $50,000 per year. The dataset contains 14 attributes, with the target fields being income and number of attributes. The data can be downloaded from `https://www.kaggle.com/wenruliu/adult-income-dataset?select=adult.csv`. However, BigQuery contains a repository of popular public datasets, so we will use that instead. Let's get started:

1. As we did previously, click on **Create new dataset** in the **Tables** tab and click on the **CREATE DATASET** button:

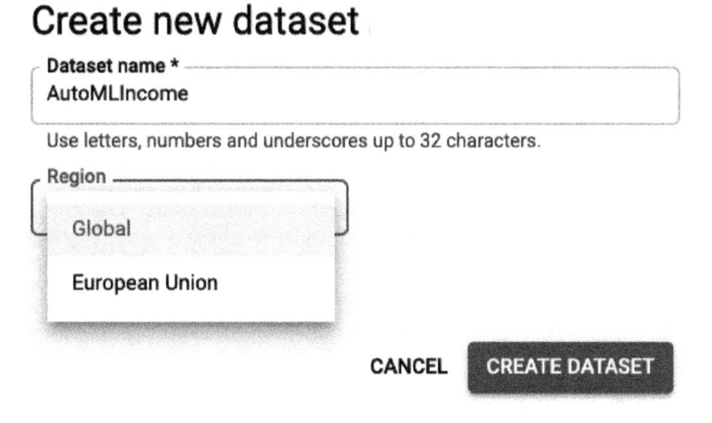

Figure 9.29 – AutoML Tables – Create new dataset prompt

2. Now, to add to the dataset, select the third option – that is, **Select a table or view from BigQuery** – as shown in the following screenshot:

Add data to your dataset

Before you begin, read the data guide to learn how to prepare your data. Then choose a data source:

- **CSV file**: Can be uploaded from your computer or on Cloud Storage. Learn more
- **Bigquery**: Select a table or view from BigQuery. Learn more

Select a data source

- ○ Upload CSV files from your computer
- ○ Select CSV files from Cloud Storage
- ◉ Select a table or view from BigQuery

Figure 9.30 – AutoML Tables – selecting the BigQuery data

3. BigQuery can be accessed at `https://console.cloud.google.com/bigquery`. This is where you can view the datasets it contains. You can do this by calling the following query:

```
SELECT  * FROM  `bigquery-public-data.ml_datasets.census_
adult_income`
```

You will see the following output. Our goal is to export this data to a bucket where it can be used for our experiment. Set the destination table for the query result as a dataset and click on **Run**:

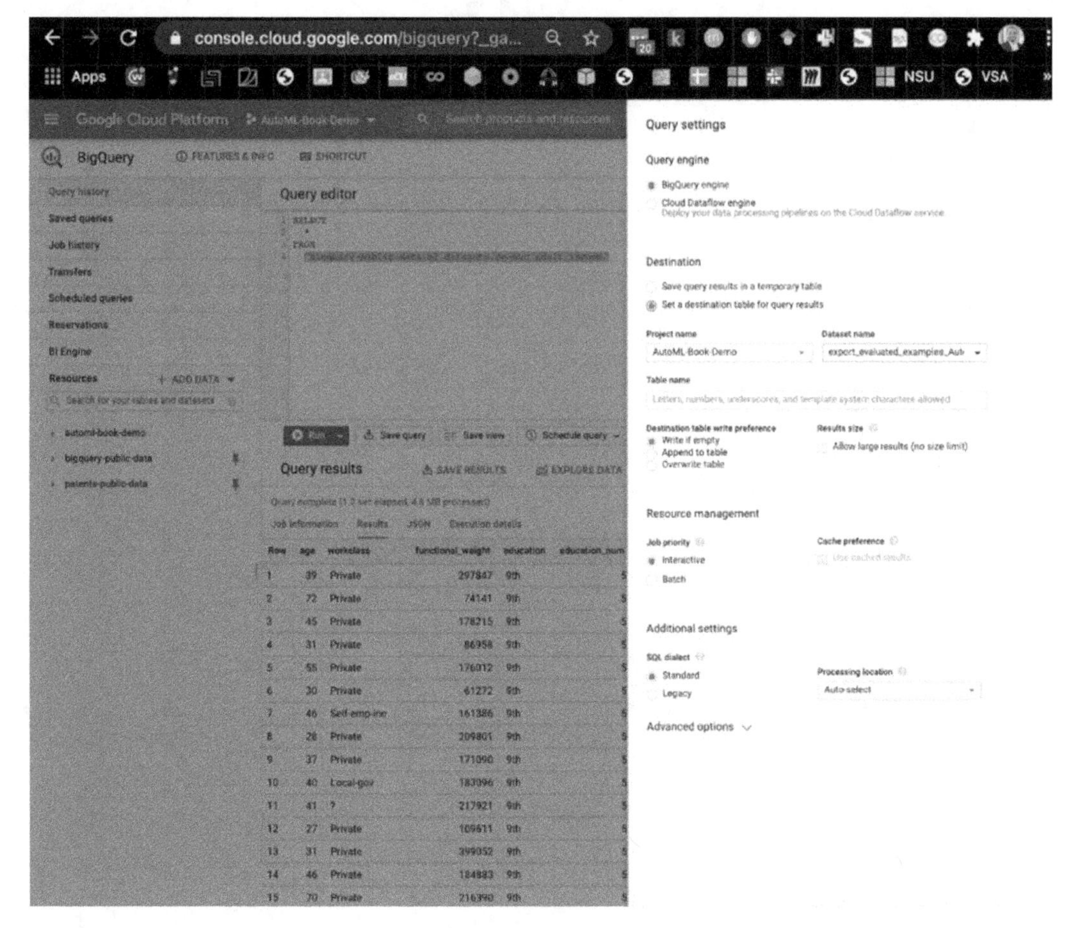

Figure 9.31 – AutoML Tables – BigQuery search results from the census adult income dataset

The following is a brief list of BigQuery public datasets. This makes using these curated datasets quite accessible and easy to use across the entire GCP suite of products:

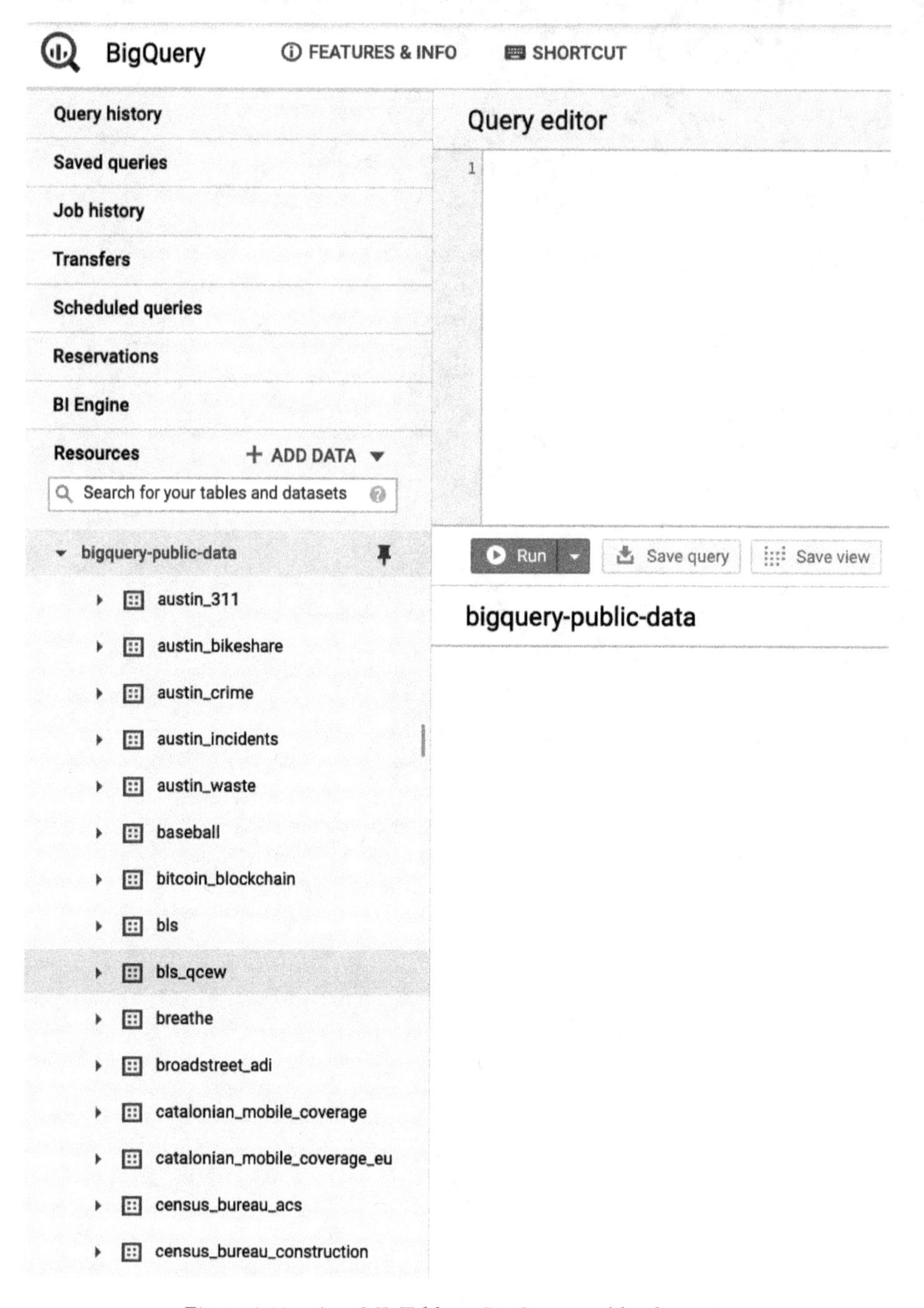

Figure 9.32 – AutoML Tables – BigQuery public datasets

In the previous step, you ran the query, which is now completed, and created a dataset, as shown in the following screenshot:

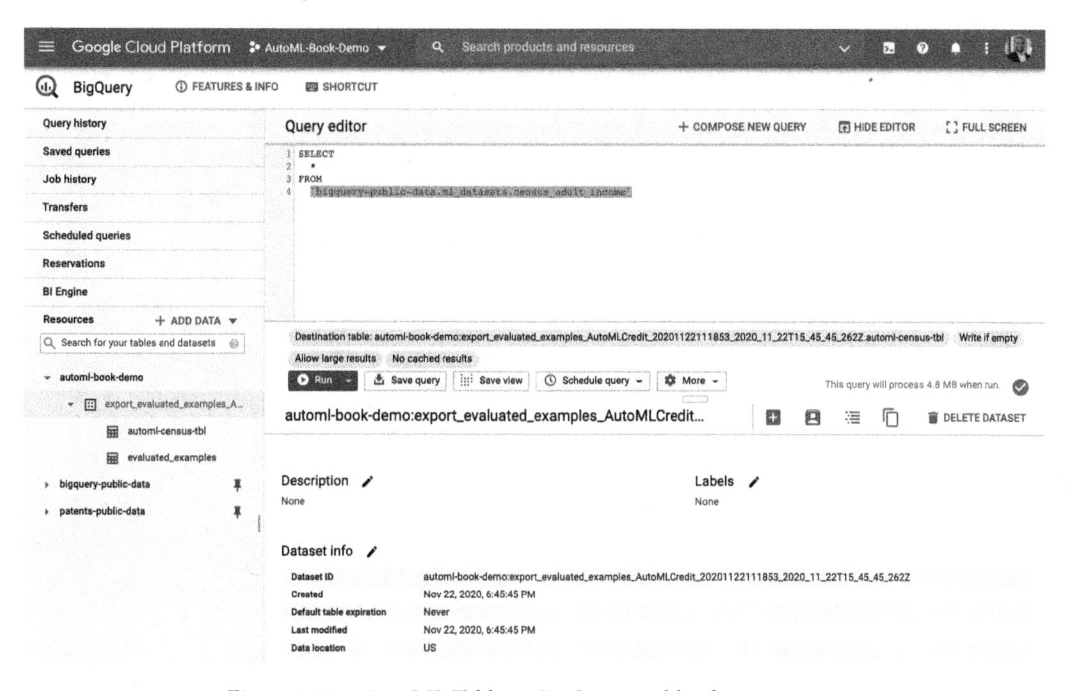

Figure 9.33 – AutoML Tables – BigQuery public dataset export

Now that the data has been exported to a bucket, you can use it to experiment in AutoML Tables.

Automated machine learning for price prediction

So far, you have seen how AutoML Tables can be used for classification problems; that is, finding classes in a dataset. Now, let's do some regression; that is, predicting values. To do this, we will use the house sales prediction dataset. The King County house sales dataset contains prices for King County, which includes Seattle. The dataset can be downloaded from Kaggle at `https://www.kaggle.com/harlfoxem/housesalesprediction`.

For this experiment, our goal is to predict a house's sale value (price) by using 21 features and 21,613 observations or data points:

1. Let's start in AI Platform by clicking on the **CREATE DATASET** button on the main page:

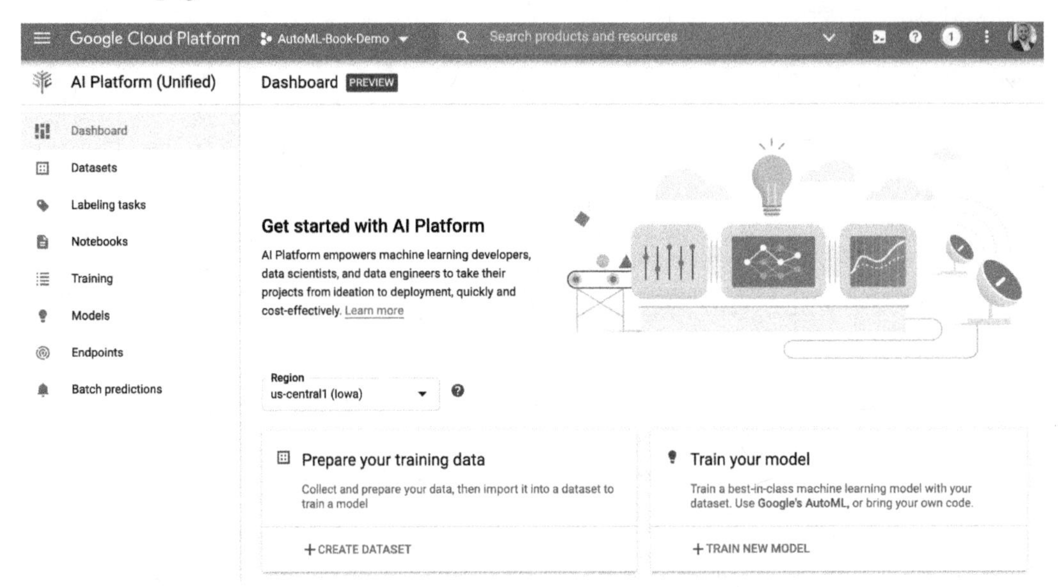

Figure 9.34 – AutoML Tables – getting started with the AI Platform home page

Here, you must choose a dataset name and region, as shown in the following screenshot. Set the dataset's type to tabular since it currently has classification and regression automated machine learning capabilities and click **CREATE**:

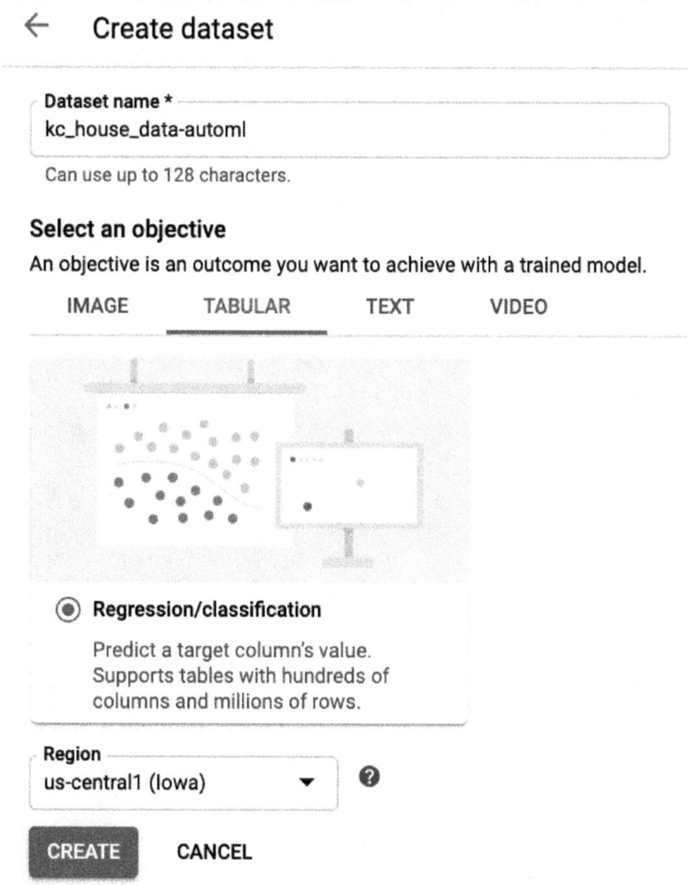

Figure 9.35 – AutoML Tables – selecting the automated machine learning objective

2. Upon clicking **CREATE**, you will see the following screen. Select the **Upload CSV files from your computer** option and upload it to cloud storage by pressing **CONTINUE**:

kc_house_data-automl
SOURCE
ANALYZE
Add data to your dataset
Before you begin, read the data guide to learn how to prepare your data. Then choose a data source:
• CSV file: Can be uploaded from your computer or on Cloud Storage. Learn more
• Bigquery: Select a table or view from BigQuery. Learn more
Select a data source
Upload CSV files from your computer
Select CSV files from Cloud Storage
Select a table or view from BigQuery
Upload CSV files from your computer
Add up to 500 CSV files per upload. The files will be stored in a new Cloud Storage bucket (charges apply). Data from multiple files will be referenced as one dataset.
kc_house_data.csv
1 file
SELECT FILES
Select a Cloud Storage path
Choose where your uploaded CSV files will be stored (charges apply)
Cloud Storage path
gs:// automl-zillow-pricing-ds
BROWSE
What happens next?
The CSV file data will be uploaded to Cloud Storage and associated with your dataset. Making changes to the referenced CSV files will affect the dataset before training.
CONTINUE
$625,000
$975,000
You can build two model types with tabular data. The model type is automatically chosen based on the data type of your target column.
• Regression models predict a numeric value. For example, predicting home prices or consumer spending.
• Classification models predict a category from a fixed number of categories. Examples include predicting whether an email is spam or not, or classes a student might be interested in attending.

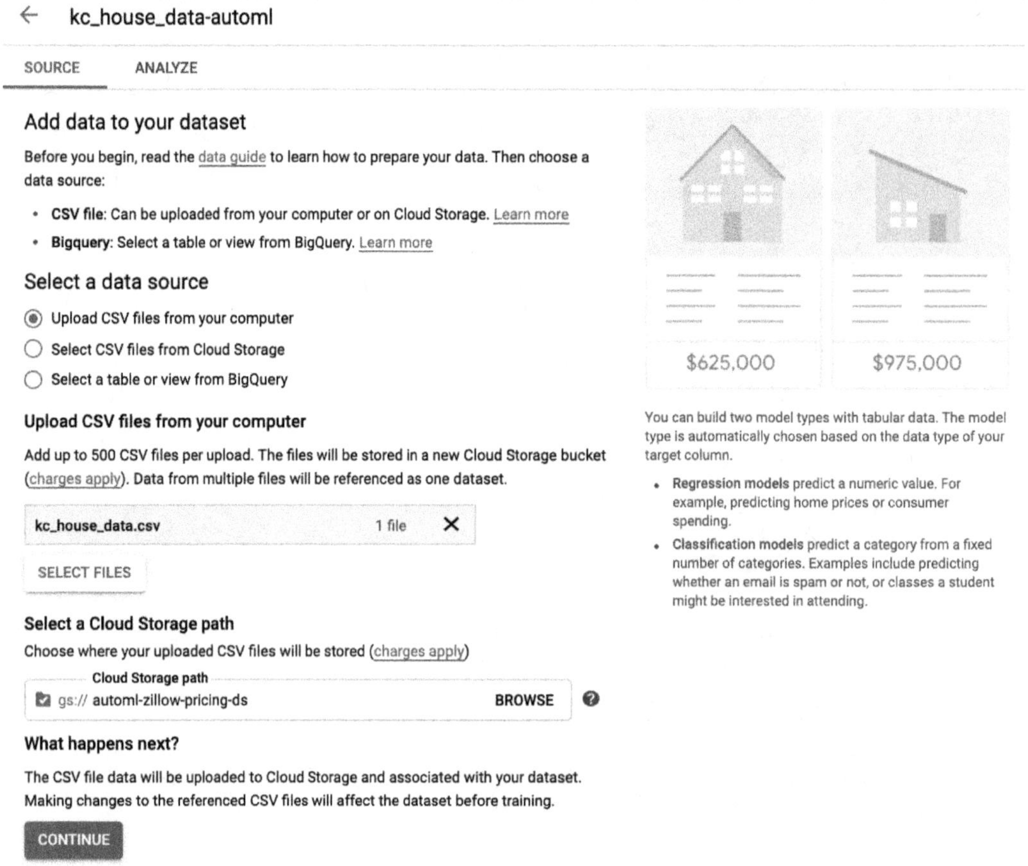

Figure 9.36 – AutoML Tables – choosing storage parameters for your data

3. Upon clicking **CONTINUE**, the dataset will be uploaded and you will see the following screen, which shows a description of the data. Click on **TRAIN NEW MODEL**:

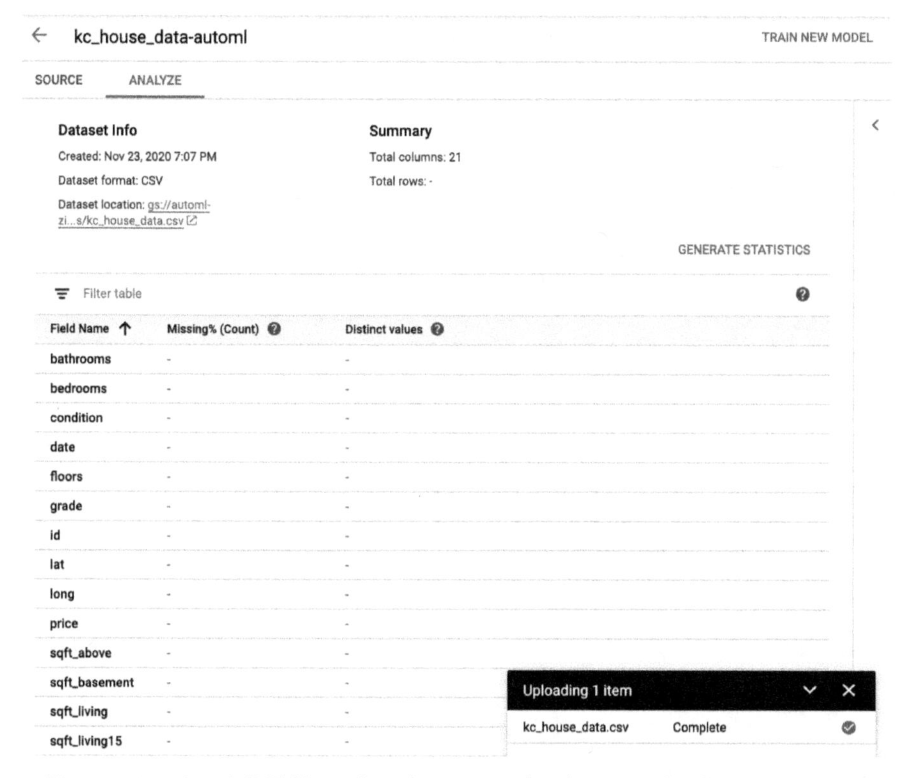

Figure 9.37 – AutoML Tables – data description details once uploading is completed

4. At this point, you are at the training the new model workflow. Here, set the objective to `Regression` and the method to **AutoML**. Then, press **CONTINUE**:

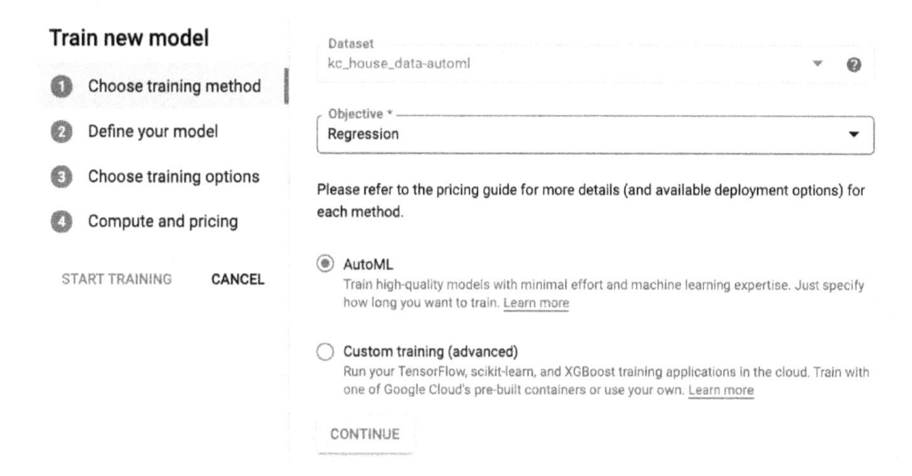

Figure 9.38 – AutoML Tables – Train new model steps

5. Next, we must edit the model definition. Here, choose the target column (the price to be predicted) and the data split; that is, how you would like the test and training data to be split. The default option of random assignment is a good choice unless you have a specific need to do manual or chronological stratification. Click **CONTINUE** to proceed:

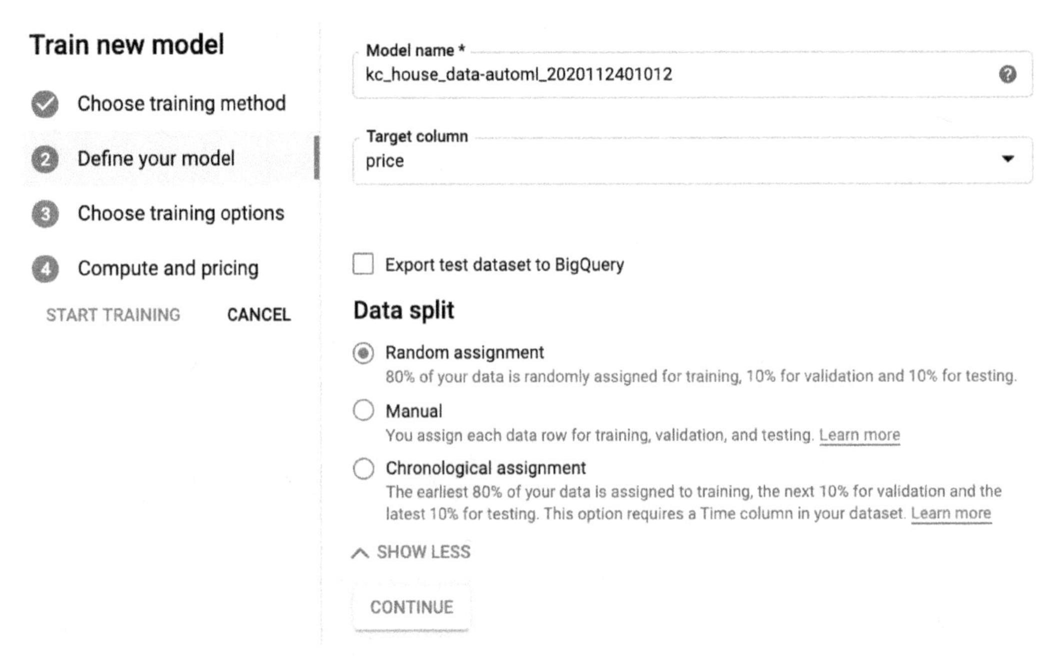

Figure 9.39 – AutoML Tables – Train new model steps

The following screenshot gives you the option to perform granular actions with data, such as removing or filtering columns, applying transformations, and more:

Figure 9.40 – AutoML Tables – description of the dataset upon being uploaded

6. You can also choose optimization objectives. The choices are **root mean square error** (RMSE), **mean absolute error** (MAE), or **root mean square log error** (RMSLE), which is robust against the outliers. Select **RMSE (Default)** and click on **CONTINUE** to proceed:

Optimization objective

- (●) RMSE (Default)
  Capture more extreme values accurately
- ( ) MAE
  View extreme values as outliers with less impact on the model
- ( ) RMSLE
  Penalize error on relative size rather than absolute value. Especially helpful when both predicted and actual values can be quite large.

^ SHOW LESS

CONTINUE

Figure 9.41 – AutoML Tables – description of the optimization objectives

7. The final thing we must look at before we start training is the training budget. This step should be familiar to you from our earlier experiments. Set a budget of 5 hours and click on **START TRAINING**. Do not forget to toggle on **Enable early stopping** – we don't want to exhaust the budget if the results are reached earlier:

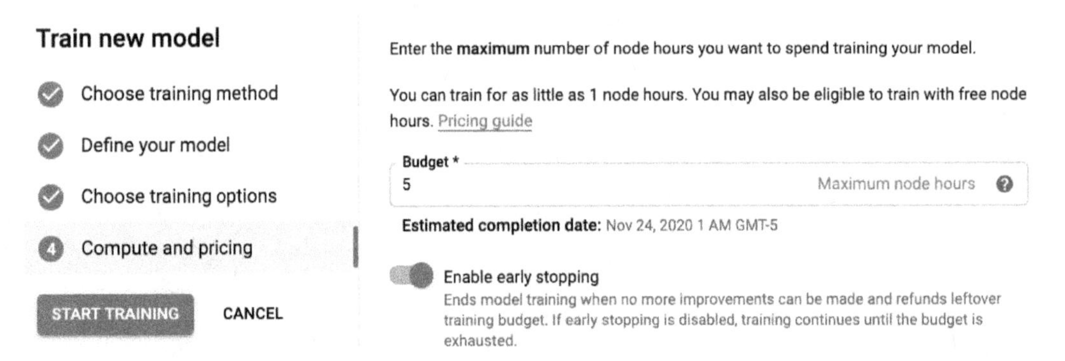

Figure 9.42 – AutoML Tables – training the new model's compute and price step

The model will start training. You will be able to see its progress in the **Training jobs and models** side panel:

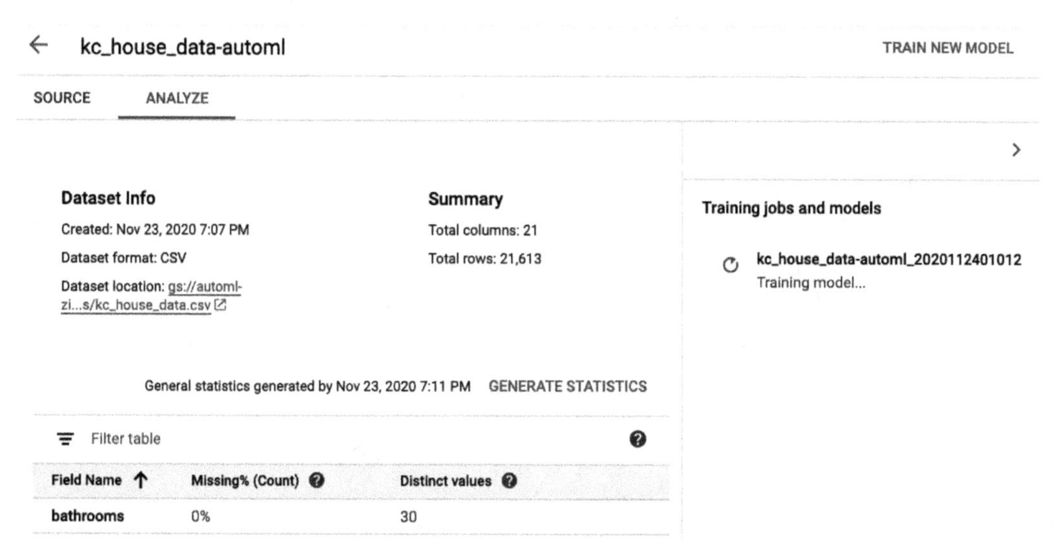

Figure 9.43 – AutoML Tables – new model training started

This specific model took 1 hour and 35 minutes to train. The following screen will appear once it's completed. This screen will show you the status attributes and training performance of the model:

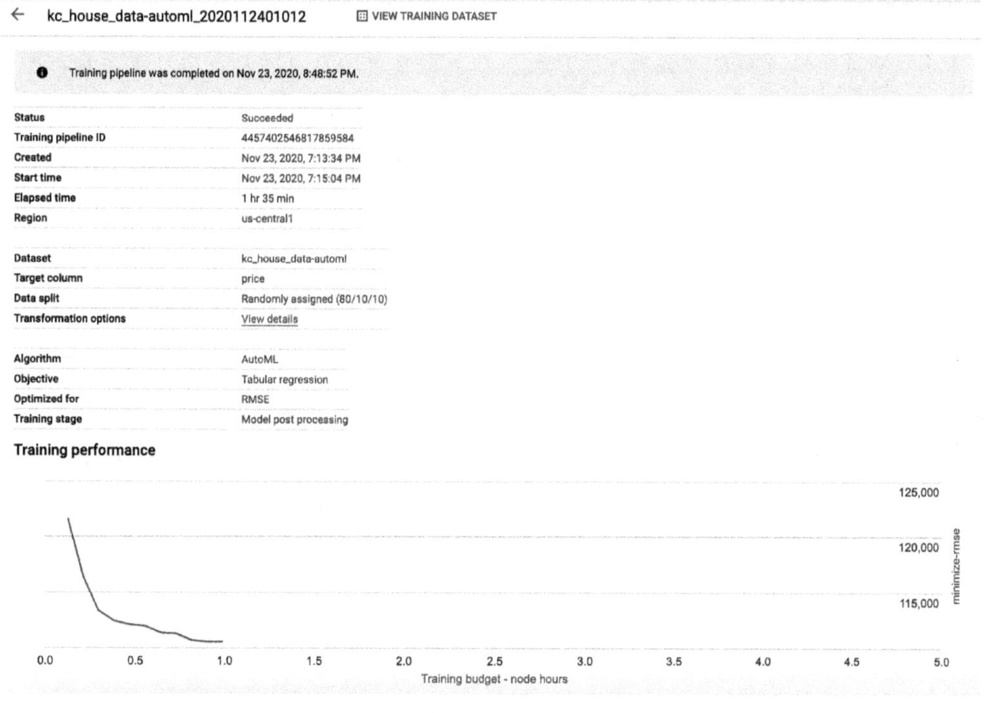

Figure 9.44 – AutoML Tables – training performance and results

Scroll down the **Training performance and results** page to view the feature importance chart for this model. This chart proves the age-old adage of real estate – location, location, location – to be correct. Also, the price of the property and the square feet of living space are closely related. This is not surprising either:

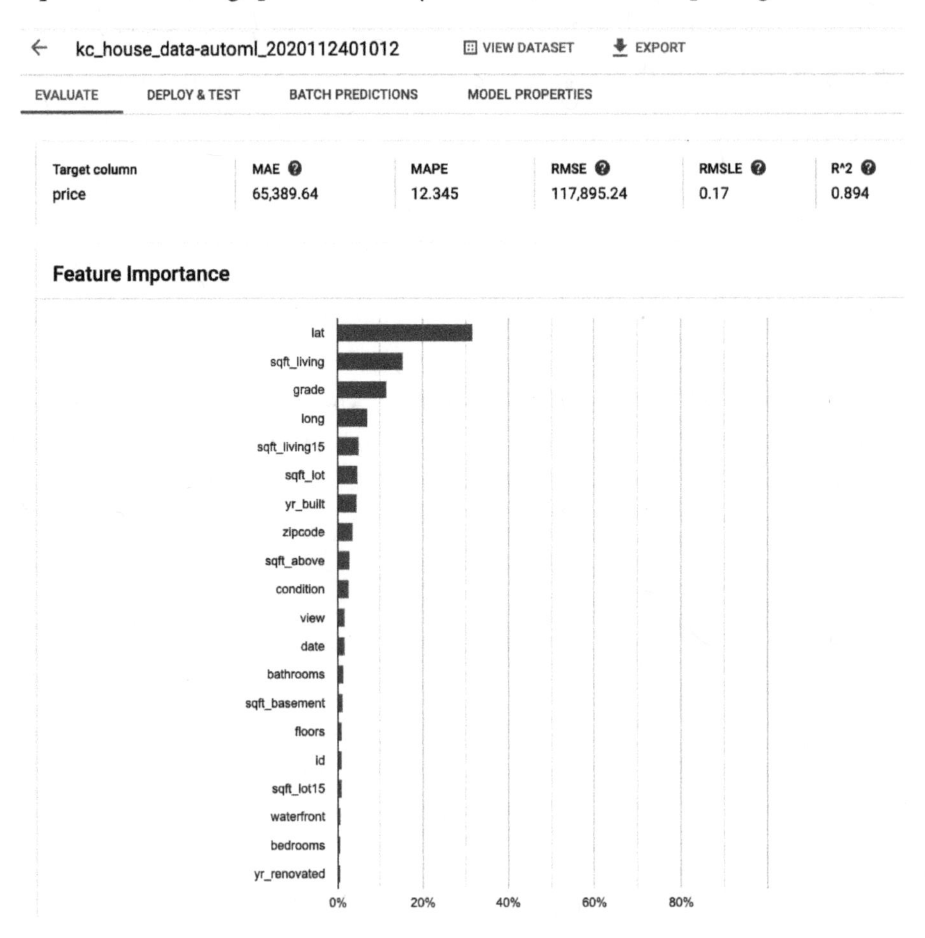

Figure 9.45 – AutoML Tables – results and their feature importance

8. At this point, you can deploy and test the model by clicking on **DEPLOY & TEST**, as shown in the following screenshot:

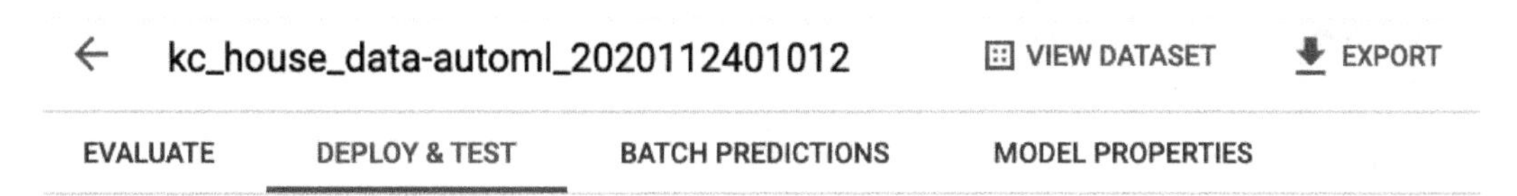

Use your edge-optimized model
Container
Export your model as a TF Saved Model to run on a Docker container.
Deploy your model
Endpoints are machine learning models made available for online prediction requests. Endpoints are useful for timely predictions from many users (for example, in response to an application request). You can also request batch predictions if you don't need immediate results.
DEPLOY TO ENDPOINT

Figure 9.46 – AutoML Tables – deploying the model to an endpoint

In the several experiments we've conducted in this section, we have found that the size of the data was a substantial factor for improved accuracy. As the number of observations in a dataset increases, automated machine learning can perform a better neural architecture search and hyperparameter optimization to get the best results.

Summary

In this chapter, you learned how to perform automated machine learning using AutoML Tables. We started by setting up a Cloud AutoML Tables-based experiment and then demonstrated how the AutoML Tables model is trained and deployed. Using multiple data sources, we explored AutoML Tables with BigQuery public datasets, as well as both classification and regression. We hope that this chapter has made you familiar with working with GCP AutoML so that you can apply it to your automated machine learning experiments.

In the next chapter, we will explore an enterprise use case for automated machine learning.

Further reading

For more information regarding what was covered in this chapter, please refer to the following links:

- AutoML Tables beginner's guide:

  `https://cloud.google.com/automl-tables/docs/beginners-guide`

- AutoML Tables notebooks:

  `https://cloud.google.com/automl-tables/docs/notebooks`

Section 3: Applied Automated Machine Learning

This final section deals with real-world usage scenarios of automated machine learning, examining problems and case studies. In this section, the applied enterprise needs of data cleaning, feature engineering, and automation are discussed in detail with AutoML's transparency and explainability offerings.

This section comprises the following chapter:

- *Chapter 10, AutoML in the Enterprise*

10
AutoML in the Enterprise

"Harnessing machine learning can be transformational, but for it to be successful, enterprises need leadership from the top. This means understanding that when machine learning changes one part of the business — the product mix, for example — then other parts must also change. This can include everything from marketing and production to supply chain, and even hiring and incentive systems."

- Erik Brynjolfsson, Director of the MIT Initiative on the Digital Economy

Automated **Machine Learning** (**ML**) is an enabler and an accelerator that unleashes the promise for organizations to expedite the analytics life cycle without having data scientists as a bottleneck. In earlier chapters, you learned how to perform automated ML using multiple hyperscalers, including open source tools, AWS, Azure, and GCP.

This chapter, however, is quite different since here we will explore the enterprise use of automated ML. We will look into applying automated ML in real-world applications and discuss the pros and cons of such approaches. Model interpretability and transparency are areas of great interest in automated ML. We will explore models of trust in ML, providing a playbook for applying automated ML in an enterprise.

In this chapter, we will cover the following topics:

- Does my organization need automated ML?
- Automated ML – an accelerator for enterprise advanced analytics
- Automated ML challenges and opportunities
- Establishing trust – model interpretability and transparency in automated ML
- Introducing automated ML in an organization
- Call to action – where do I go next?

Does my organization need automated ML?

Technology decision-makers and stakeholders don't like fads, and you probably shouldn't either. Building and using technology for the sake of technology has limited business value in a vertical enterprise; the technology has to solve a business problem or provide an innovative differentiation to be relevant. Therefore, this inquiry becomes very significant: does an organization really need automated ML or is it just one of those steps in the AI and ML maturity cycle that we can live without? Would this investment result in **Return on Investment** (**ROI**), or would it become one of those unused platforms that sounded like a good idea at the time?

Let's try to answer these questions by looking at the value proposition of automated ML and see whether it makes a good fit for your organization. As a technology stakeholder, envision yourself as someone trying to build an enterprise AI playbook and deciding whether to invest in and utilize or disregard the utility of automated ML.

Clash of the titans – automated ML versus data scientists

In an organization, large or small, you would first need to communicate this idea to your own data science team. It could be a large group of people with a mix of PhDs, ML engineers, and data scientists, or it could be a close-knit group of startup bootstrappers, whose lead data science expert is an engineer with Jupyter notebooks installed on their machine. In either case, you would need convincing arguments to present a case for automated ML.

We have said this before but will gladly repeat it: automated ML is not going to kill data scientist jobs any time soon. Having said that, you would be hard-pressed to find a data scientist who acknowledges the efficiency of using feature engineering, hyperparameter optimization, or neural architecture search using automated ML. As data scientists, we tend to think, sometimes rightfully so, that data science is an art form that cannot be brute-forced. There is a lot of subject matter expertise and knowledge that goes into model tuning, and therefore it is better left with humans who know what they are doing. The problem is that this model doesn't scale:

Figure 10.1 – Life of a data scientist

There is a perpetual shortage of qualified data scientists and other qualified individuals who can build, tune, monitor, and deploy models at scale. Our organizations are data-rich but starving for insights. We have seen several critical business intelligence, experimentation, and insight projects pushed down the priority stack for revenue-centric programs. Organizations need automated ML to let **subject matter experts** (**SMEs**) and citizen data scientists build data experiments and test their hypotheses without needing a PhD in ML. Would they get things wrong? I am sure. However, automated ML will enable and empower them to build advanced ML models and test their hypotheses.

Therefore, we must support this democratization of AI. This is not to say that a mission-critical credit risk model trained by automated ML should be rolled out to production without proper due diligence and testing – that wouldn't be the case even with handcrafted models. An organization must make sure that performance, robustness, model decay, adversarial attacks, outlier parameters, and accuracy matrices and KPIs apply to all models equally.

In a nutshell, ML progress in an enterprise is slow-paced. It is tough to scale and not very well automated. Collaboration among business teams and data scientists is difficult and actual operationalized models delivering business value are few and far between. Automated ML brings the promise of solving these problems and gives additional tools to data scientists to ease the manual drudgery of feature engineering, hyperparameter optimization, and neural architecture search.

Automated ML – an accelerator for enterprise advanced analytics

While building your AI playbook and reimagining the AI talent strategy for your organization, you should consider automated ML as an accelerator. The following are some of the reasons why you would want to consider using automated ML for your organization.

The democratization of AI with human-friendly insights

Automated ML is rapidly becoming an inherent part of all major ML and deep learning platforms and will play an important part in democratizing advanced analytics. All major platforms tout these capabilities, but for it to be an accelerator for an enterprise, automated ML must play an important role in the democratization of AI. The toolset should enable a citizen data scientist to perform daunting ML tasks with ease and get human-friendly insights. Anything short of explainable, transparent, and repeatable AI and automated ML would not be the advanced analytics accelerator you had hoped for.

Augmented intelligence

Automated ML is becoming ingrained in most modern data science platforms, and therefore will be commoditized as part of the MLOps life cycle. For a data scientist, the biggest value proposition is the ease of feature engineering, data preprocessing, algorithmic selection, and hyperparameter tuning, that is, augmenting the data scientist's skillset. MLOps platforms with built-in automated ML also provide training and tuning, model monitoring and management, and head-to-head model comparison capabilities for A/B testing. This excellent suite of tools becomes extremely helpful to strengthen a data scientist's skills, hence automated ML proves to be an augmented intelligence platform for data scientists and domain SMEs alike.

Automated ML challenges and opportunities

We have discussed the benefits of automated ML, but all these advantages are not without their fair share of challenges. Automated ML is not a silver bullet and there are several scenarios where it would not work. The following are some challenges and scenarios where automated ML may not be the best fit.

Not having enough data

The size of the dataset is a critical component for automated ML to work well. When feature engineering, hyperparameter optimization, and neural architectural search are used on small datasets, they do not yield good results. The dataset has to be significantly large for automated ML tools to do their job effectively. If this is not the case with your dataset, you might want to try the alternative approach of building models manually.

Model performance

In a small number of cases, the performance you get from out-of-the-box models may not work – you may have to hand-tune the model for performance or apply custom heuristics for improvement.

Domain expertise and special use cases

In a case where your model requires significant subject matter expertise and rules built into it, the gains from automated ML models may not provide good returns.

Computational costs

Automated ML is inherently computationally expensive. If the datasets are extremely large, you might consider using local compute resources (which are arguably cheaper) to avoid the expensive costs associated with using cloud machines. In these cases, costs incurred in training the model may outweigh the optimization benefits – caveat emptor.

Embracing the learning curve

Anything worth doing is never easy, such as exercise or ML. Automated ML takes the brunt of the work out of repetitive and tedious tasks, but there is still a learning curve. Most platforms dub their automated ML products as zero-code, low-code, or no-code approaches; however, you would still need to get yourself familiar with the tool. What happens when your results don't match your intuition or hypothesis based on years of subject matter expertise? How do you fine-tune the model based on identified important features? Which models perform well on training data but poorly on production datasets and why? These are practical considerations your citizen data scientists and stakeholders would need to go through as part of this learning curve. A lot of this learning and adaption would depend on the tool you select and how easy it makes life for you.

Stakeholder adaption

Every new technology faces the challenge of adaption – and automated ML would be dead on arrival due to its very nature. Your enterprise AI strategy should include training and incorporate the learning curve and potential disruption associated with the introduction of new technology such as automated ML. Building example templates and starter kits would help get stakeholders up to speed. We have seen in practice that getting junior data scientists and developers on board helps socialize new technology.

Let's proceed toward the next section, where we will discuss various techniques that can help in building trust in models trained by automated ML.

Establishing trust - model interpretability and transparency in automated ML

Establishing trust in the model trained by automated ML can appear to be a challenging value proposition. Explaining to the business leaders, auditors, and stakeholders responsible for automated decision management that they can trust an algorithm to train and build a model that will be used for a potentially mission-critical system requires that you don't treat it any different from a "man-made" ML model. Model monitoring and observability requirements do not change based on the technique used to build the model. Reproducible model training and quality measurements, such as validating data, component integration, model quality, bias, and fairness, are also required as part of any ML development life cycle.

Let's explore some of the approaches and techniques we can use to build trust in automated ML models and ensure governance measures.

Feature importance

Feature importance, or how much a specific attribute contributes to the result of a prediction either positively or negatively, is a model inspection technique that is offered by most, if not all, automated ML frameworks. In earlier chapters, you have seen how AWS, GCP, and Azure all offer a view of feature importance scores for their trained models. This information can be used by domain SMEs and citizen data scientists to ensure the model is accurate.

Feature importance is not only helpful in validating the hypothesis but may also provide insights into data that was previously unknown. Data scientists, with the help of domain SMEs, can use feature importance to ensure that it does not exhibit bias against any protected class and see whether any of the preferences are against the law. For instance, if a loan decision algorithm has a bias toward a specific gender, ethnicity, or race, that would be illegal and unethical in most scenarios. On the contrary, if a breast cancer database shows a significant gender bias, that is due to a biological construct and hence is not a societal or implicit bias to be mitigated or addressed. Testing an automated ML model, or any model for that matter, for feature importance with perturbation makes a good sanity check for correctness, robustness, and bias.

Counterfactual analysis

In algorithmic explainability, counterfactual analysis belongs to the class of example-based explanations. In simple terms, it uses causal analysis to show what would have happened if an event has or has not occurred. For example, in a biased loan model, counterfactual analysis will reveal that while keeping all the other factors unchanged, modifying the ZIP code of a loan applicant has an impact on the outcome. This shows bias in the model against people in a specific geography, possibly a sign of latent racial bias manifested in the ZIP code as a proxy. Besides bias, counterfactual analysis can also reveal mistakes in model assumptions that can be quite beneficial.

Data science measures for model accuracy

Standard ML approaches for performance estimation, model selection, and algorithm comparison should be applied to ensure the accuracy and robustness of the trained model. Some standard ways of validating a ML model are shown in the following figure:

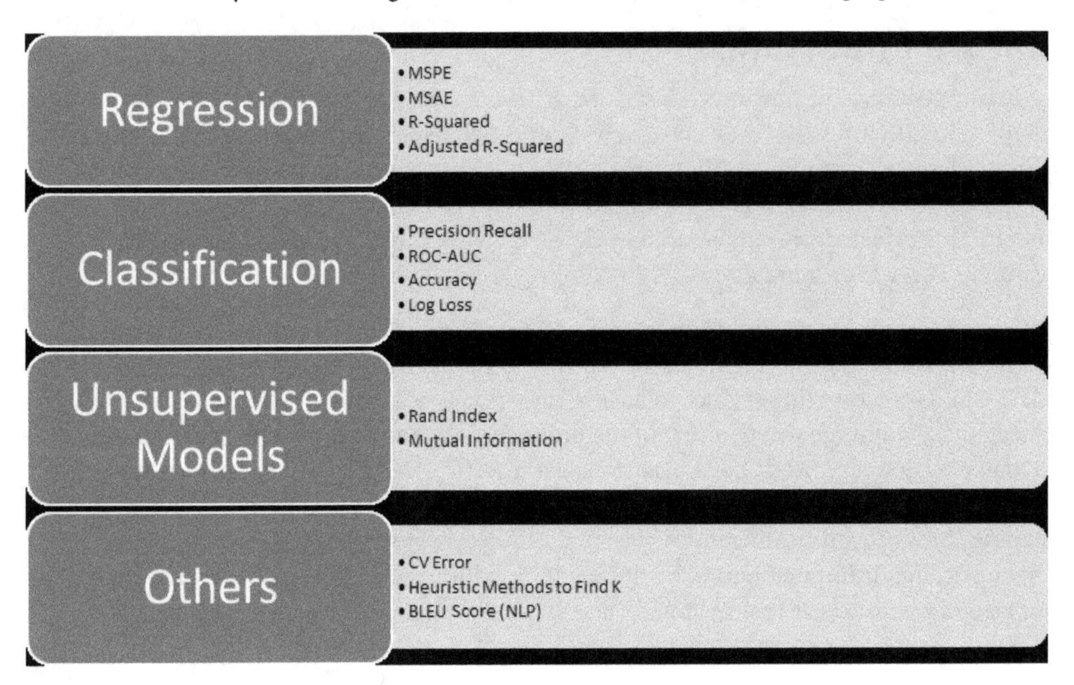

Figure 10.2 – Standard measures for data science

For performance estimation in a large dataset, recommended approaches include checking for the confidence interval via normal approximation, as well as train/test splits. For smaller datasets, repeating k-fold cross-validation, leave-one-out cross-validation, and confidence interval testing helps to ensure good estimates for performance. For model selection with large datasets, three-way holdout methods of training validation, testing split, and repeated k-fold cross-validation with independent test sets work well. To compare the model and algorithms, applying multiple independent training and test sets for algorithm comparison, McNemar's test, and Cochran's test is prescribed, while for smaller datasets, nested cross-validations work quite effectively.

For automated ML-based models to be accurate and robust, we need to ensure that we perform explainability measures across the entire life cycle. Therefore, checks need to be performed pre-modeling – that is, characterizing the input data – during the modeling – that is, designing explainable model architectures and algorithms – and post-modeling – that is, extracting explanations from outputs.

Pre-modeling explainability

The pre-modeling explainability starts with exploratory data analysis and dataset statistical description and sanitization, that is, descriptions of the variables, metadata, provenance, statistics, between variables (pair plots and heatmaps), ground truth correlations, and probabilistic models generating synthetic data. This explainability extends to explainable feature engineering followed by interpretable prototype selections and identification of meaningful outliers.

During-modeling explainability

When using automated ML algorithms, choose to adopt a more explainable model family when given the option; linear models, decision trees, rule sets, decision sets, generalized additive models, and case-based reasoning methods are more explainable than complex black-box models.

Hybrid explainable models are also used to design explainable model architectures and algorithms such as **Deep K-Nearest Neighbors** (**DKNN**), **Deep Weighted Averaging Classifier** (**DWAC**), **Self-Explaining Neural Network** (**SENN**), **Contextual Explanation Networks** (**CENs**), and **Bag-of-features Networks** (**BagNets**). The approach of joining prediction and explanation using models such as **Teaching Explanations for Decisions** (**TED**), multimodal explanations, and rationalizing neural predictions greatly helps explain the models. Visual explanations are, of course, quite effective since a picture is worth a thousand words, unless it's in a higher dimension because then it becomes quite confusing, such as postmodern art.

Explainability using architectural adjustments and regularizations as an artifact uses explainable **Convolutional Neural Networks** (**CNNs**), explainable deep architecture, and attention-based models, which are being used in natural language processing, time series analysis, and computer vision.

Post-modeling explainability

There are various built-in tools for post-modeling explainability that are part of automated ML toolkits and hyperscalers. These tools include macro explanations and input-based explanations or perturbations to see how input manipulations can potentially impact the outputs. Macro explanation-based approaches, such as importance scores, decision rules, decision trees, dependency plots, verbal explanations, and counterfactual explanations, are great resources for domain SMEs to get their head around the outcomes of a trained model.

There are also explanation estimation methods that try to probe the proverbial black box, including perturbation-based training (LIME), backward propagation, proxy model, activation optimization, and SHAP.

As described in the preceding methods, there is no singular way of establishing trust in ML models, whether they are manually trained or built via automated ML toolkits. The only way to accomplish this is by following engineering best practices; validation, reproducibility, experimental audit trail, and explainability are the best-known methods to verify and ensure it works. You may not need all these approaches as part of your ML workflow but know that these and various other approaches to validate and monitor your model throughout the life cycle are critical to ensure the success of your enterprise ML operationalization.

Introducing automated ML in an organization

Now that you have reviewed the automated ML platforms and the open source ecosystem and understand how it works under the hood, wouldn't you like to introduce automated ML in your organization? So, how do you do it? Here are some pointers to guide you through the process.

Brace for impact

Andrew Ng is the founder and CEO of Landing AI, the former VP and chief scientist of Baidu, the co-chairman and co-founder of Coursera, the former founder and leader of Google Brain, and an adjunct professor at Stanford University. He has written extensively about AI and ML and his courses are seminal for anyone starting out with ML and deep learning. In his HBR article on AI in the enterprise, he poses five key questions to validate whether an AI project would be successful. We believe that this applies equally well to automated ML projects. The questions you should ask are as follows:

- Does the project give you a quick win?
- Is the project either too trivial or too unwieldy in size?
- Is your project specific to your industry?
- Are you accelerating your pilot project with credible partners?
- Is your project creating value?

These questions help not only to choose a meaningful project but also to seek business value and technical diligence. Building a small team and appointing a leader as an automated ML champion would help further the cause.

Choosing the right automated ML platform

Open source, AWS, GCP, Azure, or something else – what automated ML platform should you use?

There are few considerations you should keep in mind when choosing an automated ML platform. First of all, if you are cloud-native, that is, your data and compute reside in the cloud, it would make way more sense to use the automated ML offered by that specific cloud provider for all practical purposes. If you have a hybrid model, try to keep your automated ML compute close to the data you plan to use, wherever it resides. The feature parity might not make such a significant difference; however, not having access to the right data definitely would.

Generally speaking, cloud compute and storage resources for automated ML can get expensive quickly, providing you are working with large models and doing multiple simultaneous experiments. If you have compute resources and data available in an on-prem scenario, that is an ideal playground without increasing your hyperscaler's bill. However, this comes with the responsibility of setting up the infrastructure and doing the setup for an on-prem toolkit. So, if cost is not a major concern and you want to quickly explore what automated ML can do for you, cloud-based toolkits would make an ideal companion. Depending on your relationship with your cloud provider, you might also be able to get a discount.

The importance of data

From the examples in previous chapters, it must have become painfully obvious that data, and a lot of it, is the single most important thing needed for a successful automated ML project. Smaller datasets do not provide good accuracy and do not make a good use case for automated ML. If you do not have large enough datasets to work with, or your use case does not lend itself well to automated ML, maybe it's not the right tool for the job.

The right messaging for your audience

Automated ML promises to deliver immense value to both data scientists and SMEs; therefore, you should carefully craft your pitch.

For business leaders and stakeholders, it is a business enablement tool that will help citizen data scientists to expedite development; business users can test their hypotheses and execute experiments in a faster manner.

For your data science allies, you should introduce automated ML as a tool that helps augment their capabilities by taking away the repetitive and mundane tasks from their day-to-day job. Among many things, automated ML can help take away the drudgery of sifting through a dataset for important features and can help identify the right parameters and models through a large search space. It will expedite the training of new data scientists and will help other engineers who like to dabble in data science experiments gain a firm understanding of the fundamentals by doing it. If your fellow data scientists and ML engineers see value in this technology and do not feel threatened by it, they will be willing to adapt. This would not be the first time they have heard of this amazing technology and would love to use it, albeit with a healthy dose of skepticism.

Call to action - where do I go next?

All good things must end, and so does this book. Whew! We covered a lot of ground here. Automated ML is an active area of research and development, and in this book, we tried to give you a breadth-first overview of the fundamentals, key benefits, and platforms. We explained the underlying techniques of automated feature engineering, model and hyperparameter learning, and neural architecture search with examples from open source toolkits and cloud platforms. We covered a detailed walkthrough of three major cloud platforms, namely Microsoft Azure, AWS, and GCP. With the step-by-step walkthroughs, you saw the automated ML feature set offered in each platform by building ML models and trying them out.

The learning journey does not end here. There are several great references provided in the book where you can further do a deep dive to learn more about the topic. Automated ML platforms, especially cloud platforms, are always in flux, so by the time you read this, some of the screens and functionality might already have changed. The best way to keep up with these hyperscalers is to keep pace with the change, the only constant.

As a parting thought, we strongly believe what Sir Richard Branson said, that *"you don't learn to walk by following rules. You learn by doing, and by falling over."* The best way to learn something is by doing it – execution beats knowledge that doesn't get applied. Go ahead and try it out, and you will be glad to have the metaphorical coding battle scars.

Happy coding!

References and further reading

For more information on the topics covered in this chapter, you can refer to the given links:

- *How to Choose Your First AI Project* by Andrew Ng: `https://hbr.org/2019/02/how-to-choose-your-first-ai-project`
- *Explainable Artificial Intelligence for Neuroscience: Behavioral Neurostimulation*, December 2019, Frontiers in Neuroscience, 13:1346, DOI: 10.3389/fnins.2019.01346
- *Interpretable ML – A Guide for Making Black Box Models Explainable* by Christoph Molnar: `https://christophm.github.io/interpretable-ml-book/`

`Packt.com`

Subscribe to our online digital library for full access to over 7,000 books and videos, as well as industry leading tools to help you plan your personal development and advance your career. For more information, please visit our website.

Why subscribe?

- Spend less time learning and more time coding with practical eBooks and Videos from over 4,000 industry professionals
- Improve your learning with Skill Plans built especially for you
- Get a free eBook or video every month
- Fully searchable for easy access to vital information
- Copy and paste, print, and bookmark content

Did you know that Packt offers eBook versions of every book published, with PDF and ePub files available? You can upgrade to the eBook version at `packt.com` and as a print book customer, you are entitled to a discount on the eBook copy. Get in touch with us at `customercare@packtpub.com` for more details.

At `www.packt.com`, you can also read a collection of free technical articles, sign up for a range of free newsletters, and receive exclusive discounts and offers on Packt books and eBooks.

Other Books You May Enjoy

If you enjoyed this book, you may be interested in these other books by Packt:

Mastering Azure Machine Learning

Christoph Körner, Kaijisse Waaijer

ISBN: 978-1-78980-755-4

- Setup your Azure Machine Learning workspace for data experimentation and visualization
- Perform ETL, data preparation, and feature extraction using Azure best practices
- Implement advanced feature extraction using NLP and word embeddings
- Train gradient boosted tree-ensembles, recommendation engines and deep neural networks on Azure Machine Learning
- Use hyperparameter tuning and Azure Automated Machine Learning to optimize your ML models
- Employ distributed ML on GPU clusters using Horovod in Azure Machine Learning
- Deploy, operate and manage your ML models at scale
- Automated your end-to-end ML process as CI/CD pipelines for MLOps

Learn Amazon SageMaker

Julien Simon

ISBN: 978-1-80020-891-9

- Create and automate end-to-end machine learning workflows on Amazon Web Services (AWS)
- Become well-versed with data annotation and preparation techniques
- Use AutoML features to build and train machine learning models with Autopilot
- Create models using built-in algorithms and frameworks and your own code
- Train computer vision and NLP models using real-world examples
- Cover training techniques for scaling, model optimization, model debugging, and cost optimization

Hands-On Artificial Intelligence on Google Cloud Platform

Anand Deshpande, Manish Kumar, Vikram Chaudhari

ISBN: 978-1-78953-846-5

- Understand the basics of cloud computing and explore GCP components
- Work with the data ingestion and preprocessing techniques in GCP for machine learning
- Implement machine learning algorithms with Google Cloud AutoML
- Optimize TensorFlow machine learning with Google Cloud TPUs
- Get to grips with operationalizing AI on GCP
- Build an end-to-end machine learning pipeline using Cloud Storage, Cloud Dataflow, and Cloud Datalab

Packt is searching for authors like you

If you're interested in becoming an author for Packt, please visit `authors.packtpub.com` and apply today. We have worked with thousands of developers and tech professionals, just like you, to help them share their insight with the global tech community. You can make a general application, apply for a specific hot topic that we are recruiting an author for, or submit your own idea.

Leave a review - let other readers know what you think

Please share your thoughts on this book with others by leaving a review on the site that you bought it from. If you purchased the book from Amazon, please leave us an honest review on this book's Amazon page. This is vital so that other potential readers can see and use your unbiased opinion to make purchasing decisions, we can understand what our customers think about our products, and our authors can see your feedback on the title that they have worked with Packt to create. It will only take a few minutes of your time, but is valuable to other potential customers, our authors, and Packt. Thank you!

Index

A

I

K

L

M

N

O

P

R

S

T

W

www.ingramcontent.com/pod-product-compliance
Lightning Source LLC
LaVergne TN
LVHW081334050326
832903LV00024B/1159

* 9 7 8 1 8 0 0 5 6 7 6 8 9 *